Historic Preservation Technology

Historic
Preservation
Technology

Robert A. Young, PE, LEED AP

WILEY

John Wiley & Sons, Inc.

Library of Congress Cataloging-in-Publication Data:

Young, Robert A., PE.
Historic preservation technology : a primer / Robert A. Young
    p.   cm.
Includes bibliographical references and index.
ISBN 978-0-471-78836-2 (cloth : alk. paper)
1. Buildings—Repair and reconstruction. 2. Historic preservation. 3. Architecture—Conservation and restoration. I. Title.
TH3401.Y68   2008
690'.24—dc22   2008003741

Printed in the United States of America
10 9 8 7 6 5 4 3 2 1

To Deborah

Contents

Acknowledgments

In the early stages of writing this book, I described the process to a colleague as one of revisiting old friends. At that time, I thought that meant reviewing the many books, articles, and papers that I have enjoyed reading and have drawn from throughout my career. Now it has deepened to include reflecting on the impact that so many people (some of whom I have only met through their writings) have had in shaping this book. Five of these are, sadly, no longer with us, and I must acknowledge them first of all for the profound effect they have had on my career.

Although I did not meet Harley J. McKee in person, his *Introduction to Early American Masonry* encouraged me to enter professional historic preservation practice. I took my first course in historic preservation technology from David Evans at Eastern Michigan University, which laid the cornerstone for what has turned out to be the foundation of my career. I also had the good fortune to take courses at the Campbell Center from Martin Weaver, who showed me the way historic preservation should be done. To these three, I offer wholehearted thanks for their pioneering work in historic preservation. In addition, I acknowledge and thank my father, Raymond W. Young, and my father-in-law, Edward Gagnon, for the preservation values and respect for the built environment that I have learned from them.

I thank Patrick Tripeny, Associate Dean and Architecture Department Chair of the University of Utah College of Architecture + Planning, for his support in obtaining the travel funding and release time that enabled me to complete this work. His insights concerning all things structural and how to wend my way through the publishing process are particularly appreciated.

I thank Brenda Scheer, Dean of the University of Utah College of Architecture + Planning for her support and encouragement. I likewise thank the University of Utah Teaching Committee for awarding to me the John R. Park Fellowship and several teaching grants that have enabled me to deepen my understanding of historic preservation.

I also thank my colleagues and friends who have provided innumerable insights into the many aspects of writing this book. These include (from the Association for Preservation Technology) David Woodcock, Hugh Miller, Andrew Ferrell, Thomas Jester, Richard Ortega, and Barbara Campagna; (from the Traditional Building Skills Institute) Willie Littig, Mike Jackson, Russ Mendenhall, John Lambert, Keith MacKay, Joe Gallagher, and Gina Gardner; (from the Utah State Historic Preservation Office) Wilson Martin, Don Hartley, Nelson Knight, Cory Jensen, and Barbara Murphy; (from Eastern Michigan University) Robert Schweitzer and Marshall McClennan; (from the University of Utah) Peter Goss, Martha Bradley, Cathay Ericson, Derick Bingman, Peter Atherton, Thomas Carter, William Miller, and Barbara Brown; (from Richardson Engineering) Chuck Richardson; (from the Utah Capitol Preservation Board) David Hart; (from the Campbell Center for Historic Preservation Studies) David Flaharty, Brian Powell, and Norman Weiss; (from the Historic New England Program in New England Studies) Abbott Lowell Cummings, James Garvin, and Ken Torino; and (from the National Park Service) Greg Dugan and Pam Holtman.

I offer special thanks to my editors and the staff at John Wiley & Sons: Paul Drougas, who saw the potential of the book, and Lauren LaFrance, Donna Conte, Raheli Millman, and Helen Greenberg who helped get the book (and me) through the production process in one piece. I also thank the proposal and manuscript reviewers Ilene Tyler, Jonathan Spodek, David Mertz, William Murtagh, and, especially, Ann Milkovich McKee, whose comments helped me to see this book from a different perspective.

Most of all, I acknowledge and dedicate this book to Deborah Young, my wife and constant companion of the past thirty-plus years, whose love, support, and motivation continue to enable me to extend myself in new directions and into new adventures.

Introduction

Historic Preservation Technology is intended to be the opening chapter in a career in historic preservation practice centered on issues of construction and remediation technology of older buildings. *Historic Preservation Technology*, however, is also about more than just preserving old buildings. It provides part of the foundation for using the sustainable design philosophy of stewardship of the built environment. This philosophy examines relationships between the built and natural environments and pursues practices that enhance the best qualities of each of them. From my professional experience, I understand the opportunities that exist when buildings are reused rather than simply torn down. Likewise, I recognize that not all buildings have historic significance and that many face physical and economic challenges that may prevent their continued use. In between, however, many buildings get destroyed simply because owners, architects, engineers, designers, contractors, and public officials do not fully understand how to successfully rehabilitate them.

Historic preservation and adaptive use of buildings represent the highest form of recycling available. For this reason, I wrote this book as an introductory tool to enhance understanding of the treatment requirements used in design review guidelines, as well as understanding construction and remediation issues of older buildings in general. Although not every project will come under design review, this book uses historic preservation technology practices, methods, and strategies that are considered appropriate by the *Secretary of the Interior's Standards for the Treatment of Historic Properties* as its central theme.

Historic Preservation Technology examines the construction practices and materials used from the late sixteenth century to the early 1960s. To provide a historic context, each chapter describes when materials or practices were introduced and how they were used. While not meant as a comprehensive history, this overview provides information to place products and materials in their historic context and can assist in defining a construction chronology during building investigations. The dates are as precise as possible, but in some cases there are conflicting accounts or the actual introduction evolved through undocumented vernacular building traditions and lack a specific date. The chapters then provide an overview of common problems and remediation strategies. Each chapter concludes with "References and Suggested Readings" that includes references cited and readings that can extend the depth of knowledge of specific subject matter as it relates to the reader's needs.

Historic Preservation Technology has five parts, each of which builds upon the earlier parts. Part I, "Process Overview," describes processes that guide decisions regarding remediation approaches. Specifically, it provides an overview of the *Secretary of the Interior's Standards*, preservation condition assessment reports, building codes, and inspection methods. Part II, "Building Materials," describes construction materials and how they were historically used. These materials include wood, masonry, concrete, and metals. Primarily, this part covers historic construction methods used for load-bearing walls and skeletal framing systems. Each chapter describes fabrication methods, decay mechanisms, and appropriate remediation techniques. These form the foundation of all remediation treatments throughout the book. Part III, "Building Fabric," describes exterior features, such as roofing, siding, windows, entrances, porches, and storefronts. Part IV, "Building Ornamentation and Finishes," describes features, such as flooring, walls, ceilings, art and stained glass, woodwork, plaster, and decorative finishes. Part V, "Special Topics," describes heating, ventilating, air-conditioning, lighting, building service systems, and sustainability.

Historic Preservation Technology concludes with two appendices. Appendix A outlines the fundamental requirements of the *Secretary of the Interior's Standards for the Treatment of Historic Properties*, which form the foundation of the premise of this book. Appendix B identifies technical resources available on the Internet that can assist readers in obtaining information from historic preservation practitioners and other allied groups. A glossary of terms used in the book follows the appendices. Accompanying this book is an instructor's manual and a website supported by John Wiley & Sons.

ROBERT A. YOUNG, PE, LEED AP
Salt Lake City, UT

PART I

Process Overview

CHAPTER 1 Overview

Stewardship of the built environment balances the needs of contemporary society and their impact on the built environment with its ultimate effects on the natural environment. Merging historic preservation and environmental conservation can create innumerable opportunities for reuse of the built environment, which fosters a more sustainable environment overall.

Historic preservation is not a recent phenomenon in the United States. The historic preservation of buildings began when the city of Philadelphia purchased Independence Hall to save it from demolition in 1816, and the first notable restoration was completed at the Truro Synagogue in Newport, Rhode Island, in 1828 (Murtagh, 2006, 12). The preservation movement made increasing progress throughout the nineteenth and twentieth centuries, leading to the National Historic Preservation Act (NHPA) of 1966. The NHPA mandates that federally funded projects affecting historic buildings must undergo what is known as a "Section 106 review" to ensure that proposed work follows the *Secretary of the Interior Standards for the Treatment of Historic Properties (Standards)*. When the NHPA first went into effect, federally funded projects included highway construction and urban renewal programs that were ravaging older and historic business districts and residential neighborhoods.

Many historic preservation projects were inspired by the American Bicentennial and other personal motivations during the early part of this period. Later projects were completed to earn tax incentives developed to assist historic property owners rehabilitate their properties. The Historic Preservation Tax Credit program continues to provide financial incentives that encourage the use of historic buildings. Tax credits and other financial incentive programs at the state and national levels have helped make historic preservation a multi-billion-dollar industry.

Environmental Conservation in the Late Twentieth Century

Since the creation of Yellowstone as the first national park in 1872, conservation has made steady advances and has become prominent in the public eye. Issues and concerns raised by the conservation movement led to the National Environmental Policy Act (NEPA) of 1969 that fostered an array of stewardship-oriented activities. One of these activities is the requirement for an Environmental Impact Statement (EIS) for large projects (e.g., highway construction, mass transit) that, in part, addresses the adverse effects that the project will have on historic resources.

The two energy crises and the revitalization efforts of the 1970s led to the recognition that the construction and operation of buildings consumes large amounts of natural resources and that growing suburban sprawl was contributing significantly to lower environmental quality. Two observations are needed to understand the sustainability aspects of revitalization: (1) "reuse of existing structures has conserved land, raw materials, and energy," and (2) "cities contain extensive infrastructure of buildings, pipes, reservoirs, conduits, streets, and parks whose reproduction would be formidably expensive" (Jakle and Wilson 1992, 232). Reusing buildings reduces the impact of demand for new land by reusing previously developed land. Reuse recycles a significant number of existing buildings in place and reduces material resource streams (e.g., construction and waste).

The conservation aspects of preserving historic buildings have since become known in much greater detail. Efforts to formalize environmentally sensitive practices have become more coherent with the formation of such programs as the United States Green Buildings Council's Leadership in Energy and Environmental Design (LEED) program and the United States Environmental Protection Agency's Energy Star initiatives (see Chapter 22 for further details).

Emergence of Stewardship of the Built Environment

The social, environmental, and economic benefits of preservation and adaptive use enhance sustainability by promoting reuse of buildings, which in turn has brought sustainability and historic preservation together as stewardship of the built environment. As such, preservation has increasingly become viewed as a tool for increasing sustainability (see Figure 1-1).

Figure 1-1 Stewardship of the built environment blends the need for new buildings with the opportunities presented by reusing existing buildings. This is true for areas within the central city (as shown here) as well as in twentieth-century suburbs.

The successful rehabilitation and adaptive use projects of the past forty years have promoted the acceptance of historic preservation. This acceptance has spurred continued interest in older buildings that has created a permanent market segment and has honed historic preservation practice into an industry served by a multitude of trained professionals.

Today, the *Standards* form the basis of the review processes used by federal agencies to confirm that work proposed and later completed using federal funds (e.g., preservation tax credits, grants) does not have an adverse effect on buildings on or eligible for the National Register of Historic Places (NRHP). As a result, state and local governments and many private organizations administer their review process based on the federal process and the *Standards*.

Thus, typically, when an individual building (or small group of buildings) is subject to a historic preservation review process, an application is made to the agency or organization (e.g., the Landmark Commission or State Historic Preservation Office [SHPO]) that will be administering the review to confirm that any proposed work conforms to their guidelines. This process may involve administrative approval or require a hearing before a review board. Once the design is approved, a building permit can be issued. Upon completion, the work is inspected to verify compliance. Proceeding without an appropriate permit may result in the owner's being fined and required to reverse any changes made.

These reviews are not required for buildings that are not protected by preservation ordinances. When not seeking federal funding or other preservation-related funding incentives, the property owner only has to ensure that the work meets local codes and zoning ordinances.

HISTORIC PRESERVATION PROCESS OVERVIEW

The Secretary of the Interior recognizes four treatment standards related to historic buildings: preservation, rehabilitation, restoration, and reconstruction. Within these standards are guidelines to determine potential directions for appropriate treatment of historic buildings.

The term "historic preservation" is used broadly to describe the efforts to retain the historic character of a building and the historic context of the place where the building is located. A confusing aspect of historic preservation concerns the differences between preservation, rehabilitation, restoration, and reconstruction.

To reduce confusion, the Secretary of the Interior developed *Standards* (see Appendix A) that broadly define the processes involved in each of these four treatments. Each standard has varying degrees of freedom and restrictions that affect the selection of technologies used in the investigation and construction phases. These treatments are defined in the *Standards* as follows:

> *Preservation* is defined as the act or process of applying measures necessary to sustain the existing form, integrity, and materials of an historic property. Work, including preliminary measures to protect and stabilize the property, generally focuses upon the ongoing maintenance and repair of historic materials and features rather than extensive replacement and new construction. New exterior additions are not within the scope of this treatment; however, the limited and sensitive upgrading of mechanical, electrical, and plumbing systems and other code-required work to make properties functional is appropriate within a preservation project. (Weeks and Grimmer 1995, 17)

SECRETARY OF THE INTERIOR'S STANDARDS FOR THE TREATMENT OF HISTORIC PROPERTIES

Standards

Rehabilitation is defined as the act or process of making possible a compatible use for a property through repair, alterations, and additions while preserving those portions or features which convey its historical, cultural, or architectural values. (Ibid., 61)

Restoration is defined as the act or process of accurately depicting the form, features, and character of a property as it appeared at a particular period of time by means of the removal of features from other periods in its history and reconstruction of missing features from the restoration period. The limited and sensitive upgrading of mechanical, electrical, and plumbing systems and other code-required work to make properties functional is appropriate within a restoration project. (Ibid., 117)

Reconstruction is defined as the act or process of depicting, by means of new construction, the form, features, and detailing of a non-surviving site, landscape, building, structure, or object for the purpose of replicating its appearance at a specific period of time and in its historic location. (Ibid., 165)

Guidelines

The treatments are accompanied by guidelines describing the level of historic sensitivity and preservation technology required for compliance. A recommended construction process is described for each treatment and includes these aspects of the building:

- Exterior Materials
  - Masonry
  - Wood
  - Architectural metals

- Exterior Features
  - Roofs
  - Windows
  - Entrances and porches
  - Storefronts

- Interior Features
  - Structural system
  - Spaces/features/finishes
  - Mechanical systems

- Site

- Setting

- Special Requirements
  - Energy efficiency
  - New additions to historic buildings
  - Accessibility
  - Health and safety

While preservation sensitively retains historic features and ensures their retention, restoration allows removal of features not within the historic period being sought. Rehabilitation has the broadest range of treatments affecting an existing building, and reconstruction may have few or no existing historic features available. Treatment guidelines provide an overview on the expected areas of concern, which may include:

- Identification, retention, and preservation of existing historic features
- Stabilization
- Protection and maintenance
- Repair
- Replacement/limited replacement of existing features
- Replacement of missing historic features
- Removal of features from other periods
- Energy efficiency/accessibility/health and safety code requirements
- Alteration/additions for the new use
- Re-creation of missing features from the historic period
- Research and documentation of historic significance
- Investigation of archeological resources
- Reconstruction of nonsurviving buildings and sites

The guidelines should be reviewed to confirm which construction processes are applicable specifically to the treatment being pursued.

The guidelines are intended to ensure retention of character-defining features. A major concern is the reversibility of any chosen method so that the changes made can be reversed at a later date. If the process used proves unsatisfactory, the historic fabric of the building is still available for the later, more sensitive procedure. While the guidelines describe "recommended" and "not recommended" processes, they do not make specific reference to product names or actual products. Examples of selected masonry rehabilitation treatments (shown below) illustrate differences between recommended and not recommended processes:

RECOMMENDED	NOT RECOMMENDED
Identifying, retaining, and preserving masonry features that are important in defining the overall historic character of the building such as walls, brackets, railings, cornices, window architraves, door pediments, steps, and columns; and details such as tooling and bonding.	Removing or radically changing masonry features which are important in defining the overall historic character of the building so that, as a result, the character is diminished.
Protecting and maintaining masonry by providing proper drainage so that water does not stand on flat, horizontal surfaces or accumulate in curved decorative features.	Failing to evaluate and treat the various causes of mortar joint deterioration such as leaking roofs or gutters, differential settlement of the building, capillary action or extreme weather.
Repairing masonry walls and other masonry features by repointing the mortar joints where there is evidence of deterioration such as disintegrating mortar, cracks in mortar joints, loose bricks, damp walls or damaged plasterwork.	Removing nondeteriorated mortar from sound joints, then repointing the entire building to achieve a uniform appearance.

RECOMMENDED	NOT RECOMMENDED
Replacing in kind an entire masonry feature that is too deteriorated to repair if the overall form and detailing are still evident—using the physical evidence as a model to reproduce the feature. Examples can include large sections of a wall, a cornice, balustrade, column, or stairway.	Removing a masonry feature that is unrepairable and not replacing it; or replacing it with a new feature that does not convey the same visual appearance.
Designing and installing a new masonry feature such as steps or a door pediment when the historic feature is completely missing. It may be an accurate restoration using historical, pictorial, and physical documentation; or be a new design that is compatible with the size, scale, material, and color of the historic building.	Creating a false historical appearance because the replaced masonry feature is based on insufficient historical, pictorial, and physical documentation. Introducing a new masonry feature that is incompatible in size, scale, material and color.

Source: Weeks and Grimmer, 1995, 67–70.

Violation of specific limitations, such as sandblasting to remove paint, can jeopardize compliance with the guidelines. Project participants who are new to historic preservation practices may inadvertently violate these limitations. Therefore, selection of appropriate preservation technology for all project phases is critical in meeting the guidelines.

Application of *Standards* and Guidelines

Buildings are considered historic when they are listed on, or are eligible for, the NRHP or are listed on a state or local historic register. A building may also be listed on both the NRHP and a state or local historic register. The *Standards* and their guidelines apply to historic buildings affected by federal funds (e.g., tax credits, highway construction, grants, and revitalization programs). In jurisdictions that have adopted the International Existing Building Code, buildings on the NRHP are eligible for exemptions from the International Building Code.

Buildings on state or local historic registers are also considered historically significant and may be controlled by local or state design review agencies and their own guidelines. These guidelines may have been adapted from the federal *Standards* or may be more restrictive. These buildings come under the control of state and local historic preservation ordinances that may make them eligible for specific protections against demolition and inappropriate alterations; state and local preservation financial incentives; and building code exceptions allowed for historic buildings. These buildings may be listed concurrently on the NRHP and become eligible for federal tax credits and preservation incentives. Review agencies overseeing the state or local register where the property is located should be consulted directly for their guidelines. When attempting to use a modern process or product, check with those agencies to obtain clarification and a "certificate of appropriateness" (written approval of the proposed work). Local, state, and federal reviewing groups annually evaluate numerous products and processes to determine if they are, regardless of the manufacturer's claims, compatible with the intent of the *Standards*.

Property owners are free to choose the treatment that suits their budget and goals for those buildings not on any recognized state or local historic register or those buildings that are on the NRHP but are not covered by provisions of state or local preservation ordinances. The choices made can lead to three outcomes with varying levels of historic sensitivity. The first outcome develops from the lack of sensitivity toward historic character-defining features that is a common occurrence in "modernization" projects and has led to insensitive alterations and removals (e.g., window replacement) that are the exact opposite of any aspect of preservation. The second outcome stems from the lack of appropriate preservation technology awareness of the property owner, architect, contractor, or code official in the use of incompatible modern products and processes and has led to the same unfortunate result, despite all their good preservation intentions. For these types of projects, the materials and processes (e.g., cleaning and resurfacing) used often can also cause irreversible harm to the remaining building fabric. A third outcome is created when the *Standards* or other formal guidelines are voluntarily used as a model and often produces a more historically sensitive result than the previous two outcomes.

Historic preservation technology combines investigation methods, materials, and construction methods used to preserve, rehabilitate, restore, or reconstruct a building. Historic preservation technology plays an important role from initial project investigations through final construction. Therefore, understanding how preservation technology choices made in a project can affect the success of the project outcome is critically important. These choices not only include the early identification of character-defining features of a historic building at the earliest stages but also their retention, protection, remediation, and maintenance needs at subsequent stages of the project. Inadequate investigations or improper remediation techniques that endanger the retention of historic building fabric can jeopardize efforts to meet the expected guidelines.

Whether the building is of the highest significance or is simply an older building, the most successful preservation-oriented projects have often included these aspects:

HISTORIC PRESERVATION PLANNING, DESIGN DEVELOPMENT, AND CONSTRUCTION

- Preservation Planning
  - Protection and stabilization of the property
  - Code and regulation verification
  - On-site investigation
  - Off-site archival research
  - Condition assessments/historic structures reports
- Design Development
  - Select treatment
  - Project programming
  - Design review and approval
  - Construction drawings and specifications
- Construction
  - Construction monitoring for compliance with project goals
  - Final approval by local building officials
  - Commissioning

With the evolution of design and construction systems, this outline should be adapted to the particular system being used. Innovations of emerging systems involve the timing of when the design team is formed, the use of project definition and collaborative practices, and the integration of building information management into the design and construction industry.

Historic Preservation Planning

Three aspects drive the planning phase: verifying code compliance and preservation regulations; assessing existing conditions; and collecting historical data. The condition assessment based on the on-site and archival research should provide the basis for the preservation treatment selection. Subsequently, these planning materials should be used to convey the overall appropriateness of the proposed methods and products. Contractors and consultants can then be selected based on their ability to work within the selected guidelines. Selecting contractors and consultants who understand the guidelines is important in minimizing problems related to the loss of historic character-defining features.

Protection and Stabilization of the Property

The property should be protected from unwanted trespassers. An unsecured building attracts vandals, transients, and illegal salvagers (see Figure 1-2). Inappropriate methods of stabilizing and investigating the property may result in the unnecessary destruction of historically significant building fabric. As a precaution, the early part of the investigation is done cautiously until the owner and the architect select the treatment to be used throughout the project.

Code and Regulation Verification

In this procedure, compliance with current building codes, especially those related to seismic, fire and life safety, and the Americans with Disabilities Act, is evaluated. This is especially true for buildings being considered for a change of use. When the

Figure 1-2 Unsecured and vacant buildings invite vandalism and unwanted trespassers, as the damage to this building demonstrates.

Figure 1-3 Various visual and physical clues indicate several changes in this entranceway: a porch roof and decking have been removed; windows have been added or modified; and drainage/moisture problems have occurred. These clues are given by changes in the brick, window sills, and lintel construction materials, as well as missing or peeling paint.

use will remain the same, the extent of the work proposed may determine whether the entire building will need to be brought up to current codes or whether just the portion of the building or building service system being affected may need to meet current codes. Code issues are further explored in Chapter 2.

All concurrent local and state preservation ordinances that pertain to the building must be identified. This process includes determining if the building is listed on the NRHP or other state and local historic registers and what is required to meet any applicable design guidelines.

On-Site Investigation

Using technology appropriately allows for data collection that is obtained by minimizing the destruction or removal of historic fabric of the building. Many visual inspection and simple probing methods can assess surface conditions (see Figure 1-3). In some instances, the damage and its sources lie beneath the surface. When subsurface damage is suspected or discovered, procedures and methods are available to perform either nondestructive testing where the fabric is analyzed *in situ* with minimal removal or the opposite case, where invasive measures include removing large sections of the surface materials and finishes to gain direct access to the internal portions of the construction. On-site investigation methods are further explored in Chapter 3.

Off-Site Archival Research

Data collection about the existing conditions on-site tells only part of the story of a building. Off-site research in state and local archives can provide information that gives additional clues as to when or why certain modifications were made. Identifying the original architect or builder can lead to identifying when the building was built, whether original drawings or other construction documents are available, and how the building has changed. Information obtained on the original and subsequent

Figure 1-4 Research can reveal historic photographs of a building that can be used to confirm materials or details that have changed through time. This photograph shows the appearance of a Federal period house prior to addition of a later porch that changed the original entrance.

owners and uses of the building, earlier tax photographs and data, and historic photographs (see Figure 1-4) can assist in project planning. This work is usually performed by a preservation consultant who specializes in architectural research. The SHPO will typically have a directory of preservation consultants.

Conditional Assessments and Historic Structures Reports

The ultimate goal of the planning phase is to gather all relevant information on the building so that a condition assessment can be made. The key to selecting the treatment is to obtain written information that allows for informed decisions. The assessment of what is an appropriate treatment arises from the findings of on-site and archival investigations, as well as consultations with local building code officials and any applicable review board(s).

A brief written report, in its simplest form, outlines recommended treatments on a single aspect of a project that may be acceptable. This format is often used when the owner's short-term budget or goals warrant it. When a building needs simple repairs, an estimate of their cost is sought directly from a contractor. If the owner agrees to the findings, the contractor may be hired to do the repairs. This method may lead to a piecemeal approach to preservation that may also be the most expensive over the long term, as well as shortsighted in terms of how one set of repairs may interact with the other building repair needs.

In a more comprehensive format, the condition assessment report provides a broad view of overall conditions. The condition assessments for separate subcomponents are developed by a team of consultants who specialize in the construction assemblies found in the building, and their findings are then assembled in a condition assessment report detailing the conditions and recommended treatments that views the entire building as an integrated whole. The report may include an assessment of the historic significance of an item and list alternative treatments. A condition assessment report can reveal long-term remediation needs and provide an opportunity for the owner, architect, and contractor to determine the budget, scope, and phases of the project. The report is then used to generate construction drawings and specifications.

The most integrated form of report is commonly referred to as a "historic structures report" (HSR), which integrates the historic research with the condition assessments of a variety of consultants. Originally developed by the National Park Service to document work on its historic buildings, the HSR is a combination of archival research, on-site architectural research identifying character-defining features, a condition analysis and assessment, a list of recommended treatments, and a prioritized action plan.

The design development phase takes the collected data and the intentions and goals of the owner and translates them into a set of construction documents that illustrate and specify the expected scope of construction work and the expected final construction result.

Architectural Design Development

Select Treatment

Even without the need to adhere to the *Standards*, an often confusing aspect of historic preservation practice concerns what is expected in the project. The treatment defines the types of recommended practices that will be acceptable to meet the goals of the project. Identifying the selected treatment is important when communicating between the different parties involved in the project so that the context of the work is known. This decision will minimize the inadvertent or accidental loss of historic character-defining features through the actions of those who may be unfamiliar with the differences, for example, between a restoration and a reconstruction.

Understanding what is considered an appropriate historic preservation technology within the selected project treatment is important in determining the actual construction practices that are to be employed. In this regard, the identification of appropriate technologies to be used is equally imperative in ensuring the retention of important historic features during the investigation phase *and* the construction phase of the project.

The project treatment selected should be based on a combination of historical significance, on-site conditions, budget, proposed use, any upgrades required by code, and the requirements of any federal, state, or local review board. Since much of what is considered appropriate is outlined in the *Standards*, these treatments must be carefully considered in determining the historic preservation technology to use throughout the project

Project Programming

Project programming enables the architect and the owner to determine parameters (e.g., space requirements, types of spaces) of the expected design. At this point, assessing the effect that the project will have on the existing building is critical. If the project includes an adaptive use or significant alterations to the existing building, multiple and sometime conflicting requirements of the building codes and the guidelines will need to be resolved. If the scope of work defined by the program cannot be adapted to the building, then alternative programming requirements or a different building may be needed.

Design Review

Preliminary plans are submitted for review by local code officials for compliance with local building codes. When applicable, the plans are also reviewed by the

Figure 1-5 This addition (left-center) underwent a design review by the local Landmarks Commission to ensure that it was compatible with the design guidelines for the historic district where it was located.

local Landmarks Commission, the SHPO, and any other review board with an interest in the project (see Figure 1-5). These reviews may involve revisions and repeat reviews to gain approval. Consultation with the reviewers before submitting the plans for the formal review can often streamline this process.

Construction Drawings and Specifications

The final step in this phase is the development of the final construction drawings and specifications that will be used by contractors to bid for the work. The preliminary drawings are updated to reflect the comments and revisions provided during the various reviews. These documents are submitted for final approval and are subsequently used to solicit bids from contractors.

Construction

Construction is usually the longest portion of the project. All historic character-defining features must be protected from harm for the duration of the project. Most large projects are done in several simultaneous operations and in phases coordinated by a general contractor.

Construction Monitoring

An important part of the construction process is monitoring compliance with project goals. Contract administration will include updating progress and completion status reports. Likewise, change orders (e.g., changes to the original scope or terms of the contract) will need to be evaluated in terms of how they affect the historic character-defining features. Materials and methods that do not comply with the treatment guidelines should not be allowed. Construction inspections must be accommodated throughout this phase, and any deficiencies will need to be corrected.

Final Approval

As the project nears completion, final inspections of the various construction processes will be made before the building is approved for occupancy. One emerging sustainable design practice during this phase is known as "commissioning." Commissioning is a process that verifies that all building systems work properly and educates the owner on how to operate and maintain them. Commissioning varies from providing a simple collection of operators' guides for the equipment and recommended maintenance activities to a formal training period for building operations and maintenance personnel. Here again, ensuring the use of methods and materials that enhance the retention of historic character-defining features is advised. Upon final approvals and completion of any commissioning activities, the building is ready for occupancy.

References and Suggested Readings

Biallas, Randall J. 1997. Evolution of historic structures reports at the U.S. National Park Service: An update. *APT Bulletin*, 28(1): 19–22.

Cox, Rachelle S. n.d. *Design review in historic districts*. Preservation Information Series. Washington, DC: National Trust for Historic Preservation.

Jakle, John A. and David Wilson. 1992. *Derelict landscapes: The wasting of America's built environment*. Savage, MD: Rowman & Littlefield.

Jandl, H. Ward. 1988. *Rehabilitating interiors in historic buildings*. Preservation Brief No. 18. Washington, DC: United States Department of the Interior.

McDonald, Travis C., Jr. 1994. *Understanding old buildings: The process of architectural investigation*. Preservation Brief No. 35. Washington, DC: United States Department of the Interior.

Morton, W. Brown, III, Gary L. Hume, Kay D. Weeks, and H. Ward Jandl. 1992. *The Secretary of the Interior's standards for rehabilitation and illustrated guidelines for rehabilitating historic buildings*. Washington, DC: United States Department of the Interior.

Murtagh, William J. 2006. *Keeping time: The history and theory of preservation in America*. New York: John Wiley & Sons.

Nelson, Lee H. 1988. *Architectural character: Identifying the visual aspects of historic buildings as an aid to preserving their character*. Preservation Brief No. 17. Washington, DC: United States Department of the Interior.

Park, Sharon C. 1993. *Mothballing historic buildings*. Preservation Brief No. 31. Washington, DC: United States Department of the Interior.

Rypkema, Donovan D. 2001. Preservation in the new century: Risk, relevance and reward. *Forum Journal*, 15(3): 4–10.

Slaton, Deborah and Alan W. O'Bright. 1997. Historic structures reports: Variations on a theme. *APT Bulletin*, 28(1): 3.

———. 2004. *The preparation and use of historic structures reports.* Preservation Brief No. 43. Washington, DC: United States Department of the Interior.

Stovel, H. 1997. A significance-driven approach to the development of the historic structure report. *APT Bulletin*, 28(1): 45–47.

Waite, John G., Clay S. Palazzo, and Chelle M. Jenkins. 1997. Watching the evidence: An HSR to guide the preservation of George Washington's Mount Vernon. *APT Bulletin*, 28(1): 29–35.

Weeks, Kay D. and Anne E. Grimmer. 1995. *The Secretary of the Interior's standards for the treatment of historic properties with guidelines for preserving, rehabilitating, restoring, and reconstructing historic buildings.* Washington, DC: United States Department of the Interior.

Woodcock, David G. 1997. Reading buildings instead of books: Historic structures reports as learning tools. *APT Bulletin*, 28(1): 37–38.

CHAPTER 2 Health and
Life Safety

The oldest known building code was the Babylonian Code of Hammurabi (ca. 1780 BC), which essentially stated that if the building owner died due to a collapse of a building, then the builder would be put to death; this meant that the builder was responsible for constructing a safe building. Since then, many building codes have been established to protect the health and life safety of building owners and occupants. Although Boston outlawed thatch roofs and wooden chimneys in 1630, the first known formal building code in the United States was introduced in what is now Winston-Salem, North Carolina, in 1788. Many larger cities enacted building codes in the early 1800s. New Orleans became the first city to require inspections in 1865 (Daggers 2003).

With the multiple codes written by local municipalities throughout the late nineteenth and early twentieth centuries, the need for a model code to craft state and local codes became evident. The first model code organization, Building Officials and Code Administrators International (BOCA), was established in 1915 and published the *National Building Code (NBC)*. The International Congress of Building Officials (ICBO) was established in 1922 and published the *Uniform Building Code (UBC)* in 1927. The Southern Building Code Congress International (SBCCI) was formed in 1940 and published the *Southern Building Code (SBC)* in 1945 (Ching and Winkel 2007). *NBC* was used in the north-central and northeast portions of the United States. *UBC* was used west of the Mississippi River, and *SBC* was used in the southeast. Model codes were adapted to address local issues such as earthquakes and flooding. These codes were oriented specifically to new construction. This

BUILDING CODES

orientation often left the repair and retrofit of older and historic buildings vulnerable to unnecessary demolition practices to install the systems prescribed for new construction. In recognition of the particular challenges that historic buildings pose, "smart" codes have been developed by several states (e.g., Maryland, New Jersey, Vermont, Washington, and California) to facilitate working with existing buildings.

A significant preservation tool developed by the National Institute of Building Science in the late twentieth century was the *Guideline on Fire Ratings of Archaic Materials and Assemblies* (United States Department of Housing and Urban Development 2000). This guideline was written to reduce the demolition of existing building assemblies that were no longer defined in the most recent code and has been incorporated as an appendix for a number of codes, including the National Fire Protection Association (NFPA) *NFPA 914: Code for Fire Protection of Historic Structures* and the *Uniform Code for Building Conservation (UCBC)*.

The International Code Council (ICC) was created to succeed the formerly separate code groups. In 2000, the ICC introduced a family of codes known as the "I-Codes" (e.g., *International Building Code, International Energy Conservation Code*). The NFPA, originally a partner in this process, opted out in favor of its own codes (e.g., *NFPA 70: National Electric Code, NFPA 101: Life Safety Code, NFPA 914: Code for Fire Protection of Historic Structures,* and *NFPA 5000: Building Construction and Safety Code*), which some municipalities adopted instead of the corresponding I-code version (Green and Watson 2005). Therefore, confirming which codes are enforced locally is imperative.

When codes are updated, an existing building may fall out of compliance. However, the building is "grandfathered" in compliance and does not need to be updated simply to meet the adoption of a new code. When the building is modified, regulations are applied to make it comply with the current code. So, without any updating, a building can fall further out of compliance. A common practice in the late twentieth century was the "25-50 percent rule," whereby if the project cost was less than 25 percent of the building's value, then the portion of the building that needed to meet the code was negotiable. When costs approached 50 percent of the building's value, all work had to meet the code. When the cost exceeded 50 percent of the building's value, the entire building had to be brought into compliance with the code. The recognition that this worked against rehabilitating existing buildings led to the abandonment of this rule in the early 1980s. After that, only new work was required to comply with current codes (Green and Watson 2005).

The *International Building Code (IBC)* has provisions that accommodate historic buildings. One important aspect of the I-Codes is the policy on existing historic buildings. Beyond the *IBC*, there is an *International Existing Buildings Code (IEBC)* that has specific provisions for historic buildings providing alternative means and processes to achieve building and life safety while allowing for the continued preservation of existing buildings (see Figure 2-1).

The *IEBC* defines four categories of work (*IEBC* Chapter 4): repair, alterations (divided into three levels), change of use, and additions. Repairs refer to patching, restoration, and minor repairs to elements, equipment, or fixtures to maintain them in good condition. Alteration Level 1 refers to removal and replacement of the covering of existing elements, equipment, or fixtures using new materials that serve the same purpose, as well as any structural work that does not involve reconfiguring the floor space. Alteration Level 2 refers to reconfiguring any space, the addition or elimination of any window or door, the reconfiguration or extension of any building

Figure 2-1 The *IBC* and the *IEBC* have provisions for code compliance to protect existing buildings that are on statewide, national, or local registers of historic buildings.

system, or the installation of additional building services equipment. Alteration Level 3 refers to situations where the work area exceeds 50 percent of the aggregate area of the building and involves reconfiguring hallways, corridors, or exit ways that requires the building to be closed for construction (commonly called a "gut rehab"). Change of use refers to a change in the purpose or level of activity in the building. An addition is a new construction that increases the floor area, number of stories, or building height (Green and Watson 2005; Kaplan 2007).

The *IEBC* provides three ways to reach compliance: the prescriptive compliance method, the performance compliance method, and the work area compliance method. The prescriptive compliance method (*IEBC* Chapter 3) requires compliance with the *International Fire Code* and *IEBC* requirements for structural systems, mechanical and electrical systems, fire egress, accessibility, and relocated and historic buildings. The performance compliance method (*IEBC* Chapter 13) provides a rating system for nineteen fire safety parameters based on specific characteristics of the building. The building meets compliance when the sum of those categories exceeds 100 or it can be demonstrated that the alternative measures have enabled the building to reach an acceptable level of compliance. The work area compliance method (*IEBC* Chapters 5–11) provides prescriptive requirements based on the four categories of work just described. Requirements for Repair (*IEBC* Chapter 5) or Alteration Level 1 (*IEBC* Chapter 6) are restricted to the physical work area. For Alteration Level 2 (*IEBC* Chapter 7) when the work area exceeds 50 percent of the floor area, additional requirements for items (e.g., egress) outside the work area are imposed. For Alteration Level 3 (*IEBC* Chapter 8), requirements can be extended to floors below the work area. For a change of use (*IEBC* Chapter 9) requirements are established by comparing the relative change in hazard between the former and proposed uses in these categories: Means of Egress, Heights and Area, and Exposure of Exterior Walls. The code is more stringent when the new use increases the level of hazard and less so when the hazard remains the same or decreases (Kaplan 2007). Additions (*IEBC* Chapter 10) require work to comply with *IBC* requirements for new construction.

For historic buildings (*IEBC* Chapter 11), requirements are modified for each category of work to reflect the building's historic nature. This section also adds the option of a report prepared by a registered design professional to

> identify each required safety item that is in compliance with this chapter and where compliance with other chapters would be damaging to contributing historic features. . . . In addition, the report shall describe each feature that is not in compliance with these provisions and shall demonstrate how the intent of these provisions is complied with in providing an equivalent level of safety. (International Code Council 2006, 51)

This report can be requested by the local code official when deemed necessary. The code official then determines whether the overall compliance requirements have been met.

ACCESSIBILITY

The Americans with Disabilities Act (ADA), enacted in 1990, mandates access to public buildings for persons with limited mobility, visual impairment, and other disabilities. The ADA applies to buildings that serve as public accommodations, commercial facilities, and state and local government operations. Religious entities, private clubs, and private residences (except those portions that serve as places of public accommodation) do not need to comply.

ADA compliance can be achieved by following the ADA Accessibility Guidelines (ADAAG), which provide examples of "readily achievable" means to do so (Jester and Park 1993; United States Department of Justice 1991). The readily achievable means were intended to be strategies that could be implemented at little or no cost and enhance the accessibility of the building or site. These readily achievable means include:

- Creating a convenient designated parking space
- Installing ramps
- Making curb cuts
- Retrofitting doors and door hardware
- Rearranging furnishings, shelves, and telephones
- Adding raised markings on elevator controls
- Installing offset hinges and other accessible door hardware
- Installing accessible water fountains
- Modifying bathrooms to increase maneuvering space
- Installing grab bars, higher toilet seats, and full-length mirrors
- Relocating dispensers

An unintended consequence of this act was misperception about how to make buildings accessible. The big problems in historic buildings have been the lack of access at entrances, the requirement for elevator access, and the alterations of bathroom facilities to accommodate wheelchairs. Early attempts to correct these deficiencies in historic buildings led to significant alterations to entrances to accommodate ramps, the addition of elevator towers, alterations to stairways, and the insertion of washroom facilities to meet the requirements of the ADA.

a   b

Figure 2-2   Many of the early attempts to comply with the ADA led to (a) visually intrusive solutions, while (b) other efforts were more sensitive. Many modification efforts now take advantage of entrances adjoining a parking lot rather than the sidewalk along the front of the building.

Unfortunately, many of these original attempts were based on processes for new construction and did little to explore how to integrate these features into buildings while mitigating their disruption to historic fabric (see Figure 2-2). Since these features all tended to be part of the historic character-defining features, the potential impact on the historic fabric was high. Ramps were erected along the primary facade; elevator shafts were constructed as adjuncts to the primary facades and, in some instances, were located at the rear of the building. Later solutions for these problems included more sensitive insertions of these access ways inside the building by identifying appropriate secondary spaces that could be sacrificed for the construction of fire stairs and elevator shafts. Likewise, handicapped-accessible restrooms were inserted in new vertical cores that were carved out of these vertically aligned secondary spaces.

Fire has destroyed many buildings. Through a combination of construction materials, urban congestion, and the use of open flames for heating, lighting, or cooking well into the nineteenth century, fires were able to rapidly consume individual buildings and large groups of buildings. The lessons learned from the cause and spread of these fires and the types of fire-related damage have been used to enhance building design and construction practices. Fire and life safety codes have emerged alongside the other building codes in response to concerns raised by such fire disasters as the Great London Fire of 1666, the Chicago fire of 1871, and the Triangle Shirtwaist Factory fire in New York City in 1911. Fires in entertainment clubs and other places of public assembly also have revealed shortcomings in design practices and brought about revisions to the codes. In 1896, the NFPA published its first *Life Safety Code*. While the ICC has its own fire and life safety codes, many communities instead have adopted separate codes developed by the NFPA and the American National Standards Institute (ANSI).

FIRE AND LIFE SAFETY

Fire and life safety codes usually focus on protection of life, protection of property, and continuation of operations. The protection of life is usually of highest priority. This is accomplished through the inclusion of fire detection and alarm systems combined with egress paths that allow the occupants to leave a building without encountering smoke or fire. Protection of property involves incorporating the ability of the fire department to access the building, set up fire-fighting equipment, and have adequate fire suppression resources (e.g., water and adequate water pressure). Protection of property also includes the concept of "compartmentation," which dictates the maximum floor area allowed for various types of construction and occupied use. This goal may force the installation of firewalls and fire-rated enclosures, such as fire doors and smoke barriers, which subdivide larger spaces into smaller ones. "Continuity of operations" refers to the ability to purge smoke from the building and to contain water flow from floor to floor.

Fire safety in historic buildings presents the problems of creating appropriate means of egress and inserting adequate fire protection systems. When older buildings are used for public access, they must meet fire egress requirements. Often compliance has meant the insensitive enclosure of large open lobbies and installation of fire doors. Over the past two decades, a number of sensitive installations have been completed that can serve as models. For example, door catches can be tied into the fire alarm system so that when the alarm sounds, the door catches release, allowing the door to close. Metal fire doors have also been faux finished to appear similar to the materials used on doors and wood trim elsewhere in the building.

Safe egress is of primary importance in evacuating a building. Fire codes dictate the maximum allowable distance from a fire-rated egress. They also dictate how the fire egress is configured to meet not only the evacuation needs of the occupants but also the potential simultaneous ingress requirements of fire and rescue personnel responding to the fire emergency.

Like restrooms, fire stairs can be inserted in secondary spaces that are vertically aligned to allow for the insertion of a fire-rated stairway enclosure. So long as

Figure 2-3 This sprinkler head (center-top) is in plain sight but could have also been recessed for less visual intrusion.

people leaving the building remain in a fire-rated enclosure space, the building can meet the code. Other interpretations of the fire code may be enhanced by the use of the alternate code *IEBC* or the *Guideline for the Fire Rating of Archaic Materials and Assemblies* (United States Department of Housing and Urban Development 2000), as both documents contain accepted values for fire ratings of materials that are no longer used in modern construction. This interpretation alone has forestalled the destruction of plaster, lath, and room finishes when updating a building to meet modern code requirements.

Other problems arise when installing new fire protection systems in a previously unsprinklered building. There are numerous solutions to this problem. Although the finishes may need to be disrupted, the need to conceal piping within walls and ceiling cavities can overrule their historic significance. The sprinkler head may also be recessed into the ceiling with a nominal cover or installed to minimize visual intrusions (see Figure 2-3). Like restrooms and fire stairs, standpipes that carry the water vertically in a building can be located in service areas.

LATERAL LOADS

The horizontal forces that occur during earthquakes, hurricanes, and tornadoes are known as "lateral loads." While similar in terms of effect, the lateral loads created during an earthquake initially act along the base of a building and cause a whipping action through the aboveground portions of the building. Hurricanes and tornadoes also produce lateral loads, but these occur along the aboveground portions of the building.

Earthquakes

Earthquakes are seismic events that occur as slippage along faults in the tectonic plates that form the Earth's crust. An earthquake is actually a series of multiple waves of energy through the overlying soil. When this slippage occurs, multiple shock waves radiate from the point of origin and create lateral forces that push against any resistance encountered. For a building, even though the impact on the structure occurs at the foundation, the aboveground portion of the structure moves with a whip-like reaction. While the base may be only nominally displaced, the displacement at the top of the structure can be substantially greater. Combining acceleration and deceleration of this movement at the top with the resistance to movement at the base, the elasticity or rigidity of the structural connections and materials in the framing, and the structural integrity of the building can literally tear a building apart.

Shock waves can also reduce the bearing capacity of soils beneath the building in a process known as "liquefaction," in which the soil behaves structurally like a liquid. Any weight resting on this soil sinks into the ground.

The first building codes to regulate seismic performance were the 1927 Palo Alto city code and the 1927 *UBC* (McClure 2006). High-risk seismic zones have been identified to enable architects, engineers, and owners to anticipate the seismic resistance needed to ensure that the building can withstand a seismic event. While the western United States is largely a high-risk seismic environment, two other hazardous areas exist in the Mississippi River region at St. Louis, Missouri, and the Charleston, South Carolina, area. The Great Lakes, Great Plains, and Gulf Coast states are relatively low in risk.

A deficiency of many historic buildings is that they were built before the recognized need for seismic safety precautions. As such, many unreinforced masonry (URM) buildings exist that do not have the reinforcement needed to maintain public safety during and after a seismic event. The National Earthquake Hazards

Reduction Act, passed in 1977, established the National Earthquake Hazards Reduction Program (NEHRP), which is responsible for reducing earthquake risks. Reauthorized in 1990, this program has led to a variety of investigational and educational efforts, such as *The Utah Guide for the Seismic Improvement of Unreinforced Masonry Dwellings* (Utah Division of Comprehensive Emergency Management, n.d.).

Efforts to assess potential risks have been advanced by the Federal Emergency Management Agency (FEMA) and the Structural Engineering Institute of the American Society of Civil Engineers (ASCE). FEMA has developed a process known as "Rapid Visual Screening" (RVS), which assesses the potential seismic risk that a building may incur (FEMA 2001). Intended to highlight potential problems, RVS is a first step in determining seismic upgrading needs. ASCE has developed the standard AEI/ASCE 31, *Seismic Evaluation of Existing Buildings*, which also provides an assessment process. When, however, more specific information is needed, one should hire a structural engineering consultant to investigate in detail the building being considered for a seismic upgrade. This may provide more detailed information, but it will take more time.

Seismic upgrading can have a dramatic effect on historic buildings as measures are incorporated to counteract the effect of lateral loads. Greater attention is being paid to the connections that tie roofs and foundations and other structural elements together. These strategies are relatively straightforward in new construction but cause significant challenges in upgrading historic buildings. Most problems relate to the many URM buildings that do not meet seismic codes. As a result, three methods of upgrading have evolved: base isolation, center core consolidation, and diaphragms and shear wall systems. A base isolation system eliminates transfer of energy from a seismic event to the historic structure (see Figure 2-4).

Base isolation involves the insertion of resilient supports beneath the building. To do this, the existing structural foundation is severed from its connections to the ground. The base isolators are inserted beneath the foundation. This insertion requires a matrix of isolators uniformly distributed underneath the building. In addition to severing the foundation horizontally from the ground, the perimeter of the vertical portion of the foundation is typically isolated in a seismic moat to allow the building to move laterally without being abruptly stopped by the surrounding soil. The moat is capped with a sacrificial covering that will fail quickly so as not to restrict foundation movement. In this manner, shock waves do not transfer energy into the structural system and the tendency for the building to whip around aboveground is reduced.

The second approach is the center core consolidation method, which stiffens URM load-bearing walls. Center core technologies involve drilling cores vertically downward from the top of an unreinforced masonry wall into the foundation. The coring penetrates into the foundation and reinforcement connections located at the bottom of the wall. Rods are inserted into the cored holes, and the void surrounding each rod is then filled with grout or epoxy that cures to create a continuous reinforced diaphragm.

The third approach is the installation of shear walls and the creation of diaphragms that absorb energy and allow the structure to remain sufficiently stiff to withstand forces that would otherwise rip the building apart. This approach ties the structure into a system of diaphragms and shear walls and often entails removing interior finishes to tie exterior wall cladding together or into a reinforcement system. Horizontal diaphragms are created by connecting the structure of each floor

Figure 2-4 Base isolation systems, although expensive, have enabled the retention of historic buildings without compromising their visual integrity. The Salt Lake City-County Building restoration completed in 1989 included a base isolation system.

horizontally so that the floor becomes a single continuous membrane. Shear walls may include the infill of wall openings to create a continuous surface. Diaphragms and shear walls resist the lateral forces generated in an earthquake. These improvements are typically the most visually intrusive and physically disruptive solutions.

The lateral forces generated by a hurricane or tornado occur aboveground and are compounded by the swirling winds and varying wind pressures as the tornado or hurricane passes the building. Winds may also carry airborne projectiles that can puncture or crush vulnerable windows, doors, and lightweight construction assemblies. In a hurricane, flooding from the storm surges that occur as the hurricane reaches shore create lateral forces as the water floods the area. These forces can destroy windows and doors, shear houses from their foundations, and strip roofing materials and the entire roof assembly, as well as rip roof and wall cladding, porches, and balconies from the building's exterior (see Figure 2-5). Once the building envelope is compromised, rain may enter the building and cause further damage. Preventive measures to reduce potential damage include protective storm shutters, the structural reinforcement that connects the walls to the roof and foundation, and insertion of bracing members to the interior framing of the roof (Uguccioni and Hernden 1997).

Hurricanes and Tornadoes

Figure 2-5 Damage from hurricane Katrina in 2005 left many historic portions of Louisiana and the Gulf Coast devastated.

HEALTH

Some materials and chemicals used well into the twentieth century are now known to be health hazards: The more potentially hazardous ones include lead, asbestos, and polychlorinated biphenyls (PCBs) that were used in construction and toxic chemicals that were used in cleaning activities. The illicit production of methamphetamines has generated increasing concerns about contamination in buildings that have formerly housed "meth labs." Some naturally occurring hazards include mold and radon.

Lead and Asbestos

Lead and asbestos are the two best-known hazards (see Figure 2-6). Their specific uses are described in the rest of this book. Their presence immediately signals the need for a licensed abatement specialist. When abatement is done professionally, the waste material becomes identified as hazardous waste and is handled as such. Specific procedures such as containing the waste in specific locations, filtering air, and then capturing all effluents and waste materials and placing them in specific containers must be used. The waste is then taken to a hazardous waste disposal site. The fees charged at these sites add to the cost of demolition.

Chemical Hazards

Chemical hazards must be disposed of in accordance with applicable local, state, and federal regulations. PCBs and other chemicals may be present on-site in electrical equipment as well as in the form of contaminants in the ground. PCBs were used in insulating electrical transformers. They have since been banned in the United States but are present in industrial locations around the country. Industrial cleaning products, such as carbon tetrachloride, used by dry cleaning establishments, were sometimes illegally disposed of on-site, thus contaminating soil. Likewise, leaking fuel tanks from gas stations allow gasoline and its additives, such as methyl tertiary-butyl ether (MBTE), to leak into the ground.

Contaminants created in the production of methamphetamines can remain after the "meth lab" has been removed. While law enforcement officials typically

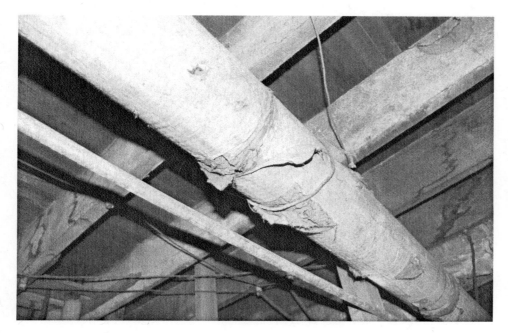

Figure 2-6 Asbestos was formerly used in a broad range of building products, including pipe insulation, as shown here.

remove the equipment and any remaining bulk chemicals, the process of making methamphetamine may have contaminated the ventilation system, the plumbing systems, and any exposed surfaces. Environmental cleaning companies are available to test for contamination and can then remove any residual contamination.

Natural Hazards

Mold presents a major health risk. Mold is attracted by moisture and a good air supply and thrives in temperatures above 40°F. Whether the moisture is due to water vapor trapped in a wall or the result of a flood or hurricane, health problems can arise from contact with mold, including asthma and hay fever-like symptoms. Mold can also cause skin and mucous membrane infections and irritation. Removal guidelines are available from the Occupational Safety and Health Administration (OSHA) (United States Department of Labor, n.d.).

Radon also poses a health risk. Radon, a radioactive gas generated by decaying uranium in soil, is found nationwide, generally collecting in basements and crawl-spaces with poor ventilation. Simple tests are available to detect radon. The problem is usually corrected by installing a radon reduction system that typically exhausts the fouled air. It is also common to seal any cracks in foundation walls or floors where radon has been detected.

JOB SITE SAFETY

Safety practices must conform to the regulations set forth by OSHA. OSHA enforces strict rules for protecting workers that include parameters for visual, auditory, respiratory, and personal injuries. Vertical access must be done in accordance with their standards, including the proper use of ladders, scaffolding, and vertical lift equipment (see Figure 2-7).

All chemicals used on-site should be used in accordance with the manufacturer's Material Safety Data Sheet (MSDS), which provides guidelines in their use and the emergency first aid measures that should be followed. The MSDS must be readily accessible in the event of an emergency.

Figure 2-7 Vertical access must protect both workers and passersby from falls and dropped tools or materials. This scaffolding includes stable work platforms as well as netting to trap falling objects.

Many older buildings also are habitats for a variety of vermin, including birds, insects (e.g., spiders, bees, and wasps), snakes, and small mammals (e.g., rodents, bats, raccoons). These pose potential health and safety problems, particularly in the early inspection and demolition phases, when contact is made directly with them or indirectly through their scat.

References and Suggested Readings

Advisory Council on Historic Preservation. 1989. *Fire safety retrofitting in historic buildings.* Washington, DC: Government Printing Office.

Alderson, Caroline and Nick Artim. 2000. Fire-safety retrofitting: Innovative solutions for ornamental building interiors. *APT Bulletin*, 31(2/3): 26–32.

Andrews, Carol, Charles Barsch, and Deborah Cooney. 1993. *Coping with contamination: A primer for preservationists.* Washington, DC: National Trust for Historic Preservation.

Artrim, Nick. 2003. Automatic fire suppression for historic structures: Options and applications. *APT Bulletin*, 34(4): 35–40.

Bailey, James S. and Edmund W. Allen. 1988. Seismic isolation retrofitting: Salt Lake City and County Building. *APT Bulletin*, 20(2): 32–44.

Battaglia, David H. 1991. *The impact of the Americans with Disabilities Act on historic structures.* Washington, DC: National Trust for Historic Preservation.

Bonneville, David R. 1988. Seismic retrofit of the Campton Place Hotel. *APT Bulletin*, 20(2): 28–31.

Ching, Francis D. K. and Steven R. Winkel. 2007. *Building codes illustrated: A guide to understanding the 2006 International Building Code.* New York: John Wiley & Sons.

Connolly, William M. 2003. *The New Jersey rehabilitation subcode and historic preservation.* *APT Bulletin*, 34(4): 19–21.

Cox, Rachelle. 2001. *Controlling disaster: Earthquake-hazard reduction for historic buildings.* Washington, DC: National Trust for Historic Preservation.

Daggers, Steve. 2003. ICC consolidation benefits building safety and public. *The Alliance Review* (March/April): 4–5.

Federal Emergency Management Agency (FEMA). 2001. *Rapid visual screening of buildings for potential seismic hazards: A handbook (FEMA 154).* 2nd ed. Washington, DC: FEMA.

Friedman, Donald. 2003. Structural triage of historic buildings: Combining safety and preservation interests after disasters. *APT Bulletin,* 34(1): 31–35.

Green, Melvyn. 1981. Seismic rehabilitation of historic buildings: A damage analysis approach. *Bulletin of the Association for Preservation Technology,* 13(2): 23–26.

Green, Melvyn and Anne L. Watson. 1988. Building codes: Evaluating buildings in seismic zones. *APT Bulletin,* 20(2): 13–17.

———. 2005. *Building codes and historic buildings.* Washington, DC: National Trust for Historic Preservation.

International Code Council. 2006. *International Existing Building Code 2006.* Falls Church, VA: International Codes Council.

Jackson, Mike. 2003. Main Street and building codes: The "tin ceiling" challenge. *APT Bulletin,* 34(4): 29–34.

Jandl, H. Ward. 1998. *Rehabilitating interiors in historic buildings.* Preservation Brief No. 18. Washington, DC: United States Department of the Interior.

Jester, Thomas C. and Sharon C. Park. 1993. *Making historic properties accessible.* Preservation Brief No. 32. Washington, DC: United States Department of the Interior.

Kaplan, Marilyn. 1992. *Safety, building codes and historic buildings.* Washington, DC: National Trust for Historic Preservation.

———. 2003a. Considering fire-safety improvements in historic buildings. *APT Bulletin,* 34(4): 10–17.

———. 2003b. Rehabilitation codes come of age: A search for alternative approaches. *APT Bulletin,* 34(4): 5–8.

———. 2007. Adopting 21st century codes for historic buildings. *Model Public Policies* (A Public Policy Report published by National Trust Forum), (May/June): 1–8.

Look, David W., Terry Wong, and Sylvia Rose Augustus. 1997. *The seismic retrofit of historic buildings: Keeping preservation at the forefront.* Preservation Brief No. 41. Washington, DC: United States Department of the Interior.

Lynch, Michael F. 2003. Planning projects to prevent damage during construction: A property owner's primer. *APT Bulletin,* 34(4): 43–46.

McLure, Frank. 2006. Modern earthquake codes: History and development. Berkeley, CA: Computers and Structures (accessed from www.csiberkeley.com/Tech_Info/McClure_book_smll.pdf).

Park, Sharon C. and Douglas C. Hicks. 1995. *Appropriate methods for reducing lead paint hazards in historic housing.* Preservation Brief No. 37. Washington, DC: United States Department of the Interior.

Pianca, Elizabeth G. 2001. Smart codes: A new approach to building codes. *Forum News,* 3(5): 1–2, 6.

Sewall, Jim, and Claudette Hanks Reischel. 2005. *Treatment of flood-damaged older and historic buildings.* Washington, DC: National Trust for Historic Preservation.

Staley, Ronald D. 2002. Scaffolding historic structures: Planning and implementation of fixed scaffolding systems. *APT Bulletin,* 33(2/3): 33–38.

Stein, Benjamin, John Reynolds, Walter Grondzik, and Alison Kwok. 2006. *Mechanical and electrical equipment for buildings.* New York: John Willey & Sons.

Stockbridge, Jerry G. and Robert A. Crist. 1988. Pre-quake seismic diagnostic techniques. *APT Bulletin,* 20(2): 10–12.

Uguccioni, Ellen and Joseph Herndon. 1997. *Hurricane readiness guide for owners and managers of historic resources.* Washington, DC: National Trust for Historic Preservation.

United States Department of Housing and Urban Development. 2000. *Guideline on fire ratings of archaic materials and assemblies.* Washington, DC: United States Department of Housing and Urban Development.

United States Department of Justice. 1991. *Standards for accessible design: ADA accessibility guidelines (ADAAG).* Washington, DC: United States Department of Justice.

United States Department of Labor. n.d. *A brief guide to mold in the workplace: Safety and health information bulletin*. Washington, DC: Occupational Safety and Health Administration. http://www.osha.gov/dts/shib/ shib101003.html

United States Environmental Protection Agency. 1998. *Lead in your home: A parent's reference guide*. Washington, DC: United States Environmental Protection Agency.

———. 2007. *A citizen's guide to radon: The guide to protecting yourself and your family from radon*. http://www.epa.gov/ radon/pubs/citguide.html#lower

Utah Division of Comprehensive Emergency Management. n.d. *The Utah guide for seismic improvement of unreinforced masonry dwellings*. Salt Lake City: Utah Division of Comprehensive Emergency Management

Watts, John M., Jr. 2003. Fire-risk indexing: A systemic approach to building-code "equivalency" for historic buildings. *APT Bulletin*, 34(4): 23–28.

Winter, Thomas A. 2003. A pioneering effort: The California State Historical Building Code. *APT Bulletin*, 34(4): 17–21.

CHAPTER *3* Building Pathology: Investigation, Analysis, and Assessment

OVERVIEW

Building pathology, the study of decay mechanisms in buildings, is an established specialty in historic preservation technology. Practitioners integrate knowledge of modern building sciences and both contemporary and historic construction technologies into a system of identification, diagnosis, and remediation approaches that can correct building system deficiencies with particular sensitivity to historic preservation standards. In some instances, someone with extensive expertise in building pathology may be able to diagnose the sources of building decay problems using what seem to be intuitive processes. In actuality, this intuition is a knowledge-based skill gained from years of practice and is invaluable in reducing the time needed to do certain aspects of the survey work. This does not mean that the expertise eliminates the investigation process in its entirety.

Practitioners who are inexperienced in dealing with older buildings may not realize that a thorough investigation process is critical. Projects that skimp on this step to save time in the early stages often see delays and additional expenses incurred in the construction phase when "unforeseen" but perhaps predictable problems occur. The desire to move the project into the design and construction phases must be tempered by the need to accommodate competing demands for safety and sensitivity to historic character-defining features (e.g., those that contribute to the historic nature of the building or interior space). So, in adapting the woodworker's adage "measure twice, cut once" to the investigation phase, spending appropriate time investigating the building will reveal issues that, left undetected, may later cause delay and waste money.

Historic preservation technology is an important part of the investigation process. For smaller projects, the investigations are still informally done. In many cities, concerns for public safety have led to ordinances that define how investigations and testing should be performed, especially on multistory buildings. Accordingly, the American Society of Testing and Materials (ASTM) and the American National Standards Institute (ANSI), have developed standards that are used nationwide to ensure that findings are made using recognized methods. Using appropriate investigation methods and their associated technologies allows the gathering of information while protecting the historic building fabric. The construction technologies and material fabrication methods used over time have generated a substantial number of inspection and testing methods that could (and do) fill numerous volumes by themselves. Therefore, only an overview will be presented here to provide a context for the analytical and research aspects of historic preservation technology used to assist in selecting the treatment for the project.

IDENTIFYING FEATURES AND CONDITIONS

The goals of the building investigation are to identify character-defining features (elements that enhance the historic qualities) of the building, document existing conditions, and identify sources of decay and deterioration. Through a combination of on-site and archival research, missing features and changes that have occurred through time (see Figure 3-1) can also be identified. Based on the findings from these investigations, the desired rehabilitation, preservation, or restoration treatment can be selected.

Figure 3-1 These two originally similar houses demonstrate how buildings change over time. Of particular note are the differences around the windows.

Without a firm understanding of existing conditions, sources of decay, and the effect of their remediation on the historic character-defining features of the building, a project may not be considered to be in compliance with the guidelines. As the historical sensitivity and significance of a building increase or when the design team's familiarity with historic construction technologies is minimal, a more preservation-sensitive assessment process must be used. Even when as-built documentation is available, an investigation is imperative to determine that the as-built drawings still match the existing configuration of interior spaces and exterior facades.

Many design guidelines call for the retention of exterior and interior features that are significant to historic, architectural, and cultural values. Two key criteria for a successful project thus become the *identification* and *retention* of character-defining features. The remaining character-defining features of the original construction or even some later constructions typically define the level of historic integrity left in the building. Later remodeling and updates may have left these features intact, altered them, or removed them completely. These later updates may have also contributed their own set of character-defining features.

Character-Defining Features

While the building investigation focuses on building condition assessment, an important aspect of the earliest investigations is to identify these character-defining features of the building and site; this identification is essential in selecting the treatment approach to the project. As noted environmental activist Stewart Brand stated:

> Buildings loom over us and persist beyond us. They have the perfect memory of materiality. When we deal with buildings we deal with decisions taken long ago for remote reasons. We argue with anonymous predecessors and lose. . . . From the first drawings to the final demolition, buildings are shaped and reshaped by changing cultural currents, changing real-estate value, and changing usage. (Brand 1994, 2)

Essentially, the building is the accumulation of its previous usages and reflects the needs of its occupants through time. Character-defining features are identified by how a feature or group of features affects the presence of the building on the site and in its neighborhood context; how one's awareness of surface, detail, and form changes as one approaches the building; and how the interior space is perceived due to both surface ornamentation and spatial relationships.

Exterior character-defining features may include the overall building height, shape, and massing; roof and roof features, such as chimneys, roof monitors, or cupolas; projections, such as porches, oriel windows, or bay windows; recesses, such as galleries, arcades, or recessed porches and balconies; openings, such as windows and doorways; and various exterior materials that contribute to the building's uniqueness. The exterior character-defining features also communicate how the building relates to its site and the context of the neighborhood where it is located (both in its historic origins and contemporary form). Context is important when considering exterior changes that will affect how the building relates to neighboring buildings.

Although the exterior of a building is often its most prominent visible aspect, its interior is frequently more important in conveying the building's history and development. Interior features to be considered for preservation may include the overall building plan (space sequence and circulation patterns), spaces (rooms and volumes), individual architectural features, and the finishes and materials of the walls, floors, and ceilings (Jandl 1988). They may also include intangibles such as daylighting or acoustical qualities. All of these features together form the historic integrity that becomes jeopardized as more of these features are lost or compromised.

If the subject building is listed on the NRHP or on a state or local historic register, then its character-defining features may already have been identified in preparing nominations to these registers. As part of the off-site archival research process, it is important to verify these features with the local, state, and federal preservation organizations administering these registers.

The identification of character-defining features is actually an ongoing process that occurs simultaneously with all other on-site activities. Removal of a later update may reveal previously unknown features and modify the direction of the work to be done. Much of the on-site investigation process involves an evaluation of the remaining historic integrity of the building; a determination of how subsequent modifications have affected the integrity; and an evaluation of whether those modifications can be appropriately reversed. Understanding how these character-defining features came into being and how they should be incorporated in the future use of the building is a necessary goal of the investigation. A *rehabilitation* treatment would evaluate the historic nature of the features and balance decisions on the needs for contemporary use with the desire to retain character-defining features as sensitively as is practical. A *preservation* treatment would retain everything together as the physical document of all character-defining features that existed immediately prior to the time of the preservation project. However, the need to identify character-defining features becomes critical when a *restoration* treatment to a specific time period is desired. A true restoration often raises the question "How does the historic value of historic character-defining features from subsequent time periods compare in significance with those remaining from the time period targeted for the restoration?" This issue must be considered with care.

The decision processes that arise in a restoration project are revealed in the Paul Revere House in Boston, Massachusetts. The house was originally constructed in 1680 after a fire destroyed the previous building on the site. Originally owned by a merchant, it was eventually purchased by Paul Revere in 1770. Throughout the nineteenth century, the house was used for a variety of purposes, including a candy store, a cigar factory, a bank, and a grocery store. In 1902, it was purchased with the intent of saving it as a landmark. After much deliberation, the eighteenth- and nineteenth-century changes were removed, even though several of these changes reflected important historic trends of those periods, and the house was returned to its c. 1700 appearance (see Figure 3-2).

According to the Paul Revere Memorial Association, which administers the house as a museum, " . . . the restored dwelling, with its third story front extension removed, resembles its late seventeenth century appearance. Ninety percent of the structure, two doors, three window frames, and portions of the flooring, foundation, inner wall material and raftering, are original" (Paul Revere Memorial Association 2006). While this is an extreme example, this same choice has been repeated numerous times nationwide whenever a true restoration is desired based on physical evidence discovered on-site and documentation found off-site.

Preliminary Walk-through

The building's condition directly affects the scope and cost of the work. Obtaining a valid condition assessment before starting design or construction is critical. Two indicators that contribute to the immediate perception of potential costs are structural integrity and moisture damage. Initially, preliminary assessments are made of

Figure 3-2 The Paul Revere House (center) underwent a significant restoration that removed a third-floor addition to return it to its c. 1700 appearance. This "scrape approach" to restoration has caused lengthy debate among preservationists.

the overall building envelope, structural construction, and other architectural features to determine if they are in suitable condition to proceed with the proposed work or if they need major alterations.

A common approach used to initiate a project is a simple walk-through of the building and its site. In this approach, a preliminary assessment of the conditions and the extent of deterioration form the initial decision to either pursue or stop further plans. Depending on the complexity of the building and the expertise of the parties involved, this phase may be completed in as little as a few hours. With current digital technology, it is possible to take notes and photographs as needed to summarize both overall and specific problems and opportunities. The walk-through enables the design team to form initial opinions about proceeding with the project and often allows the identification of systems and building elements in need of stabilization or temporary protection before a more formal and detailed inspection can safely occur.

Reconnaissance Survey

If the findings from the preliminary walk-through are positive, then the next step is the reconnaissance survey, in which the conditions are studied more closely. Note-taking and photography become more systematic and formal at this point since they will be used to prepare more detailed reports that will be used in the overall condition assessment. This phase includes both a thorough exterior envelope and structural assessment and an interior room-by-room analysis. Any number of consultants may be used for this work as need for their specific expertise is identified. The complexity of the work proposed and the historic significance of the building will dictate the length of time and number of disciplines involved. This phase may take as little as a few days or as much as several months or more.

This survey can be a combination of visual and hands-on inspection methods that include surface mapping, nondestructive testing, and material sampling for laboratory analysis. The goal is to assess physical conditions and sources of decay. Treatments addressing only the symptoms of the problem ultimately fail if the cause of decay is not corrected.

With these aspects in mind, the reconnaissance survey should include a focused assessment of these components of the building and its site and their influence on the project goals:

- *Site/Soil Conditions*: verify bearing strength of soils; recent and historic groundwater fluctuations; presence of expansive clay soils; drainage patterns; existence of subsurface structures on-site; and possible archaeological resources.

- *Structure*: verify the physical integrity and load capacity of structural members (both foundation and structural components above grade); occurrences of cracking, sagging, bulging, and displacement; and seismic and lateral load capacity as required by local codes.

- *Envelope and Enclosure Systems*: verify the structural and physical integrity of the roof, walls, ceilings, floors, windows, and doors and identify sources of moisture penetration or vermin damage.

- *Building Service Systems*: verify capabilities of mechanical, electrical, plumbing, and other building systems.

- *Environmental Hazards*: confirm the need for mitigation/abatement of health hazards due to mold, asbestos, radon, lead, and contamination from solvents, polychlorinated biphenyls (PCBs), and petroleum-based fuels and their additives.

The reconnaissance survey is largely a visual inspection to gather preliminary information needed to make a decision on how to proceed and whom to contact to conduct further detailed inspections. The term "visual inspection" is somewhat of a misnomer since it implies that it means to look but not touch. In actuality, it is a process of probing and poking that relies on the inspector's senses of touch, hearing, sight, and smell. This method is used to gain a better sense of the condition of an object using these senses to confirm suspicions aroused by its appearance. It also relies on the experience of the inspector to recognize where potential problems lie beneath the surface of something that appears to be intact.

Other aspects of this approach use the physical appearance to suggest the potential for underlying problems. Such techniques as shining a high-intensity light across a surface may locate surface irregularities invisible to the naked eye alone that can indicate alterations to the space or finishes; tapping lightly or pressing gently on plaster surfaces can give an indication of their soundness; and using small probes, such as a screwdriver, to carefully push on surfaces to determine their structural integrity, can assist in reaching a preliminary opinion of conditions. Many architectural inspectors use a small kit of tools and objects to assist them with the inspection. These items can include, but are not limited to, a flashlight, assorted screwdrivers and blunt probes, pliers, cutters, small automobile mechanic mirrors, ball bearings/marbles/ rubber balls (to check floor levelness), a 6-inch torpedo level, 10-foot and 25-foot tape measures, resealable plastic bags, envelopes, or containers for samples, labels for sample identification, a hammer, a wooden or rubber mallet, pry bar(s), a metal ruler, pencils and marking pens, a notepad, clipboard, voice recorder, camera, binoculars, magnifying glass, and a first-aid kit. An outline of a reconnaissance survey method is shown in Sidebars 3-1 and 3-2.

Beyond providing an initial condition assessment and identifying the character-defining features of the building, the reconnaissance survey should identify conditions that may need further and more specific inspections by a certified consultant; identify conditions that require immediate stabilization; and identify conditions that need to be monitored. Next, these issues are referred to the appropriate consultants who will do the formal inspections. In these more detailed inspections, several

SIDEBAR 3-1: An Exterior Reconnaissance Survey Method

GOAL

- Compile a summary of the building/site/setting.
  - Photograph overall views of building(s), facades, and site.
  - Photograph each problem and all character-defining features.
  - Record all findings.
  - Identify conditions that need further assessment by consultants.

PRELIMINARY PREPARATION

- Review existing construction plans or other building documentation.
- Develop simple sketched elevations and site plans if none are available.

EXTERIOR

- Look at the building's exterior in general.
  - Note sagging, broken, or displaced structural elements/confirm source.
  - Note foundation and soil conditions along the perimeter of the building.
  - Note character-defining features and materials.
  - Note the general level of repair or missing features.
  - Identify significant changes.

- Check the condition of the roof.
  - Note sagging ridge or structural members.
  - Note missing/damaged materials.

- Proceed to each facade.
  - Identify problems and sources.
  - Identify historic features.
  - Identify alterations.
  - Note all findings for each facade.
  - Trace continuity of defects.

- Walk the site along the perimeter and then explore the site.
  - Sketch a site plan/identify site features.
  - Locate and note overgrown elements or suspicious landscaping.
  - Observe how neighboring buildings are similar or different.

SIDEBAR 3-2: An Interior Reconnaissance Survey Method

GOAL

- Compile a summary of the building's interior.
  - Photograph interior spaces.
  - Photograph each problem and all character-defining features.
  - Record all findings.
  - Identify conditions that need further assessment by consultants.

- Preliminary preparation
  - Review floor plans or other building documentation.
  - Develop simple sketched floor plans if none are available.

- Interior
  - Verify the operational status of building systems.
  - Note potentially hazardous conditions.
  - Go to the lowest level (basement/crawl space).

  Check for structural problems/confirm their source.
  Check for water problems/confirm their source.
  Check for signs of alterations.

- Go to the highest level (attic/crawl space/roof).
  - Check for structural problems/confirm their source.
  - Check for water problems/confirm their source.
  - Check for signs of alterations.

- Proceed room by room through the building.
  - Sketch or photograph interior elevations, floors, and ceilings.
  - Use the floor plan to indicate problems/locations.
  - Identify problems and their sources.
  - Identify historic features.
  - Identify alterations.
  - Note all findings for each space.
  - Trace continuity of defects.

historic preservation technologies for nondestructive testing and sample analysis may be integral to the completion of inspections and assessments.

FIELD MONITORING

Most commonly, field monitoring entails a question that cannot be answered immediately during the course of the various inspections. The two specific areas of concern generally relate to structural integrity and moisture migration. Field monitoring involves a variety of recording instruments either installed temporarily or maintained on-site indefinitely to gain a long-term perspective on the conditions and whether they fluctuate or are static. Long-term monitoring is often used for moisture monitoring, where a computer data logger and other sensing devices can track a number of parameters throughout the building simultaneously. For structural assessment, several devices are available to track the displacement activity of structural members. These devices range from simple strain gauges placed across cracks to indicate movement to regular and periodic displacement data recovery using remote sensing devices.

An important aspect of the investigation process is the stabilization and protection of assemblies and materials based on concerns for the imminent loss of historic fabric or for life safety. In some instances, installing a durable protective covering over the surfaces may be all that is needed to ensure that later activities will not damage sensitive materials. However, concerns for structural stability and safe access during construction may require measures to immobilize structural and envelope assemblies to prevent collapse (see Figure 3-3).

TEMPORARY PROTECTION AND STABILIZATION

Figure 3-3 The stone facades (left and center) of this building are being stabilized to protect them from damage during construction.

INSPECTION METHODS

The process of completing a rehabilitation, preservation, or restoration project successfully begins with a thorough assessment of the building and its site. The early investigations may raise further questions that lead to the use of historic preservation technologies in assessment techniques such as nondestructive testing and various sampling methods. These concerns should be noted and addressed in the reconnaissance survey and verified by later, more detailed inspections as needed.

Nondestructive Testing

Nondestructive testing is an in situ process that can be performed by persons in a wide variety of disciplines. As the name implies, the goal of nondestructive testing is to make an assessment of conditions without significantly harming the material being tested. ASTM and ANSI, among other organizations, have developed standards for each material testing process. The primary uses for nondestructive testing are to investigate subsurface conditions that are not visible to the naked eye and to test various aspects of structural integrity, capacity, and movement. The essential feature of these tests is that they produce minimal or no loss or removal of historic material and its subsequent replacement or reconstruction when compared to the alternative of dismantling or demolishing portions of historic structures to permit firsthand visual inspection of concealed assemblies or removal of historic materials to a lab for destructive testing.

The primary testing strategies employ a variety of techniques ranging from simple sounding of materials (e.g., rapping, tapping, chaining) to the use of sophisticated instruments that use light waves, electromagnetic radiation, x-rays, electrical conductivity, fiber optics, and mechanical pressure to assess conditions. A representative sample of these building pathology historic preservation technologies is the following:

- *Infrared Thermography*: used to document radiant energy as a thermal indicator. The infrared photograph of energy emanating from a surface is used to identify moisture, heat leakage, and dissimilar materials within the assembly.

- *Impact Echo*: used to apply a mechanical pulse to a structure and then measure reflected waves (echoes). The method detects cracks, flaws, delaminations, cold joints, and voids in concrete materials.

- *Impulse Radar or Ground-Penetrating Radar*: used to assess a wide variety of materials to a depth of several feet. This method detects changes in material density within a wall (e.g., metals in concrete) or within the ground (e.g., a buried foundation wall).

- *X-Ray Fluoroscopy*: used to investigate subsurface conditions within an assembly that is accessible from both sides. This method enables an x-ray analysis to reveal varying materials, connection types, and abnormalities in the assembly and, in certain instances, can be used to detect lead.

- *Tomographic Imaging*: used to develop three-dimensional models of an assembly using x-rays. The method, based on medical computerized axial tomography (CAT) scan technology, allows an evaluation of different materials or an assessment of physical displacement of materials in a structural assembly.

- *Ultrasonic Pulse Velocity Measurement*: used to assess continuity and physical integrity in concrete, masonry, and metal structures.

- *Closed-Circuit Television (Fiber Optic Probes)*: used to examine voids inside a wall assembly or to explore chimneys and cavities that are inaccessible for direct visual inspection. The method allows for a nominal loss of material where the probe is inserted but provides rapid visual assessment.

■ *Borescope*: used as a visual inspection device for linear voids, such as boring holes. This method allows inspection in a fashion similar to that of fiber optic probes.

■ *Pulse Induction Metal Detection*: used to detect the presence of metal beneath the surface, such as nails, rebar, or other metal objects.

■ *Moisture Meter*: used to calculate the presence of water in a material. The method allows for a preliminary assessment of moisture migration (e.g., rising damp).

■ *Bore Hole Dilatometer*: used to determine the modulus of deformability. This method is often used to define the difference between masonry layers (wythes) on the surfaces of a wall and those behind it.

■ *Probe Penetrometer*: used to provide an estimate of surface hardness for a masonry unit or the mortar joint around it. This method provides a means to assess structural stability.

■ *Rebound Hammer*: used to determine strength based on surface hardness. This is an in situ test used on concrete structural elements.

When nondestructive testing is inadequate in answering specific questions, selected elements may need to be tested in situ until failure or samples may need to be taken to a laboratory for testing.

Sampling and Laboratory Analysis

Sampling involves obtaining small portions of the building fabric that are then tested for a variety of purposes. Testing may be done in the field (on-site) or at a laboratory (off-site). For example, sampling is frequently done to identify paint layers and their respective colors; detect the presence of hazards, such as asbestos; assess the components and structural characteristics of mortar; or perform core testing of specific structural samples. Similar to the nondestructive testing standards, ASTM and ANSI standard testing procedures must be followed to attain certifiably accurate results. Refer to Appendix B for further information on these standards and procedures.

Samples must be labeled and catalogued to permit correlation of findings to the location from which they were obtained (see Figure 3-4). Care must be taken to select inconspicuous locations for samples that still provide the needed information. Lastly, the removed samples must be properly handled so as not to diminish their physical integrity or overall composition.

CONDITION ASSESSMENTS

These individual and specific inspections have three goals. The first is to assess the existing conditions and identify the causes of deterioration. The second is to provide a recommended course of remediation. The third is to assess the continued or future performance of the item being inspected. For example, early structural systems may be oversized by modern standards and early mechanical and electrical systems may not comply with modern codes or meet users' needs.

Inspection methods involve gathering data that are reviewed as a whole and then sorted into prioritized problem areas. As the reconnaissance survey, off-site research, and other specific inspections draw to a close, findings from the various investigations

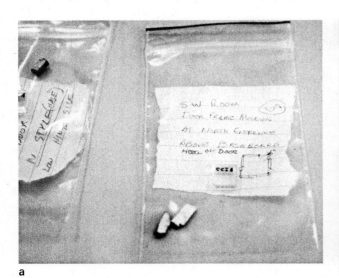
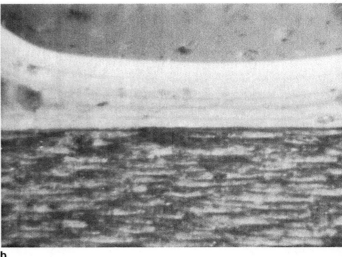

a   b

Figure 3-4   A paint sample shown in its collection sample bag (a) and through a microscope (b). The microscopic view reveals several layers of paint (top). The bottom portion shows the wood to which the paint had been applied.

can be categorized based on the priority level for a recommended action that forms the basis for the final condition assessment report. The condition assessment report, in turn, provides the information needed for selecting the appropriate treatment. A suggested range of priorities (Martin 1988) includes the following information:

- *Immediate*: Conditions that pose a considerable threat to life safety or the loss of historic features and require short-term stabilization.

- *Urgent*: Conditions that pose a threat to life safety or building integrity that may become an immediate threat if not stabilized or corrected within the next year.

- *Important*: Work required to correct water ingress, dry rot, security problems, and so on within two years.

- *Necessary*: Work required to correct problems related to structural movement and accelerating decay within five years.

- *Desirable*: Work desired by the client, including improvements in use and aesthetic changes..

- *Monitor*: No action needed at present, but monitor for future condition changes.

This system addresses the building's needs before concerns related to aesthetics. It relates directly to correcting sources of problems for the sake of retaining the building for its long-term benefits to the environment rather than ignoring problems for short-term gain.

HISTORIC STRUCTURES REPORTS

The investigation generates a wide range of information describing the history and condition of the building and the remediation work needed to complete the project. This information is drawn from the various members of the investigation team and is compiled into a historic structures report (HSR). The HSR starts with an executive summary of the administrative data, overall findings, and recommendations. The main body of the HSR consists of two sections. The first part provides a narrative that traces the chronological evolution, the physical description, the existing condition, and a determination of the historic significance of the building. The second part discusses the historic preservation objectives and recommendations for the treatment selection. The HSR concludes with a bibliography and appendices containing the historical documentation and technical data used to develop the HSR.

In certain instances, adding a postproject supplement documenting all work performed is recommended. This supplement provides a record of what actually was constructed and the specific processes that may have varied from the original recommendations of the HSR. Some organizations, government agencies, and incentive programs (e.g., tax credit programs) consider this to be the final part of the report and not just a supplement. Physical evidence discovered during the construction work or new documentary evidence discovered after completion of the report should be recorded as an appendix to the report. An important goal of the report process is to maintain the report so that it can be used for future reference in all potential design and construction activities occurring within the building (Slaton 2004).

References and Suggested Readings

Anthony, Ronald W. 2004. Condition assessment of timber resistance drilling and digital radioscopy. *APT Bulletin*, 35(4): 21–26.

Association for Preservation Technology International. 1997. *Historic concrete investigation and repair*. Chicago: Association for Preservation Technology International.

———. 1997. *Metals in historic buildings: Investigation and rehabilitation*. Chicago: Association for Preservation Technology International.

——— 2004. *Workshop in preservation engineering: Diagnostics: Nondestructive testing for the evaluation of historic structures*. Galveston, TX: Association for Preservation Technology International.

Becker, Norman. 2001. *Inspecting a house*. 2nd ed. Newtown, CT: Taunton Press.

Brand, Stewart. 1994. *How buildings learn: What happens after they're built*. New York: Viking Press.

Cauldwell, Rex. 2001. *Inspecting a house*. Newtown, CT: Taunton Press.

Erdley, Jeffrey L. and Thomas A. Schwartz, eds. 2004. *Building facade maintenance, repair, and inspection STP 1444*. West Conshohocken, PA: ASTM International.

Foulks, William G., ed. 1997. *Historic building facades: The manual for maintenance and rehabilitation*. New York: John Wiley & Sons.

Friedman, Donald. 2000. *The investigation of buildings: A guide for architects, engineers, and owners*. New York: W. W. Norton & Company.

Harris, Samuel Y. 2001. *Building pathology: Deterioration, diagnostics, and intervention*. New York: John Wiley & Sons.

Jandl, H. Ward. 1988. *Rehabilitating interiors in historic buildings*. Preservation Brief No. 18. Washington, DC: United States Department of the Interior.

Laefer, Debra, Ashley Evans, and Jon Frazier. 2006. Forensic-investigation methodology for structures experiencing settlement. *APT Bulletin*, 37(2/3): 23–31.

Martin, Wilson G. 1988. *Building inspections: A model methodology (master's thesis)*. University of York, Institute of Advanced Architectural Studies.

Mason, James A. 2005. Nondestructive analysis and strengthening design for a historic church tower in Kentucky. *APT Bulletin*, 36(1): 45–52.

McDonald, Roxanna. 2003. *Introduction to natural and man-made disasters and their effects on buildings*. Burlington, MA: Architectural Press.

McDonald, Travis C., Jr. 1994. *Understanding old buildings: The process of architectural investigation*. Preservation Brief No. 35. Washington, DC: United States Department of the Interior.

Nelson, Lee H. 1988. *Architectural character: Identifying the visual aspects of historic buildings as an aid to preserving their character*. Preservation Brief No. 17. Washington, DC: United States Department of the Interior.

Park, Sharon C. 1993. *Mothballing historic buildings*. Preservation Brief No. 31. Washington, DC: United States Department of the Interior.

Paul Revere Memorial Association. 2006. The Paul Revere Home. http://www.paulreverehouse.org/about/paulreverehouse.shtml

Piper, James E. 2004. *Handbook of facility assessment*. Lilburn, GA: Fairmont Press.

Poore, Patricia, ed. 1992. *The old house journal guide to restoration*. New York: Dutton.

Rabun, J. Stanley. 2000. *Structural analysis of historic buildings: Restoration, preservation and adaptive reuse for architects and engineers*. New York: John Wiley & Sons.

Rose, William B. 2005. *Water in buildings: An architect's guide to moisture and mold*. New York: John Wiley & Sons.

Rosina, Elisabetta and Jonathan Spodek. 2003. Using infrared thermography to detect moisture in historic masonry: A case study in Indiana. *APT Bulletin*, 34(1): 11–16.

Silman, Robert. 1996. Applications of non-destructive evaluation techniques in historic buildings. *APT Bulletin*, 27(1/2): 69–73.

Slaton, Deborah. 2004. *The preparation and use of historic structures reports*. Preservation Brief No. 43. Washington, DC: United States Department of the Interior.

Swanke, Hayden, Connell Architects. 2000. *Historic preservation: Project planning and estimating*. Kingston, MA: R. S. Means.

PART II

Building Materials

CHAPTER 4 Wood

This chapter is an overview of wood used as a structural element in a building and describes the types of wood, the historical practices of using wood as structural framing, sources of decay, and methods of remediation. The uses of wood in other nonstructural applications will be discussed in later chapters. Wood decay and remediation will be discussed here to reduce redundancies in other chapters.

Trees are categorized as hardwoods or softwoods. Hardwoods are deciduous (leaf-bearing), while softwoods are coniferous (needle-bearing). Deciduous trees include oak, maple, hickory, walnut, and chestnut. Coniferous trees include pine, fir, hemlock, and cedar. The proper species identification is important in determining suitable replacements for deteriorated or missing sections of existing wood. The use of many species of wood in early North American construction was influenced by their durability and accessibility. Unless a suitable means of transportation (e.g., water and, later, railroads) was readily available to bring in more durable wood from a distance, locally available species were used.

In the original settlements of the eastern seaboard, especially in the Northeast, the size, height, and girth of the trees in the pristine forests amazed the first settlers who came from Europe, where timber was extremely limited and expensive. As the population outgrew the settlement hearths of the eastern seaboard and moved westward, they encountered numerous varieties of species not found earlier. Today, the more durable woods are recognized to be chestnut, eastern white cedar, red cedar, cypress, eastern hemlock, black locust, white oak, burr oak, white pine, redwood, and walnut. Less durable species include aspen, white birch, black oak, red oak, sugar maple, poplar, and white spruce. Even so, the wood cut from these early trees is viewed as much stronger than wood taken from more recent trees of the same species due to growth management practices of the past three centuries (Weaver 1997).

Structural Composition

Cutting a cross section through a tree trunk (see Figure 4-1) will reveal three major components of the tree structure: the bark, the bast, and the cambium. The bark is the outermost layer that protects the inner components. The bast is the portion of the tree immediately below the bark where sap flows. The remaining portion is the cambium, which is composed of sapwood and heartwood. Sapwood is the outermost portion of the cambium and relies on the migrating sap for nourishment. The innermost portion of the cambium is the heartwood, which no longer relies on sap for growth but provides structural support for the tree (Edlin 1969). The heartwood is identifiable by its darker color than the sapwood. Besides water, wood is primarily composed of cellulose, hemicellulose, and lignin, with nominal amounts of minerals and other miscellaneous materials that vary throughout the year.

Seasonal changes in sunlight and moisture cause growth in the cambium layer located just beneath the bark to vary. This growth is faster in the spring and early summer and then slows during the late summer. As a result, the cell structure creates alternating darker and lighter concentric growth rings to form outward from the heartwood of the tree. In nontropical tree species, these alternating bands are called "annual rings." In tropical species there may be more than one cycle of growth in any given year, so the term "annual ring" is not used.

These growth rings provide different types of information that are important to historic preservation. First, when the tree is cut into smaller pieces, such as timbers, studs, and other boards, these tree rings form what is known as the "grain" of the wood. The grain provides clues to the strength of the wood, particularly when comparing old-growth wood (i.e., trees originally found in the original pristine forests) to modern samples of the same species. Old-growth wood typically grew more slowly

Figure 4-1 This section of tree trunk shows the major components of the trunk. The cambium is composed of sapwood (lighter color) just inside the bark and heartwood (darker color), which in this case has decayed substantially.

and is stronger than newer wood that may either be from a succeeding generation of the same forest or cultivated using contemporary growth management strategies. Second, the orientation of the end grain is important when selecting wood for its ability to resist warping and cupping. Quarter-sawn wood that is cut along a radius from the center is most stable, while back-sawn wood that is cut tangentially to the tree rings is least stable and can cup and warp more readily. A third aspect of the grain is the pattern or figure found along the exposed face of a piece of wood. Highly figured woods have long been used as ornamentation.

The rings also aid in determining the date when a house was built. This process is called "dendrochronology" and consists of removing a cored sample from framing members in the house in question. The growth rings are then measured and compared to an existing baseline reference sample for the specific region where the house is located. By aligning the growth rings with the known sample, the period when the wood was cut can be relatively determined and the date of construction estimated. The method is limited by the lack of widespread baseline samples in North America. Dendrochronology has been used to determine approximate construction dates for buildings in the Northeast, mid-Atlantic region, and Southwest that were built before written records were kept. One difficulty in using this approach is that it cannot determine specifically whether the framing member that was sampled was in its original location or has been reused.

STRUCTURAL CONSTRUCTION METHODS

The earliest European immigrants to North America in the late sixteenth and early seventeenth centuries adapted the traditional construction practices of their native countries to local climatic conditions and material resources. Examples of such vernacular architecture abound throughout North America, where these early practices have left a rich heritage of variations and adaptations to local conditions and the then contemporary aesthetic mindset. The surviving buildings of the earliest English, French, Dutch, and Spanish colonial settlements display the strongest concentration of these vernacular practices.

By the early seventeenth century, many of the primeval forests of Europe had been depleted and access to timber for housing was controlled by the government. A significant attraction of settling in the New World was access to pristine natural resources. While specific construction practices varied largely by country of origin, early settlers from England, France, and Holland constructed buildings using timber framing techniques that had been passed down for generations. Later immigrants from Germany and Scandinavia brought their own practices based on their heritages of log-based and timber framing construction.

Upon arrival in North America, early settlers constructed temporary shelters that often emulated structures used by local natives such as wigwams. These structures used bark, animal hides, branches, and thatch. They then set about constructing more permanent housing. In the Northeast, early buildings were constructed using timber framing traditions while in the South a combination of traditional timber framing and brick construction was used. The Dutch and Spanish used stone, timber, and masonry in their construction practices. The French used timber, log, and masonry in their settlements in Canada and along the Mississippi River valley. Climatic adaptations and locally available materials for construction largely account for variations in construction techniques used throughout the eastern seaboard.

After the Revolutionary War, settlement expanded west of the Appalachian Mountains. The Monroe Purchase in 1803, the 1846 Treaty of Washington, and the defeat of Mexico in 1848 formed the essential borders of what eventually became the contiguous forty-eight United States. With the expanding railroad networks and technological innovations of the Industrial Revolution of the early nineteenth century, American building traditions emerged. While vernacular practices and their adaptations frequently reappeared in subsequent frontier settlements and often persisted into the late nineteenth and early twentieth centuries in certain areas, log construction and timber framing for nonindustrial uses gave way to stick framing practices that continue to this day. In the twentieth century, wood framing systems were complemented by the introduction in 1905 of plywood (Jester 1995b), which was used in structural and sheathing applications and glued laminated timbers in 1934 (McNall and Fischetti 1995) that could be used instead of long spans of timber. Regional variations still occur based on those original vernacular traditions, but modern building codes and the now widespread marketing of new materials and techniques tend to homogenize new construction practices.

Timber Framing

Building construction started with tree harvesting, which usually occurred in the winter to eliminate problems with sap and to allow easier transportation over frozen ground. Before the increasing availability of sawmills by the mid-seventeenth century, trees were felled and limbed by hand using a felling axe. The logs were then cut to a length predetermined by the intended use. If a timber was to be made, the bark was removed and a scoring axe was used to roughly hew the log into a square cross section. The timber was then hewn to its approximate final dimensions using a broad axe and dressed to its final dimensions using an adze. If planks were to be made, the process was similar, except that the planks were cut using a whip saw or, as they became available, frame saws. In the earliest method, logs were moved to and secured over a pit, and the sawyer would stand atop the debarked log while the pitman held the other end of the saw in a pit beneath the log. Planks were cut lengthwise from the log or timber. This was a time-consuming and laborious process and resulted in the predominant use of timber framing in early colonial housing. The timbers or planks could then be transported on a sledge to the building site (Wilbur 1992). In many instances, the frame was assembled using unseasoned wood since the need for housing often could not be postponed to allow the wood to season. As the wood dried inside the finished construction, checks or splits occurred, which led to the misperception that checks were an indicator of a decay mechanism when in fact they were present soon after or from nearly the beginning of a dwelling's habitation.

Timber framing methods (see Figure 4-2) were brought to North America by English, French, and Dutch colonists. The principles of these methods relied upon the ability to select appropriate timber and then to create the fastening system that held it together. Builders were adept at making connections to form a "bent," a structural frame that consisted of vertical posts, horizontal beams (also known as "summer beams" and "girts"), and diagonal bracing.

Although methods varied by settlement and by the vernacular traditions of the builder, the typical timber-framed building was started by constructing the stone foundation, along the top of which sill plates were placed. The various members were assembled into individual bents that were then stood vertically and fit into cavities mortised into the sill plate. These bents were then connected to one another by girts, beams, and other horizontal members, such as floor joists. A common hall and parlor house would contain at least four bents: one at each end and a pair on

Figure 4-2 Timber framing traditions, such as the mortise-and-tenon construction shown here, were brought to North America by the English, French, and Dutch.

either face of the chimney stack. Bents were then capped by a connecting top plate running along the entire length of the building. Next, the rafters were joined to the top plates and stabilized using a series of horizontal purlins, tie beams, and collar ties. The frames were often reinforced with diagonal bracing at the corners. Numerous variations on these practices correspond to the vernacular styles of the period (e.g., "square houses" and other regional adaptations).

The spaces in the exterior walls between timber-frame bents were often filled with wattle and daub. Wattle and daub was a combination of the thin branches of the tree that were then covered with a clay-like soil that formed the traditional exterior finish that is common to half-timbered houses in England. An alternative method was to use unfired (dried in the sun) brick covered with a stucco finish on the exposed surfaces. Due to the extremely harsh conditions in New England, it was soon found to be necessary to cover these materials with clapboards on the exterior to protect them from moisture penetration and subsequent decay. A German construction practice used fired bricks. In all instances, this infill material was not designed to be load-bearing and was simply an enclosure method.

Mortise-and-Tenon Joints

Joining of these members was accomplished through a series of joints created specifically to interlock members together. To do this, the ends of the members were cut away to form either a mortised hole or a protruding tenon that would mate together. Once mated, an auger was used to drill through the two members where they connected. Since hand-wrought nails were expensive or perhaps not locally available, a treenail (also known as a "trunnel") usually consisting of a wooden peg was inserted into each hole. The pegs were left protruding from the connection or were cut off flush with the surface (see Figure 4-3).

The greatest skills involved the ability to cut both the mortise and the matching tenon. The simplest connections were made where two members met at right

Figure 4-3 Mortise-and-tenon joints were commonly used before the widespread distribution of cut and wire nails. Notice the pegs holding the mortise and tenons together at each joint.

angles. However, complexity increased significantly when angle bracing or roof framing connections were included at the same joint. The joiner also had to know how to compensate in the cutting of the joint for anticipated shrinkage related to the use of unseasoned wood.

The growing availability of water-powered sawmills in the eighteenth century and steam-powered sawmills in the nineteenth century dramatically increased lumber and timber production. Ironically, as these means of producing timber became more readily available, they enabled the introduction of light-framing systems that hastened the decline of widespread mortise-and-tenon timber framing practices. Concurrently, the introduction of new fastening technologies using threaded bolts, gusset plates, and steel reinforcement systems in the nineteenth century reduced the demand for the craft of creating mortise-and-tenon connections. As demand dwindled and the masters of this craft aged and died, this trade nearly became extinct. Despite these factors, the traditional mortise-and-tenon form of timber framing lasted well into the late nineteenth and early twentieth centuries, particularly for small buildings, barns, and outbuildings. A renewed interest in this craft emerged in the late twentieth century.

Plates, Bolts, and Other Hybrids

In 1798 in the United States, a lathe for mechanically threading screws was patented, which opened the way for the eventual mass production of threaded fasteners such as screws, nuts, and bolts. By the mid-nineteenth century, machine-threaded nuts and bolts were being introduced as an alternative fastening system in timber frame and cast iron construction. As emerging steel technologies improved in the second half of the nineteenth century, use of threaded rods, through-bolts, and gusset plate fastening systems began to replace the mortise-and-tenon joint. With the industrialization that occurred at the turn of the twentieth century, more factories and mills were built using this hybrid framing system. This transition can also be seen as the precursor to steel framing systems introduced at the end of the nineteenth century that have evolved into the multitude of variations used today.

Contrary to the romantic American myths generated in the mid-nineteenth century surrounding the use of log cabins by the earliest settlers along the Atlantic seaboard, log cabins and buildings were actually introduced to North America in the seventeenth century by the French in Canada and the Mississippi River valley and by the Swedish in the New Sweden settlement in Delaware. The traditional log cabin as we perceive it today did not appear until the early eighteenth century, when it was introduced by Germans, Scots-Irish, and Scandinavians who first settled along the mid-Atlantic coast and then moved inland. Due to the lack of timber craftsmen along the expanding frontier, the log cabin found its way into the American landscape. While now viewed as a primitive building form, log buildings provided a relatively quick means of creating a durable shelter.

Log Construction

A log building can be categorized as either a cabin or a house. Log cabins were more primitive and rougher in construction materials, especially in their use of round logs, and were often used as temporary construction. Log houses were more finely crafted and often had logs hewn with flat surfaces on the inside and outside. The interior often included a wooden floor and finished walls made of plaster or wood. The exterior may have been left as exposed logs or covered in stucco. Shingles and wood siding were common additions as they became available. While some log buildings were originally meant as temporary structures, a great many remained as the primary home for the first generation of settlers in an area. Later prosperity could mean the move to a larger, more contemporary house or the start of a series of additions to the original structure. Due to their construction method and size, log buildings could be moved to make way for new construction or disassembled and reassembled elsewhere. Many original log buildings were later used as outbuildings.

Log building construction required skill in cutting the end notches to secure the structure at each corner. The traditional craft of the builders' original homeland can be often identified by the types of notching used in a log cabin (see Figure 4-4). As notching skills began to fade, many early-twentieth-century log cabins were constructed simply using a half-lap connection at each end of the log through which spikes were driven to secure them to the log below.

Early log cabins and houses were limited by how readily large logs could be moved and lifted into place. After the logs were notched, they were moved into place and erected to form walls. Some vernacular construction practices included hewing the bottom of the upper log to rest on the log beneath, while others simply alternated narrower and thicker ends. In most instances, gaps or joints between two

Figure 4-4   Notching techniques: (a) steeple notch (Danish), (b) dovetail notch (Finnish), (c) square notch (Anglo-American), and (d) saddle notch (Swedish).

logs were daubed and/or chinked. The process of daubing filled in the gaps with clay or mud mixed with moss or straw as a binder. The process of chinking that is especially common in the western United States involves attaching poles split into lengthwise quarters to the space between the logs. In some instances, mortar may have been used where available, particularly when Scots-Irish construction practices may have been used. Log buildings were typically one or two stories tall and had either a shed roof or a pitched roof. The roof was framed using rafters and purlins shaped from smaller logs taken from thinner portions of the felled tree. Typically, a heat source of some type (e.g., a fireplace or stove) was located at the end wall or in the center of the cabin, as determined by local climate conditions.

In describing log cabins, rooms are denoted as "pens." In general, most buildings were constructed as a single pen or a series of connected pens. A single pen would be at most 14 to 16 feet in its longest dimension, although these dimensions could be shorter or longer as space needs and construction resources warranted. Many but not all log buildings were constructed as a single pen. Other variations included a double pen with an interior door connecting the two pens; a saddlebag where two attached pens had doors leading only to the outside; a dogtrot with both pens opening onto a "breezeway" created by a continuous roof overhead; and multiple variations of these combinations.

As the frontier closed at the end of the nineteenth century, an ever-increasing demand for recreational activities fostered the introduction of "rustic" architecture in the form of stylized log buildings with much larger proportions and greater complexity than had generally existed in earlier vernacular settlements of the nineteenth century. Although many of these appeared to be built only of logs, some also had integrated other framing and connection technologies, such as steel rods or girders encased in a log shell. Homesteading acts and railroad expansion in the West in the late nineteenth and early twentieth centuries further exposed these remote regions to the demand for access and accommodation by the emerging recreational tourism industry. The railroad industry built numerous large, multifloor, multiroom hotels in and adjoining the national parks. In addition, the incipient recreation industry built lodges catering to easterners desiring a rustic experience. This resulted in the emergence of vacation homes in the mid-twentieth century that accelerated after World War II.

Stick Framing

After the War of 1812, the United States continued expanding westward. The rapid rate of growth around such gateway cities as Chicago revealed that timber framing and log cabin technologies were inadequate for providing quick housing. The growing population led to a critical shortage of housing. Due to the lack of skilled builders familiar with the earlier framing systems, stick framing became popular.

Adoption of stick framing, so called because of the stick-like nature of the framing members (e.g., 2" × 4", 2" × 6") when compared to log or timber framing (e.g., typically a minimum thickness of 5"), was encouraged by three emerging industrial technologies of this era. First, the circular saw was introduced in 1814 in the United States, replacing the earlier vertical frame sawing and manual pit sawing methods. The use of this cutting technology can be identified by the saw marks left on the cut member. Pit sawing and frame sawing are indicated by parallel vertical marks, while circular saws leave concentric circular marks. Second, the increasing use of steam- and water-powered technologies to provide cutting power as well as to enhance railroad and riverboat transportation increased the capacity of saw mills to mass-produce precut lumber. Third, in 1790, industrialization led to processes that could cut flat iron bar stock along alternating diagonals into nails known as "cut nails." Following the development of wire nails in France in 1835, American manufacturers devised wire nail fabricating machines and wire nails became widely used in America by 1855. The invention of new nail types did not immediately eliminate those previously in use, but by 1890 the use of wire nails finally outnumbered that of cut nails. Cut nails remained in use for specific purposes, such as securing subflooring or attaching wood to cement or plaster, until they were supplanted in the 1950s by other fasteners, such as cement-coated nails. With these technologies in place, saw mill owners were soon able to cut and transport lumber to locations where as few as two men could quickly build a house using simple carpentry methods.

Balloon Framing

Although stick framing methods can be found in earlier buildings in Australia and England (Weaver 1997), the first stick framing methods used in the United States developed around Chicago in 1832 and became known as "balloon framing." In this process, the vertical members or "studs" were erected on top of the foundation in parallel with one another in each wall (see Figure 4-5). The floors were then constructed by attaching the horizontal framing members or "joists" to the studs and laying the subflooring across these joists. In the early forms, the lack of diagonal bracing caused the construction to wrack out of shape and in some instances fail. Later variations included inserting diagonal framing members in the walls or attaching exterior sheathing on the diagonal and laying subflooring diagonally across the floor joists. Another drawback was recognized after numerous fires had occurred. The framing system formed a continuous vertical cavity that ran from the basement to the attic space and formed a chimney-like chase between the studs. When a fire started in the basement, the smoke and flames quickly spread to the attic and roof. Later variants of balloon framing incorporated horizontal fire stops in the cavities using leftover scrap lumber to close off the cavity.

Platform Framing

Nearly seventy years after balloon framing became popular, a second framing method now known as "platform framing" emerged. Unlike balloon framing, platform framing built one floor (or platform) at a time, erected framing up to the next floor, and then constructed the next floor (see Figure 4-6). This process could be repeated up to several stories in height.

By the early twentieth century, constructing small wood-framed buildings had become a national industry serving local, regional, and national markets. Unlike the acceptance of balloon framing, which occurred quickly in fast-growing areas but only moderately in slowly growing areas or where traditional timber framing building practices persisted into the late nineteenth century, platform framing was well accepted nationwide. At this point, platform framing had become the method of choice for a segment of the construction industry that sold "kits" of housing plans and materials through catalogs and shelter magazines, which came into vogue after the passage of the Rural Free Delivery Act of 1896 opened mail delivery to rural areas. National catalog-based retailers, such as Sears & Roebuck, along with a number of kit home manufacturers, had largely adopted the platform framing method

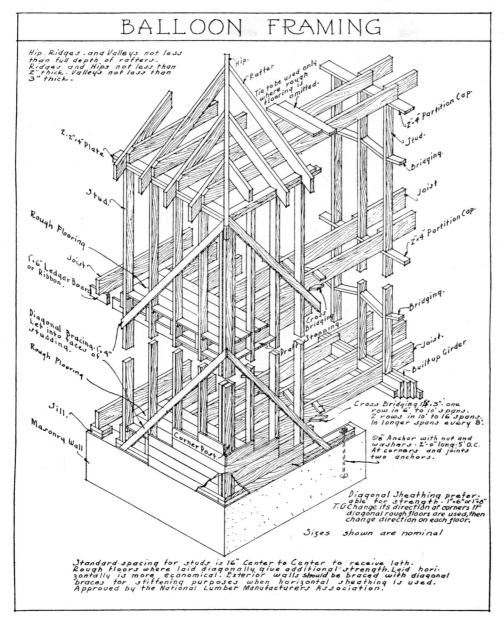

BALLOON FRAMING

Hip, Ridges, and Valleys not less than full depth of rafters. Ridges and Hips not less than 2" thick. Valleys not less than 3" thick.

Hip.

Rafter

Tie to be used only where rough flooring omitted.

2"×4" Partition Cap.

Stud.

2·2"·4" Plate

Bridging.

Joist.

Stud.

2"×4" Partition Cap.

Rough Flooring

Bridging.

Joist.

1"×6" Ledger Board or Ribbon.

Cross Bridging

Draft Stopping

Diagonal Bracing · 1"×4" Let into Faces of Studding.

Joist.

Rough Flooring

Built up Girder

Sill.

Cross Bridging 1"×3" · one row in 6' to 10' spans. 2 rows in 10' to 16' spans. In longer spans every 8'.

Masonry Wall

Corner Post

5/8" Anchor with nut and washers · 2'·0" long · 5' O.C. At corners and joints two anchors.

Diagonal Sheathing preferable for strength. 1"×6" or 1"×8" T.G. Change its direction at corners If diagonal rough floors are used, then change direction on each floor.

Sizes shown are nominal

Standard spacing for studs is 16" Center to Center to receive lath. Rough floors where laid diagonally give additional strength. Laid horizontally is more economical. Exterior walls should be braced with diagonal braces for stiffening purposes when horizontal sheathing is used. Approved by the National Lumber Manufacturers Association.

Figure 4-5 Balloon framing (from Ramsey/Sleeper, *Traditional Details for Building Restoration, Renovation, and Rehabilitation,* ©1991 by John Wiley & Sons, Inc. Reprinted with permission of John Wiley & Sons, Inc.).

to facilitate transportation, handling, and construction of the shorter individual members and parts to form a building. By World War I this method was firmly entrenched in the American market, and after World War II it became and remains the predominant form of wood framing construction.

Construction practices evolved through time based on available craft skills and local resources. Certain practices used in the eighteenth and early nineteenth centuries showed up only as regional variations, while others were widely adopted nationally, particularly in the late nineteenth century as communications systems,

Derivatives and Variations

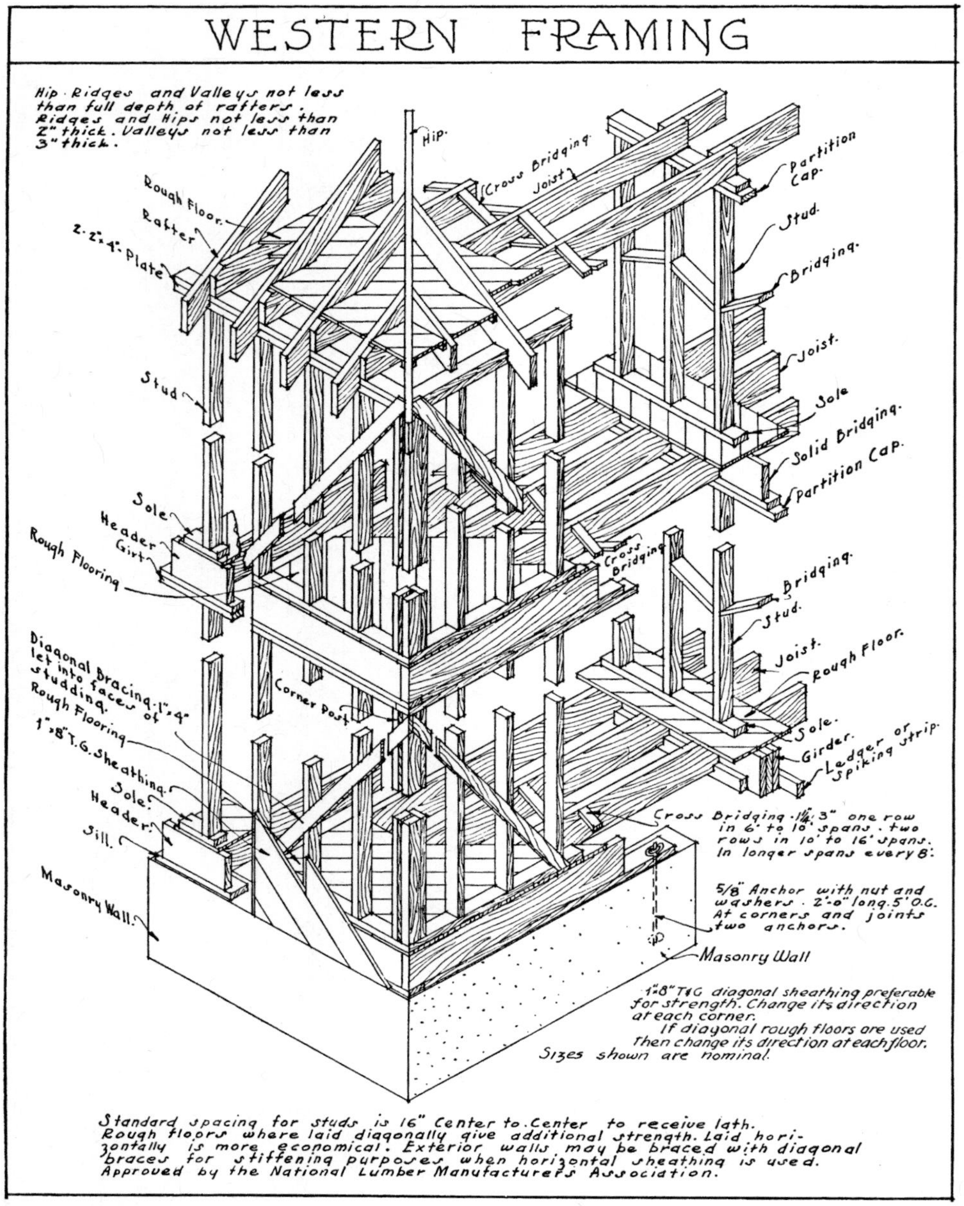

WESTERN FRAMING

Hip Ridges and Valleys not less than full depth of rafters. Ridges and Hips not less than 2" thick. Valleys not less than 3" thick.

Cross Bridging.

(Cross Bridging) Joist.

Hip.

Partition Cap.

Stud.

Bridging.

Rough Floor.

Rafter.

2. 2". 4". Plate.

Joist.

Sole.

Stud.

Solid Bridging.

Partition Cap.

Sole.

Header Girt.

Cross Bridging.

Rough Flooring

Bridging.

Stud.

Diagonal bracing. 1". 4". let into faces of studding.

Rough Flooring

Joist.

Corner Post.

Rough Floor.

Sole.

1". 8". T.G. Sheathing.

Girder.

Sole.

Ledger or Spiking Strip.

Header.

Cross Bridging. 1½. 3" one row in 6' to 10' spans. two rows in 10' to 16' spans. In longer spans every 8'.

Sill.

5/8" Anchor with nut and washers. 2'-0" long. 5' O.C. At corners and joints two anchors.

Masonry Wall.

Masonry Wall

1"-8" T&G diagonal sheathing preferable for strength. Change its direction at each corner.
If diagonal rough floors are used then change its direction at each floor.
Sizes shown are nominal.

Standard spacing for studs is 16" Center to Center to receive lath.
Rough floors where laid diagonally give additional strength. Laid horizontally is more economical. Exterior walls may be braced with diagonal braces for stiffening purposes when horizontal sheathing is used.
Approved by the National Lumber Manufacturers Association.

Figure 4-6 Platform framing (from Ramsey/Sleeper, *Traditional Details for Building Restoration, Renovation, and Rehabilitation*, ©1991 by John Wiley & Sons, Inc. Reprinted with permission of John Wiley & Sons, Inc.).

transportation networks, and product distribution channels were developed and promoted a more nationally based market. A few of the more significant practices are discussed below.

Braced Framing

In New England, as craftsmen from the early colonial period grew older, methods for timber framing began to evolve as their apprentices matured in the trade. Variations in framing systems appeared that no longer could strictly be termed "timber framing" in the traditional sense. While some features such as posts and girts

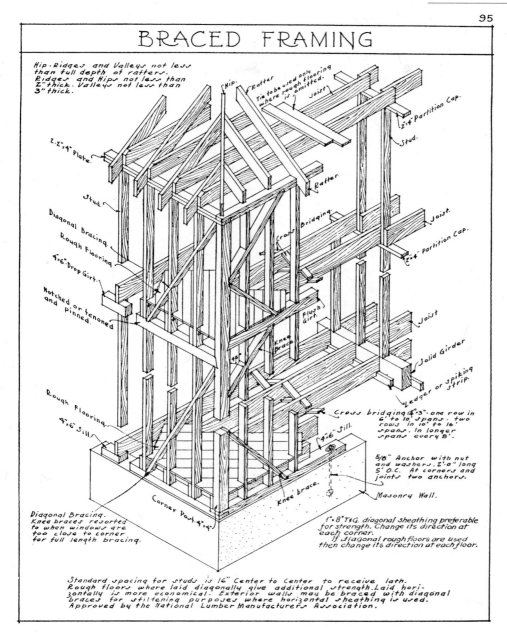

95

BRACED FRAMING

Hip-Ridges and Valleys not less than full depth of rafters. Ridges and Hips not less than 2" thick. Valleys not less than 3" thick.

Tie to be used only where rough flooring is omitted.

Hip.

Rafter

Joist

2.4 Partition Cap.

Stud.

2·2·4 Plate.

Stud.

Rafter

Diagonal Bracing.

Rough Flooring.

4·6 Drop Girt.

Cross Bridging.

Joist.

2·4 Partition Cap.

Notched or tenoned and pinned.

Joist

Flush Girt.

Knee Brace.

45°

Solid Girder

Rough Flooring.

4·6 Sill.

Ledger or spiking strip.

Cross bridging 1"·3" one row in 6' to 10' spans. two rows in 10' to 16' spans. In longer spans every 8'.

4·6 Sill.

5/8" Anchor with nut and washers. 2'-0" long 5' O.C. At corners and joints two anchors.

Masonry Wall.

Corner Post 4·4.

Knee brace.

Diagonal Bracing. Knee braces resorted to when windows are too close to corner for full length bracing.

1"·8" T&G. diagonal sheathing preferable for strength. Change its direction at each corner. If diagonal rough floors are used then change its direction at each floor.

Standard spacing for studs is 16" Center to Center to receive lath. Rough floors where laid diagonally give additional strength. Laid horizontally is more economical. Exterior walls may be braced with diagonal braces for stiffening purposes where horizontal sheathing is used. Approved by the National Lumber Manufacturers Association.

Figure 4-7 Braced frame (from Ramsey/Sleeper, *Traditional Details for Building Restoration, Renovation, and Rehabilitation*, ©1991 by John Wiley & Sons, Inc. Reprinted with permission of John Wiley & Sons, Inc.).

remained, certain framing members were reduced in size and bracing was added at the cap and sole of the corner posts (see Figure 4-7). This method combined the heavy structural system of the timber frame method with lighter framing techniques that approach those of the stick framing methods (Garvin 2001).

Plank Frame

Another variation in framing is the "plank frame." In this method, planks were typically stood on end and fitted directly into the sill plate using a trench cut along the middle of the sill plate or a series of mortise-and-tenon joints.

Plank on Plank

Examples of the plank-on-plank framing method can be found nationwide. In this method, planks were stacked in alternating courses atop one another with an offset at each corner of the building. Spaces between the planks were filled with planks of similar thickness. The planks were secured at the corners with a spike driven through the plank and into the plank below. In buildings with living spaces, the planks were offset in alternating layers to provide a key for plastering the interior wall. Exterior surfaces were left exposed or covered with stucco or other cladding. While not commonly used for houses, the plank-on-plank method can most readily be seen in outbuildings on farmsteads throughout the Midwest, the intermountain West, and parts of Canada.

Laminated Wood

The decreasing quality of lumber after the consumption of the original old-growth forests fostered the use of laminated wood products. In the late nineteenth century, as choice wood resources became more costly, manufacturers sought ways to make use of shorter pieces of clear lumber that occurred between knots. Two separate systems evolved: plywood and glued laminated timbers. Experimentation with plywood and other laminating processes led to the introduction of plywood plated girders and glued laminated timbers that formed the beginnings of the modern engineered lumber industry.

Plywood consists of veneers of wood that are glued together with the grain of alternating layers oriented at a 90 degree angle to one another. Originally, plywood was developed for furniture making in 1865; later, it was used in door panels and other building products at the turn of the twentieth century and in structural applications after World War I. The use of plywood in making airplanes during World War I led to rapid development of adhesives that were stronger and more water resistant. In the 1930s, plywood began to be used both as a structural element and as sheathing. Stressed-skin panels and structural insulated panels were introduced but did not gain significant recognition until the energy crises of the 1970s and the resurgence of timber-framed construction in the late twentieth century. Other advances in the mid-twentieth century included the use of plywood for plating and built-up girders that were composed of plywood sheets fastened together to form custom structural shapes. Ironically, the use of plywood as a gusset plate in wooden truss systems declined as the development of aluminum gusset plates forming a quicker-fastening system during World War II permanently reduced the use of plywood in this capacity. Experimentation after World War II led to several variations of plywood. Structural flake board composed of layers of strips of wood, was introduced in 1958, and oriented strand board (OSB), patented in 1965, became commercially available in the 1970s.

Glued, laminated timbers were introduced to the United States in 1934, although they had been originally developed in Europe in the late nineteenth century. They were composed of short lengths of 1- or 2-inch cross sections of wood taken from between imperfections, such as knots, in a larger piece of wood. These pieces were glued together, with the grain of the individual pieces aligned in parallel with each other. The ends were connected using a simple butt joint prior to World War II, but the stronger hooked, scarf, or fingered joints were used thereafter. The "glulam" could be shaped into free-standing arches and frames to enclose large spaces, such as gymnasiums, church sanctuaries, and other public assembly spaces requiring clear spans longer than those that could be achieved by conventional timber framing. As adhesive technology improved, these glulams were also used outdoors as well.

During an inspection, evidence that the structural integrity of the building has been compromised may be revealed. This loss of structural integrity or stability is revealed as:

IDENTIFYING DECAY MECHANISMS

- Floors and ceilings that are not level or slope in one or more directions
- Walls or framing that are out of plumb
- Structural members that sag over openings
- Doors and windows that bind when opened or closed
- Walls, roofs, or other surfaces that bulge or sag
- Structural members that are missing, rotted, cracked, split, or broken
- Structural connections that have failed
- Floors that creak or jiggle when stepped on
- Any rigid surface that moves under light to moderate pressure

Wooden buildings move in response to many causes: wind, thermal expansion and contraction, overloading, and loss of structural integrity. When investigating wooden building, be alert to the way this movement has affected the overall stability of the building. An apparently sound structure may conceal any number of failures beneath the surface. Be sensitive to sounds created when entering the building. Creaking floors usually are a good indicator of structural movement and possible loss of structural integrity. Tapping or rapping on surfaces and structural members can also help assess the soundness of the material.

The first step is to assess the original and existing construction systems to determine if sound practices were used and then look beyond the construction itself for other issues that may be contributing to the lack of integrity. These include problems created by moisture, fungi, insects and other invasive pests, fire, and other causes.

Construction Issues

The largest construction issue in wood construction is proper sizing of wood members to withstand a load. Timber framing was adapted over time to act as a system that transferred the load to its various members, which made it more durable. In lighter framing methods, a common problem is an undersized member supporting too long a span. Due to the pliable nature of wood this member may not fail immediately, but it will sag over a long period of time.

As noted above, balloon framing underwent a long series of adaptations until finally being largely replaced by platform framing. Many of these adaptations were to account for inadequate resistance to wracking (movement out of the square and plumb condition) caused by a lack of bracing (see Figure 4-8). Simple modifications to balloon framing construction practices essentially included incorporating a triangular reinforcement system. This reinforcement was done by adding diagonal bracing, installing cross bracing between the floor joists, and installing collar ties and tie beams between two opposing roof rafters.

The last construction issue is missing structural members. The time when the building was constructed and the framing techniques used may possibly be the cause

Figure 4-8 Lack of corner bracing has contributed to the wracking movement of this framing system.

of missing structural members. As just noted, the building may have been built originally without specific bracing to resist wracking. Alternatively, structural members may be missing due to alterations in the building. For instance, an original wall could have been (re)moved to enlarge smaller spaces or a new addition necessitated the inclusion of new doorways or openings. This change was common in timber frame and log cabin construction if the building was subsequently modified or moved.

Moisture

Under modern construction standards, wood is either kiln-dried or seasoned in the open air. By its very nature, wood draws water into itself through capillary action. Excessive moisture allows a number of decay mechanisms in wood to start. The most predominant forms of decay are rot due to fungi, stains due to mold, and damage caused by insects and other animals attracted by moisture in the wood. In the course of drying, moisture that nurtures the growth of the fungi and mold spores is removed. Within the building, many factors contribute to wet conditions. The major sources are leaky roofs, leaky plumbing, inadequate ventilation, inadequate vapor barrier control, poor drainage, and groundwater fluctuations. All of these introduce moisture or prevent its removal or evaporation. When the moisture content rises above 30 percent, it reaches a point that allows fungi and mold spores to begin growing.

The lack of control of water in its various phases—vapor, liquid, or ice—can often be the greatest source of problems in wooden buildings. The sources of moisture can sometimes be traced to leaking roofs and faulty plumbing. These repairs can be readily made when the point of the leakage is identified. Less obvious are moisture sources caused by inadequate ventilation and inadequate vapor barrier control. Warm air can hold more water vapor than cold air. The water vapor in air is relatively benign, and the amount of water in the air is often noted by the term "relative humidity" (RH). The RH indicates the percentage of moisture content relative to the overall moisture content that the air can contain at a given temperature. As air cools, the RH increases and approaches 100 percent. At 100 percent RH, the air can hold no more moisture at the temperature referred to as the "dew point temperature." If the air is warmed, the RH will decrease unless more moisture is also introduced along with the heat. However, when the air is cooled, the vapor condenses. This condensation can occur on any surface that is cooler than the dew point temperature for the given humidity or moisture content of the air. This phenomenon is readily demonstrated by the moisture that forms on the outside of a chilled beverage container when it is removed from a cool environment and placed in a warm, humid environment. When sufficient surface condensation has occurred, water flows down to the surface that the container is on and forms water rings (hence the invention of coasters for wood furniture).

Similarly, in the building, condensation can occur on obvious surfaces, such as glass, but it can also form on wall surfaces or penetrate into wall cavities and condense within them. Moisture, as water vapor or condensation, can be drawn into an untreated surface or any open joints, fissures, or cracks in an impermeable surface and begin the decay processes described earlier. On exposed surfaces, moisture problems can be observed directly as the presence of standing water or the failure of painted surfaces. Within the wall cavities, this insidious process may go undetected for a much longer period of time. As moisture condenses and is absorbed by the wood or pools at the base of the wall cavity, it attracts fungi, insects, and other vermin and the decay mechanisms they bring with them.

One aspect of older buildings that has caused complications in controlling moisture is that many moisture-producing amenities of modern life did not exist when these buildings were built. Other than kitchens and bathrooms, little moisture was introduced by the occupants themselves. Moisture control focused on maintaining a weather-tight living space, repairing leaky roofs, and controlling seasonal dampness or humidity based on the local climate. Cold, damp air and warm, humid air were to be avoided for both comfort and health reasons. The construction technology of many buildings allowed for the exchange of fresh outdoor air with fouled indoor air via windows, doors, and gaps in construction materials. Historically, buildings were drafty, and the air that infiltrated from outdoors allowed the buildings to "breathe." Until the late twentieth century, when concerns about energy conservation dictated tightening of buildings to eliminate infiltration and drafts, this air exchange was the primary means of flushing out accumulated seasonal moisture.

Fluctuations in moisture content can lead to wood swelling and shrinking. When the construction cannot tolerate dimensional changes caused by shrinkage or swelling, displacement of connections and surface disruptions may contribute to the loss of structural integrity or subsequently admit additional moisture. This displacement can occur in exposed interior timber framing where interior moisture levels drop sufficiently to cause "checking" (rupturing along the grain) of timbers. Exposed exterior logs in log construction can check as they dry in the sun. Checking that occurs where water can collect is a potentially significant source of decay.

Another potentially damaging aspect of moisture relates to how the site and soil react to changes in groundwater and surface runoff from the building and the site. Expansive clay soils can create uplift pressure on foundations and basement floors that can work to deform the structural systems aboveground. This deformation is particularly noticeable when varying subsurface conditions beneath the building cause differential or nonuniform settling and uplifting as moisture fluctuations and freeze-thaw cycles occur. An additional trigger of groundwater fluctuations is the removal of mature trees that would otherwise draw water out of the ground. Once the tree is removed, this groundwater remains and may introduce moisture problems that previously did not exist. Changes in ground surface conditions over time may occur slowly. An example of this is when soil naturally builds up over time due to decaying grasses and leaves that, in turn, shift surface drainage patterns toward the building. This buildup, coupled with any long-term settling of the building, can cause the sole plate of timber framing and the base logs of log construction to become exposed to moisture absorption. Likewise, long-term wetting conditions, such as rainfall that back splashes onto wood elements as it drops from a roof or prolonged exposure to melting snow, invariably have a long-term damaging effect on wood and other porous surfaces. A common construction guideline is that no wood surfaces, especially wood structural members, such as sill plates and beams, should be in direct contact with the ground, particularly in snowy regions. Therefore, the common practice is to eliminate this contact when repairing or replacing the decayed member by installing a new foundation if none exists or removing the soil to reduce the soil contact problem.

Lastly, moisture in combination with multiple freeze-thaw cycles causes a series of expansions and contractions that can degrade the structural integrity of wood.

Fungi

Mold, rot, and mushrooms are all mature forms of fungi that signal the presence of moisture. Fungal spores are common in nature. To flourish, the spores must have a suitable food supply, sufficient moisture for germination and growth, an air supply, and suitable temperatures. When the moisture content exceeds 30 percent and temperatures are above 40°F (and, optimally in the 75–90°F range), spores can germinate. Then, with a continued source of air, the spores grow into their mature form (see Figure 4-9).

Mold growth does not decrease the structural integrity of wood, although it does signal the presence of elevated moisture levels that may invite more harmful fungi that cause rot. Mold usually forms on the surface of wood and can be removed by brushing, sanding, or planing it away. Sapstains, also known as "bluestains" because of their blue color, affect the cell membranes in the sapwood portion of the tree. They have no harmful structural effects and can normally be ignored.

Rot is classified as brown rot, white rot, soft rot, and black rot. While the visual evidence of each type varies, the long-term result is that the fungi attack the cellulose, hemicellulose, and/or lignin that comprise the major structural components of wood, destroying its structural integrity. Brown rot is found in softwoods and consumes the cellulose and hemicellulose while leaving the brown lignin portions of the wood. Brown rot is therefore identified by the way it reduces wood to small cubic or irregular block-shaped pieces of brown wood through checking both along and across the grain. While some fungi are known to cause "dry" rot by providing their own moisture resources, some moisture must be present to activate it. This growth occurs in locations that may only be intermittently damp and causes the

Figure 4-9 Wood exposed to moisture will rot over time.

perception of "dry" rot since inspections may be done when the damp conditions needed for rot to occur are not present. White rot consumes lignin and cellulose while leaving a white, stringy residue. White rot also retains the dimensions of the wood member being attacked until loss of structural integrity causes it to collapse. White rot is primarily found in hardwoods but has been known to attack softwoods. Soft rot attacks hardwoods that are subjected to long-term wet conditions. By consuming the cellulose at the cellular level in the middle of the cell wall, this rot often severely reduces the structural integrity while producing only a slight brown or gray discoloration of the wood. Black rot, also known as "cellar rot" or "wet rot," thrives in moisture contents of 40–50 percent. The attacked wood turns blackish and appears to have been burnt. This type of rot is most commonly found in marine environments and other wet locations.

Insects and other invasive pests use wood as a food source and a nesting place. Attracted by moisture and by the wood itself, insects will bore into the wood, thereby weakening its structural integrity and perhaps admitting additional moisture. The most common problem insects are termites, beetles, carpenter ants, carpenter bees, and wood wasps. Other invasive pests include shipworms and crustaceans that are collectively known as "marine borers."

Insects and Other Invasive Pests

Termites attack both hardwoods and softwoods. Except for Alaska, various species of termites exist across the United States and southern Canada. Termites eat wood for food. Wood in contact with the ground provides direct access for termites; however, subterranean termites may also construct light brownish tube structures along the face of foundations to access wood aboveground. Subterranean termites excavate burrows along the concealed interior grain of the wood. This subsurface activity is concealed from exterior view and is only revealed when structural integrity fails. Less common nonsubterranean termites access wood directly without constructing tubes and eat the wood in all directions. Their presence is further indicated by tiny fecal pellets located outside an entry hole in the wood.

A variety of beetles exist in different regions of North America. Each has its own preferred wood type and species, but together they can inflict significant damage on any form of wood product. Most attack wood after it is cut and during its seasoning process, but a few attack the dry wood in finished products, particularly unfinished hardwood. One of the most common species is the powder post beetle, whose presence is indicated by tiny holes (e.g., 1/16–1/8 inch in diameter) in the wood surface with a light, powdery residue, referred to as "frass," below or around them. They lay their eggs in checks and open pores. When the eggs hatch, the resulting larvae burrow through the wood in search of food sources located in the sapwood. Over time, the cycle of (re)infestation and burrowing can cause significant damage to the wood.

Unlike termites, carpenter ants use the attacked wood for nesting rather than eating. Their presence can be identified by piles of sawdust outside openings to their burrows. Their nesting galleries run parallel to the grain of the wood and are free of frass and sawdust residue. The ants themselves may also be seen outside the burrows.

Carpenter bees attack softwoods like fir, pine, cedar, cypress, and redwood. Their presence is identified by $1/2$-inch-diameter holes in wood trim near eaves and gables of homes. These bees bore a chamber several inches long running in parallel with the grain. They can also be identified by coarse sawdust and frass that may leave unsightly stains beneath the entry hole. Wood wasps attack dead trees and can confuse bark-covered logs with dead trees.

The marine borer known as a "shipworm" is actually a bivalve mollusk that bores $1/2$-inch and larger tunnels in untreated wood structures such as pier pilings and wooden vessels commonly found in harbors and other marine locations. Small crustaceans known as "gribbles" or "sea mites" attack wood but make much smaller tunnels. In either case, they can completely destabilize the integrity of the item that they have infested.

Fire or Extreme Heat

Wood is a combustible material that is vulnerable to fire damage. While earlier timber frame construction was oversized, and could withstand some exposure to fire and still maintain structural integrity, stick frame construction used in balloon and platform framing methods is readily consumed and can quickly lose structural integrity. The conclusion that fires in early balloon-framed buildings were accelerated due to the inclusion of the natural flue created between the studs led to the incorporation of fire stops placed at intervals to eliminate this pathway. The later platform framing eliminated this problem.

Wood exposed to extreme heat may char but not combust. The heat causes a checking action that reduces the cross-sectional area and the strength of the wood.

Ultraviolet energy from the sun can break down lignin in exposed untreated or clear finished wood surfaces. As lignin leaches out of the wood, the integrity of the wood suffers. If the wood is left untreated, gaps created by this process may admit moisture, which can lead to rot or attract insects, further reducing the integrity of the wood. This damage usually appears in three forms. First is a color shift to silver gray, which is primarily an aesthetic concern. Second, in log buildings facades that face south or west and receive long periods of sunlight throughout the year, damage can accelerate moisture penetration as fibers rupture or as logs check. Third, on thinner members, such as exposed shingles or exposed plywood, the damage from ultraviolet light exposure can be more pronounced. Since they are thinner and have exposed end grains, the ultraviolet decay will break down the surface and allow additional moisture to collect in inclement weather. This exposure may begin a series of wetting-drying or freeze-thaw cycles that further reduce the integrity of the material.

Sunlight

When a wood surface is in frequent intermittent contact with harder materials, the wood can be worn away through abrasion. While not common for enclosed structural members, abrasion is common for interior surfaces, such as floors, woodwork, and casework, due to contact with foot traffic or other abrasive sources. Likewise, exterior wood surfaces can be damaged by swaying tree branches and cables or wind-blown sand. Wear is more readily apparent in softwoods than hardwoods. Normal wear and tear can be expected in high-traffic areas, such as floors and stair treads. However, when doors and windows get stuck in or abrade their frames, this may be a sign of structural settlement and should be investigated further.

Abrasion

Animals initiate decay mechanisms that may not damage the building's structural integrity directly. However, the damage they cause to surfaces may help initiate the decay mechanisms described above. Insects in wood attract predators, such as woodpeckers, that destroy the wood while accessing the insects. While not immediately a structural concern unless an extremely large number of holes have been made, holes can be an unsightly nuisance. Mice and other vermin can nest in cavities in walls by chewing and enlarging openings in the walls. House pets can scratch or chew woodwork and severely damage surface finishes. Farm animals can lick, gnaw, or kick wood members, causing moderate to severe damage to enclosures and structural posts.

Animals

If the growth of the tree from which logs or timbers were obtained was not uniform due to wind exposure, solar orientation, and location (e.g., a leaning tree), unrelieved stresses in the wood fibers may have caused the wood to twist, warp, check, or splinter as they seasoned in place. If unseasoned or inadequately seasoned wood was used in the original construction, these forces may cause deflections to occur as the wood in the structure ages. These decay conditions may be further accelerated if seasonal exposure to moisture, temperature, and sunlight varies significantly and frequently. Walls rack and joints open in response to shrinkage, causing any number of the other decay mechanisms described above to begin.

Other Problems

Any potential remediation method offers a variety of alternatives to consider. For example, in repairing a building whose sill plates, timbers, or logs have rotted due to inadequate separation from the ground (e.g., inadequate or no foundation; settlement or soil buildup), steps need to be taken to eliminate the direct soil contact by

REMEDIATION METHODS

either installing a new foundation or removing the built-up soil. Both of these solutions introduce a change in the physical appearance that may be unavoidable when trying to preserve the building in the long run. In any specific remediation, consider several alternatives that mitigate the visual changes to or losses of historic building fabric before commencing work.

In all treatment strategy selections, the overriding guideline is to remove or mitigate the source of the decay mechanism even when no other work is deemed necessary. Treating the symptom without removing the source is of little benefit since the problems will simply recur over time. Four basic strategies are available for the remediation of wood. The first is to leave the item as is. The remaining strategies can be described in increasing order of intervention as stabilization and conservation; infill or repair; and replication and replacement. The historic value of the component in question will often dictate the strategy selected. If the component is in suitable condition, then the best course of action may be to leave it as is and protect it during the remainder of the project. Stabilization and conservation refers to addressing the physical integrity of the item. It generally involves eliminating decay mechanisms such as insects and fungi and then using consolidants and epoxies to reinforce the remaining historic material. Infill repairs incorporate some form of reinforcing materials into the historic material that goes beyond simple epoxy injections. Lastly, if the material is beyond all hope of repair, then replacement or replication may be the only appropriate course of action (see Figure 4-10).

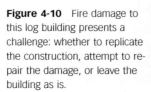

Figure 4-10 Fire damage to this log building presents a challenge: whether to replicate the construction, attempt to repair the damage, or leave the building as is.

As the sources of decay and their impact become identified, the damaged material may possibly be preserved, with only nominal repairs being made. In this manner, materials can be stabilized and conserved without significant further loss of historic fabric. Three of the more common stabilization and conservation strategies are described below.

Stabilization and Conservation

Moisture Abatement and Control

Controlling moisture migration is often the single most important remediation strategy in wooden buildings. While leaking roofs and faulty plumbing can often be repaired or replaced, moisture damage within a wall cavity is often much more time-consuming and expensive to correct. In certain instances, the moisture may have had a long-term preexisting source, such as the vapor generated in kitchens or bathrooms. In other cases, relocating or introducing these uses without attention to the moisture-generating sources may have introduced moisture to rooms that originally had none. Physical evidence, such as mold spores, mildew, rot, damp surfaces, peeling paint finishes, and efflorescing salt crystals in plaster or brick, may provide the clues needed to confirm the presence and extent of the problem. However, correcting the problem becomes more difficult as the historic significance of the building increases.

In most cases, identifying problems within the wall cavity without removing some enclosing materials is virtually impossible. The exception would be inspections using nondestructive testing techniques that may indicate the presence of moisture but possibly not the full extent of the damage. To obtain a visual assessment, the removals range from small, discrete sections that allow visual insertion of fiber optic inspection devices to larger sections for the direct physical inspection of certain areas. Similarly, the actual repair dictated by the damage may require that large sections be removed to enable repair. From the inspection process, it will be possible to develop the appropriate treatment strategies to correct the problem. If health or life safety is involved, removal of sufficient historic fabric will be needed.

Otherwise, when no health or life safety issues are present, it may be possible to simply monitor the conditions over time to see if the moisture is in fact introducing further decay. Traditionally, atmospheric temperature and moisture conditions are monitored using hygrothermographs that record the humidity and temperature on a continuous paper graph. Modern digital data loggers are also available to perform the same function and provide the means to download the data to a computer for analysis. The options for moisture monitoring within a material or wall assembly are vast and sometimes complex. Simple hand-held moisture meters can readily indicate moisture content in wood as well as devices that likewise can measure the moisture content of the air. These meters work well for initial inspections or to assess periodic conditions. When longer-term monitoring is needed and labor resources are at a premium, installing moisture monitoring data-logging devices that store data that can be downloaded to a computer for analysis may be economically feasible. At the high end of this strategy is the use of a centralized computer directly connected to these remote data loggers, which eliminates the labor costs of accessing data from each individual monitoring station. These loggers introduce a visual and physical component because, unless wireless technology is feasible for the building being monitored, all the data-logging components must be physically hard-wired back to the central computer. Careful planning is needed to conceal these cables from view for both aesthetic and security reasons. The central computer itself may also be on-site or may access the data via a dial-up modem connection to a local data monitoring device controller. Some suggested locations for the cabling and terminal devices include heat, ventilation, and air-conditioning (HVAC) chases, closets, and other nonpublic portions of the building.

The two energy crises of the 1970s resulted in a greater concern for energy consumption. Numerous construction methods and technologies were developed to minimize energy usage in new construction. Unfortunately, many of these technologies were cost-prohibitive to retrofit into older and historic buildings. This was true for both the monetary expenses and the loss of historic building fabric. Two of these processes directly impacted the control of moisture in a building. First, attempts to reduce infiltration and the unconditioned air that entered a building severely diminished a building's capacity to flush out moisture naturally. Second, new insulation methods were employed to introduce thermal resistance into cavities within the walls. As more heat was retained inside the building, exterior surfaces remained cooler and became subjected to both more condensation and freeze-thaw cycles that accelerated their decay. These modifications have created recognition of the importance of vapor barriers and the difficulty of achieving true vapor control in many historic construction assemblies. An assortment of moisture control products, including vapor barrier paints, have been used, with varying levels of success, and all introduce changes to the historic appearance or removal of historic fabric. Rather than removing substantial portions of room finishes to install a true continuous vapor barrier, an ongoing debate exists about whether "air barriers" can be installed at the points where airborne vapor can be prevented from entering wall cavities. So, in essence, attempts to control energy use must be tempered with the recognition that deviations from the long-existing thermal cycles within a wall assembly may introduce moisture control issues that did not previously exist or may require extensive removal of historic fabric.

While a moisture source in a specific space may be controlled by installing ventilation and exhaust equipment, controlling environmental or naturally occurring climatic sources is complicated. Where the moisture migration is a result of a climate- or site-related source, explore possible methods that can be used to alleviate the moisture, including the following:

- Evaluate whether changes in landscaping contribute to water infiltration into the building and, where possible, modify the landscape to provide drainage away from the building.

- Verify soil conditions and consider some form of subsurface drainage system (e.g., French drains) to mitigate or eliminate moisture at the perimeter of the building.

- Verify that moisture is not entering through the soil of unfinished basements or crawl spaces and that sufficient ventilation is present to remove the water vapor.

- Ensure that all penetrations into attics or unheated spaces from occupied spaces are sealed to prevent moisture migration, especially from bathrooms and kitchens.

- Install a vapor barrier on the warm side (toward the occupied space) of the insulation layer if the attic is to be (re)insulated.

- Verify that attic and crawl space ventilation openings are sufficiently sized and unobstructed to permit air to move freely through them.

■ Consider the use of dehumidifiers, where appropriate, to reduce moisture due to seasonal humidity and dampness when other means of mitigation are not feasible.

Pesticides and Preservatives

There are two approaches to controlling pest damage to wood. The first is through the use of pesticides, including both insecticides and fungicides. The second is through the use of preservatives. Pesticides are the primary means of effectively killing invasive pests. While pesticides are potentially hazardous to humans and pets, when used with appropriate preparation and caution they can be effective. As with all chemical agents, care must be taken to verify where, when, and how a pesticide product may be used. Local requirements for the transportation and disposal of the by-products of the process must be met. Confirm these requirements with the local building and health departments before starting the process. Contractors trained and licensed to carry out the fumigation process should be hired to mitigate any potential liability issues from the use of the particular pesticides in question.

Several common methods of pest removal are available. First are dry or wetted powders that the pest contacts directly by walking through, breathing, or ingesting them. Second are pesticides used as fumigants in a gaseous spray that penetrate the recesses of the nest or infestation site directly. In some instances, this process requires that the entire space or building be sealed or "tented" (e.g., enclosed in an airtight fabric) to ensure that the lethal dosage concentration is achieved. In this case, the tenting remains in place for a predetermined number of hours or days as needs warrant. This method has a significant drawback: it may be necessary to repeat the fumigation process several times to eliminate the pests entirely, and this introduces lethal toxins into the environment.

Conversely, rather than attacking the pests after they have become a problem, a common practice was and still is to coat or apply preservatives to the wood before or during construction. A variety of inorganic and organic pesticides have been developed and used in various building applications. Each new pesticide was more toxic than the previous ones. In the mid-nineteenth century, a process was developed that forced creosote and other chemicals into the wood with the intention of preventing a problem in the first place. In a later method that has been used since the 1940s, toxic chemicals, such as chromium copper Arsenate (CCA), are applied to and forced into the wood surface using heat and pressure. These "pressure-treated" wood products have been pretreated with preservatives to withstand attacks from fungi and other wood-consuming pests.

Since the beginning of the environmental movement in the 1960s, these products have come under scrutiny because of their carcinogenic toxicity in humans. The toxicity of a pesticide is rated by the term "LD50," which is an abbreviation for the lethal dosage needed to kill 50 percent of the pest population. This is the number of grams of poison per kilogram weight of the pest. Lower LD50 dosage levels indicate higher toxicity and can be quantified as follows:

■ LD50 <50 g/kg: extremely toxic

■ LD50 <50 g/kg to <500 g/kg: moderately toxic

■ LD50 >500 g/kg: slightly toxic

■ LD50 >5000 g/kg: nontoxic.

For example, the inorganic insecticide sodium fluoride has an LD50 rating of 0.18 g/kg and is classified as extremely toxic. The growing awareness of the health issues related to certain pesticides in the United States and Canada has led to the banning or restricted use of pentachlorophenol (PCP) and dioxins in residential applications. In the case of pressure-treated lumber, the use of CCA for residential uses in the United States has been banned since 2003 and products using alkaline copper quat (ACQ) and copper boron azole (CBA) are now available instead (Litchfield 2005). Another commonly used inorganic pesticide is sodium borate, whose LD50 rating of 5.66 g/kg qualifies it as extremely toxic to most of the fungi and insects that damage wood; however, the benefit of borate-based preservatives "is that they are effective against brown rot and white rot fungi and most wood-destroying insects while being relatively safe for both users and the environment" (Sheetz and Fisher 1993, 3).

Consolidants and Epoxies

The chemical formulations for consolidants and epoxies have varied through time. Consolidants are low-viscosity liquids that penetrate small fissures and cracks in the surface of the material being conserved, where they cure or harden to stabilize the decayed material. They are sometimes used before an epoxy treatment that is intended to fill the larger voids created by the decay mechanism (e.g., checking, rot, loss of material). Epoxies are filler materials, typically made from resins or other plastic materials, which can be used to fill gaps where original materials have been lost. Some epoxies can be sanded, planed, or carved to match the profile contours of the original material. When left to cure, both of these treatments bind the existing decayed material into a more stable condition. For the most part, due to their penetrating nature, these treatments are not reversible but are used when there is sufficient historic significance of the material being treated or when replacement or replication of the element is not available or desired (see Figure 4-11). Superficial surface damage and limited internal damage to a structural element can be repaired in this manner without destroying the remaining healthy wood. Consolidants and epoxies are appropriate where the exposed surfaces will be repainted or concealed by subsequent rehabilitation or repair processes.

Infill or Repairs

When the damage is extensive, damaged materials can be repaired or replaced. In repairing a structural member, it is crucial to consider both the visual and structural aspects of the replacement method. Certain methods can conceal the repair from public view, while others are visually intrusive. The selection of the appropriate method must balance the level of acceptable visual intrusion with the need to consider health and life safety. The selection must also consider the economic feasibility of completing a possible repair.

Structural Reinforcement

When damage is severe, consolidants and epoxies alone may not be sufficient for proper repairs. Damage at the ends or along the length of the structural member may have created more stresses that cause connections to fail or the member to sag or move. In this situation, where the damage has not totally destroyed the member, the member may be reinforced at the location of the damage without having to replace the entire member.

a   b

Figure 4-11 A penetrating consolidant and an epoxy were applied to this window sash, which was then sanded smooth, primed, repainted, and reglazed. Shown here is the sash before (a) and after (b) treatment.

Along with any number of or combination of materials, such as clamps, angle iron, gusset plates, or brackets, two common methods use epoxies in combination with other materials to complete the needed repair. The first method is the wood and epoxy reinforcement (WER) system developed in the 1970s in Canada; the second is the BETA system developed in the Netherlands. Both methods have inspired many variations in the combination of epoxies and reinforcement materials in attempts to retain historic materials. As noted by Weaver (1997, 43) in his seminal work *Conserving Buildings*, "Neither system should be used indiscriminately particularly because the epoxies are, for practical purposes, irreversible." Both methods use epoxies in combination with fiberglass, plastic, or metal to make rods that are then inserted into channels and holes that have been cut or bored into structurally weakened members. After insertion, additional epoxy is poured into the void surrounding the rod and left to cure. The epoxy may also be used in a mold in situ to replicate missing segments of structural members. In this approach, the newly molded member is held in place using fiberglass rods and the adhesive bond of the epoxy where it contacts the remaining original structural member.

Dutchman/Splicing

Beyond structural reinforcement of materials, another method is to remove only the most severely damaged portions of the wood member and then infill the cleared void space with a patch cut to match the opening created by the removal of the decayed wood. This method, known as a "dutchman," can be used to provide a continuous surface on the exterior exposed face of the wood member. The dutchman can be used to conceal an epoxy repair or any other internal structural reinforcement treatment. The dutchman can also be planed, carved, or milled to provide a continuous surface on an ornamental surface as well. The dutchman may be attached by either nailing or screwing it into place. If aesthetics require that the means of attachment be concealed, then it may be attached using water-resistant synthetic resin adhesives that are compatible with the materials and water/moisture conditions or using a combination of predrilled holes and an epoxy binder as described in the preceding section. A second method of infill repairs consists of removing a damaged section of a wood member and splicing in a new piece of wood. The connection is secured by either nailing, bolting, screwing, or in some instances using adhesives to bind the new wood with the old (see Figure 4-12).

Repairing the deteriorated crowns (ends) of a log building also involves a splicing technique. The deteriorated crown is removed and a new crown is spliced into the former location. For this application, the connection consists of drilling three

Figure 4-12 These deteriorated rafter ends were reinforced by splicing in new pieces. After the repair, the work will be concealed when the porch ceiling is replaced.

matching sets of aligned holes into the remaining existing log, matching the crown, and then securing the crown in place using fiberglass rods and epoxy (Goodall 1989).

An alternative to removing the weakened member is leaving the damaged wood in place and "sistering" one or more new pieces of wood alongside it. In either case, the spliced or sistered reinforcement member is then fastened to a healthy section of the original member using through-bolts, epoxies, or adhesives where appropriate.

Dutchman repairs can be done in a variety of locations, particularly if the surface is to be concealed by paint. Splicing and sistering are generally done where the work will not be in public view, as they alter the visual appearance of the structural members. In either case, the repair may be painted a specific color or left unfinished to contrast the repair with the original construction so that future investigators will understand that it is not the original construction.

Replication and Replacement

In a rehabilitation project, replication and replacement of failed or missing items is done when the cost of physically correcting the problem far exceeds the historic value of the item. Replacing historic fabric is sometimes necessary. In the remediation techniques described above, a key strategy is to minimize visual awareness of the repairs. The *Standards*, however, do allow replication of missing elements based on physical evidence. They also allow new replacement construction if it is sufficiently differentiated from the original. Therefore, it is possible to replicate or even replace missing features. The replacement materials must have comparable strength characteristics. The replacement wood must be adequately seasoned to minimize shrinkage once it is installed. Similarly, if wood is to be left unpainted or clear finished, then the grain and color of the wood must be matched using the same tree species. However, as much as possible, the building must remain safe to occupy and the structural repair methods must be approved by an accredited professional. Removal of historic fabric should not be taken lightly, but when it is unavoidable, care should be taken to maintain the visual integrity of the space or exterior appearance by concealing the new structural elements or, if they must be revealed, by differentiating them from the remaining original fabric. Even in concealed areas, it is suggested that new structural members be differentiated from prior ones by the use of colored paint or other markings to alert future workers to their later addition to the original construction.

A key precaution is necesary when removing any structural member. The remaining structure must be stabilized to eliminate movement. By doing so, it is possible to remove rotted sill beams or logs from the very bottom of the structure and still maintain structural integrity. This integrity can also be maintained when major repairs to the foundation and the adjoining wooden structural members are warranted.

References and Suggested Readings

Alf, Glenn Forrest and John R. Bowie. 2002. Converging disciplines during structural and environmental upgrades: The Troxell-Steckel House. *APT Bulletin*, 33(1): 9–12.

Ashurst, John and Nicola Ashurst. 1988. *Practical building conservation, Vol. 5: Wood, glass, and resins*. New York: Halsted Press.

Belle, John, John Ray Hoke, Jr., and Stephen A. Kliment, eds. 1994. *Traditional details for building restoration, renovation and rehabilitation from the 1932–1951 editions of Architectural Graphic Standards*. New York: John Wiley & Sons.

Bensen, Tedd with James Gruber. 1980. *Building the timber frame house: The revival of a forgotten craft*. New York: SImon & Schuster.

Bomberger, Bruce D. 1991. *The preservation and repair of historic log buildings*. Preservation Brief No. 26. Washington, DC: United States Department of the Interior.

Carll, C. G. and T. L. Highley. 1999. Decay of wood and wood-based products above ground in buildings. *Journal of Testing and Evaluation*, 27(2): 150–158.

Caron, Peter. 1988. Jacking techniques for log buildings. *APT Bulletin*, 20(4): 42–54.

Carroll, Orville W. 1973. Mr. Smart's circular saw mill c. 1815. *Bulletin of the Association for Preservation Technology*, 5(1): 58–64.

Condit, Carl W. 1968. *American building: Materials and techniques from the beginning of the colonial settlements to the present*. Chicago: University of Chicago Press.

Cummings, Abbott Lowell. 1979. *The framed houses of Massachusetts Bay, 1625–1725*. Cambridge, MA: Harvard University Press.

Curtis, John O. 1973. The introduction of the circular saw in the early 19th century. *Bulletin of the Association for Preservation Technology*, 5(2): 162–189.

De Mare, Eric, ed. 1951. *New ways of building*. London: Architectural Press.

Edlin, Herbert L. 1969. *What wood is that? A manual of wood identification*. New York: Viking Press.

Fidler, John. 2002. Plastic dreams: Weathering of glass-reinforced plastic facsimiles. *APT Bulletin*, 33(2/3): 5–12.

Garvin, James L. 2001. *A building history of northern New England*. Hanover, NH: University Press of New England.

Gay, Charles Merrick and Harry Parker. 1943. *Materials and methods of architectural construction*. 2nd ed. New York: John Willey & Sons.

Gelertner, Mark. 1999. *A history of American architecture: Buildings in their cultural and technological context*. Hanover, VT: University Press of New England.

Goodall, Harrison. 1989. *Log crown repair and selective replacement using epoxy and fiberglass reinforcing rebars*. Preservation Tech Notes: Exterior Woodwork Number 3. Washington, DC: National Park Service.

—— and Renee Friedman. 1980. *Log structures: Preservation and problem-solving*. Nashville, TN: American Association for State and Local History.

Hoadley, R. Bruce. 1990. *Identifying woods: Accurate results with simple tools*. Newtown, CT: Taunton Press.

——. 2000. *Understanding wood: A craftsmen's guide to wood technology*. Newtown, CT: Taunton Press.

Isham, Norman Morrison. 1928 (reprinted 2007). *Early American houses with a glossary of colonial architectural terms*. Reprint New York: Dover.

Jandl, H. Ward. 1983. *The technology of historic American buildings: Studies of the materials, craft processes, and the mechanization of building construction*. Washington, DC: Association for Preservation Technology International.

Jester, Thomas C., ed. 1995a. *Twentieth century building materials: History and conservation*. New York: McGraw-Hill Book Company.

——. 1995b. Plywood. In Jester 1995a, 132–135.

Kidder, F. E. 1909. *Building construction and superintendence: Part II: Carpenters work*. 8th ed. New York: William Comstock.

Koch, Peter and Norman C. Springate. 1983. Hardwood structural flakeboard: Development of an industry in North America. *Journal of Forestry*, 83(3): 160–161.

Lewandoski, Jan Leo. 1995. Transitional timber framing in Vermont, 1780–1850. *APT Bulletin*, 26(2/3): 42–50.

Litchfield, Michael W. 2005. *Renovation*. 3rd ed. Newtown, CT: Taunton Press.

McNall, Andrew and David C. Fischetti. 1995. Glued laminated timber. In Jester 1995a, 136–141.

McRaven, Charles. 1994. *Building and restoring the hewn log house*. 2nd ed. Cincinnati: Betterway Books.

Merritt, Frederick S., ed. 1958. *Building construction handbook*. New York: McGraw-Hill Book Company.

Nash, Stephen E., ed. 2000. *It's about time: A history of archaeological dating in North America*. Salt Lake City: University of Utah Press.

Nelson, Lee H. 1996. Early wooden truss connections vs. wood shrinkage: From mortise-and-tenon joints to bolted connections. *APT Bulletin*, 27(1/2): 11–23.

Noble, Allen G. 1984. *Wood, brick and stone: The North American settlement landscape, Volume 1: Houses.* Amherst: University of Massachusetts Press.

——. 1984. *Wood, brick and stone: The North American settlement landscape, Volume 2: Barns and farm structures.* Amherst: University of Massachusetts Press.

O'Brien, Tom. 1997. Restoring wood with epoxy. *Fine Homebuilding* (February/ March): 60–65.

Peters, Tom F. 1993. *Building the nineteenth century.* Cambridge, MA: MIT Press.

Peterson. Charles E. 1950. Burning buildings for nails. *The Journal of the Society of Architectural Historians,* 9(3): 23.

——. 1973. Sawdust trail, annals of sawmilling and the lumber trade from Virginia to Hawaii via Maine, Barbados, Sault Ste. Marie, Manchac and Seattle to the year 1860. *Bulletin of the Association for Preservation Technology,* 5(2): 84–153.

—— ed. 1976. *Building early America: Contributions toward the history of a great industry.* Radnor, PA: Chilton Book Company.

Phillips, Maureen K. 1993. "Mechanic geniuses and duckies," a revision of New England's cut nail chronology before 1820. *APT Bulletin,* 25(3/4): 4–16.

Priess, Peter. 1973. Wire nails in North America. *Bulletin of the Association for Preservation Technology,* 5(4): 87–92.

Prudon, Theodore H. M. 1979. In-situ injection of wood preservatives. *Bulletin of the Association for Preservation Technology,* 11(1): 75–80.

Rabun, J. Stanley. 2000. *Structural analysis of historic buildings: Restoration, preservation and adaptive reuse for architects and engineers.* New York: John Wiley & Sons.

Rhude, Andreas Jordahl. 1998. Structural glued laminated timber: History and early development in the United States. *APT Bulletin,* 29(1): 11–17.

Ridout, Brian. 2001. Timber decay in buildings: The conservation approach to treatment. *APT Bulletin,* 32(1): 58–60.

Robson, Patrick. 1999. *Structural repair of traditional buildings.* Dorset, UK: Donhead Publishing.

Rockhill, Dan. 1988. Structural restoration with epoxy resins. *APT Bulletin,* 20(3):29–34.

Rose, William. 2005. *Water in buildings: An architect's guide to moisture and mold.* New York: John Wiley & Sons.

Schweitzer, Robert, and Michael W. R. Davis. 1990. *America's favorite homes: Mail-order catalogues as a guide to popular early 20th century homes.* Detroit: Wayne State University Press.

Sheetz, Ron and Charles Fisher. 1993. *Protecting woodwork against decay using borate preservatives.* Preservation Tech Notes: Exterior Woodwork Number 4. Washington, DC: National Park Service.

Sobon, Jack and Roger Schroeder. 1984. *Timber frame construction: All about post and beam buildings.* Pownal, VT: Gardenway Publishing.

Szabo, T. 1977. Plywood reinforcement for structural wood members with internal defects. *Bulletin of the Association for Preservation Technology,* 9(1): 11–15.

Upton, Dell, ed. 1986. *America's architectural roots: Ethnic groups that built America.* New York: John Wiley & Sons.

Weaver, Martin E. 1997. *Conserving buildings: A guide to techniques and materials.* New York: John Wiley & Sons.

Wells, Jeremy C. 2003. Methods for the preservation of French colonial poteaux en terre (posts-in-ground) houses. *APT Bulletin,* 34(2/3): 55–61.

Weslager, C. A. 1969. *The log cabin in America: From pioneer days to the present.* New Brunswick, NJ: Rutgers University Press.

Wilbur, C. Keith. 1992. *Homebuilding and woodworking in colonial America.* Philadelphia: Chelsea House Publishers.

Williams, Lonnie H. 1996. Borate wood-protection compounds: A review of research and commercial use. *APT Bulletin,* 27(4): 46–51.

CHAPTER 5 Masonry

MATERIAL OVERVIEW

Building products classified as masonry include natural stone, bricks, and terra-cotta. Masonry can be used in both structural and nonstructural ways. This chapter discusses the use of masonry as a structural element as well as the ornamental aspects of structural masonry materials. The general overview of masonry construction is followed by a focused look at the properties and uses of stone and brick in their historic context. Nonstructural uses of masonry flooring and roofing will be discussed in later chapters.

CONSTRUCTION METHOD OVERVIEW

Load-bearing masonry construction has existed for millennia and has consisted of simply stacking individual blocks to form enclosures and load-bearing structures. While masonry units can still be used in load-bearing situations in small buildings, they typically are not considered load-bearing when used in curtain-wall applications of modern skeletal framing systems. Masonry in these applications has become simply an enclosure system where individual terra-cotta and masonry "panels" are connected to the structural frame.

Load-Bearing Masonry

Before the advent of steel and reinforced concrete skeletal framing systems, masonry construction consisted of a load-bearing wall and foundation system with the weight of upper floors supported by the walls of the floors below it. The walls of the lowest floor supported the entire weight of the building and consequently needed to be thicker to provide the needed support.

Consequently, interior spaces were smaller on lower floors, particularly in the basement and ground floor levels. Interior spaces were defined by dividing larger spaces using load-bearing masonry or wood-framed walls, depending on local building customs and the intended use of the spaces. Where continuous foundations were not required, caissons or piers could be constructed to support weight from a column resting on the pier. Window and door openings tended to be smaller when arches and timber framing for lintels were unavailable.

Arches, Vaults, and Domes

The use of masonry for load-bearing construction was significantly enhanced by the arch, vault, and dome. Developed by Roman builders, the arch was developed by the recognition that loads can be transferred through blocks placed in a continuous semicircular or segmented alignment. By constructing temporary supports, known

a   b   c

d   e   f

Figure 5-1 Common arches: (a) jack, (b) segmented, (c) Roman or semicircular, (d) Gothic, (e) elliptical, and (f) horseshoe.

as "false work," below the arch, vault, or dome, the blocks could be aligned to transfer loads down to a pier or column. A "keystone" was inserted at the top, and the false work was then removed. The arch remained in place through gravity and a combination of lateral and vertical loads pressing on the stone (see Figure 5-1).

The principles used to create the arch also resulted in the vault and the dome. A vault was an arch construction extended laterally from the arch to form a barrel. Further development led to groin vaults, where two or more barrel vaults intersected. For a dome, the construction practices for an arch were built up to the keystone in a 360-degree configuration. Once the vault or dome was completed, the area above could be filled with masonry to create the upper floors. Arches and vaults have formed the fundamental approach to a number of construction systems, including those historically used for churches, civic buildings, and mill buildings.

Throughout history, the evolution of construction practices has been motivated by the desire to build cheaper, faster, and easier. Numerous efforts were made in the eighteenth century to recover many of the earlier fabrication methods for concrete construction developed by the Romans and other ancient societies that had been lost over time. With the success of these efforts, the limitation of load-bearing masonry construction became more clearly problematic. The volume and massing required to achieve taller buildings were prohibitive in already congested cities. The growth of cities has long been directly related to the time it took to build new buildings and construct large load-bearing masonry. These limitations were further compounded by the introduction of the elevator and other service systems. The combination of these limitations hastened the development of skeletal structural framing systems that could be constructed much more quickly and offered more space than their masonry counterparts.

Load-Bearing Masonry: The Modern Zenith

The second half of the nineteenth century saw the emergence of skeletal framing systems where building loads were supported by a frame rather than by load-bearing walls. In the United States, this transfer was accomplished using two separate materials: steel or reinforced concrete. Advances in steel production in the United States and the refinement of reinforced concrete systems in Europe and the United States created new ways to construct taller buildings. By the end of the century, the shift away from load-bearing masonry construction led to the demise of masonry as a major load-bearing system in many future buildings (see Figure 5-2).

With the introduction of thin stone veneer at the turn of the twentieth century (Scheffler and Gerns 1995), masonry completed its translation to a non-load-bearing enclosure system. Terra-cotta, cast stone, or thin stone veneer products used with a skeletal framing system was the predominant construction system used on many large buildings in the United States from 1900 through the 1930s. The various finishes on the exposed face of terra-cotta were formulated to match the appearance of stone. Ceramic veneer systems, originally derived from terra-cotta manufacturing technology and ultimately refined by the 1930s, continued in use through the 1950s. Cast stone could be made to match the colors of quarried stone. Thin stone veneer was natural stone cut to a nominal thickness of 2 to 4 inches. Thus terra-cotta, cast stone, and thin stone veneer were seen as economically attractive alternatives to stone.

Masonry in the Twentieth Century

Advances in fabrication methods for aluminum and other architectural metals led to a shift in curtain wall technologies after World War II. Economics and shifting

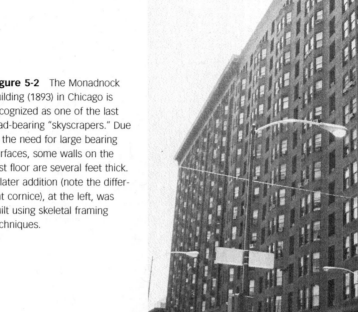

Figure 5-2 The Monadnock building (1893) in Chicago is recognized as one of the last load-bearing "skyscrapers." Due to the need for large bearing surfaces, some walls on the first floor are several feet thick. A later addition (note the different cornice), at the left, was built using skeletal framing techniques.

aesthetic tastes reduced the use of masonry products in favor of a range of architectural metals and glazing products used in panelized curtain walls. Likewise, advances in reinforced concrete systems, cast concrete shell structures, and precast fabrication systems moved construction practices away from stone, brick, and terra-cotta for large public, commercial, and institutional buildings.

In retrospect, masonry started with simple hand-held units or cast materials and evolved to reach a substantial size at the peak of its load-bearing usage in the nineteenth century. The introduction of structural framing systems in the late nineteenth century brought about a decline in masonry use for load-bearing construction in large buildings and its modern use as a non-load-bearing enclosure system. While masonry is still available for load-bearing construction, economic factors have often continued to make other structural systems more financially attractive.

STONE

Stone has long been admired for its strength and the durability it provided from the destructive forces that ravaged lesser materials. The earliest stone buildings used rough, undressed stones in crudely stacked walls. However, as the skills and technology for shaping dressed stone improved, stone became a display of wealth and power. Throughout the pre-Columbian and European settlement periods in America and the recorded history of other settlements worldwide, stone was an important building material, as the various temples, castles, fortifications, and other works of stone demonstrate. This perception continues today, whether these constructions are still in use or exist only as ruins. Despite the common use of wood and lesser construction materials historically, the important buildings today and the major remnants of buildings from earlier civilizations are often those made of stone. Stone endures.

Types

Building stone is crafted from naturally occurring rock found within the Earth's crust. Rock is classified by the way it was formed: igneous, sedimentary, and metamorphic. Igneous rock was formed from volcanic action that created molten rock and then cooled slowly into a network of tightly interconnected mineral crystals. The most common igneous rock type is granite.

Sedimentary rock formed as various minerals and materials were deposited in beds of ancient oceans. These materials include weathered rock particles captured from eroding rock and transported to the sedimentation site by water and wind. These particles are classified by grain size (from smallest to largest) as clay, silt, or sand. The weight from the accumulation of materials, along with the pressure of the water, compressed the materials into continuous rock layers. Commonly, as sea life in the water died, calcium from its remains was added to the accumulating sediment and formed fossils. A unique characteristic of sedimentary rock is a structural weakness due to the formation of bedding planes. These planes form parallel to the water surface and are caused by variations in the materials being deposited. Two common examples of sedimentary rocks are sandstone and limestone (see Figure 5-3).

Metamorphic rocks originally were igneous or sedimentary rocks that, due to the high pressures and temperatures in the Earth's crust, were reformed into denser, more crystalline structures than the original rock. Metamorphic rock is characterized by small crystals of the original mineral in the sedimentary or igneous rock. In

Figure 5-3 Stone can be worked into many ornamental forms.

some instances, the rock was also fractured by movement of the Earth's crust that allowed magma to form "veins" and "intrusions" of different mineral crystals. This veining is common in marble. Examples of metamorphic rock include marble (made from limestone), quartzite (made from sandstone), and gneiss (made from granite).

Fabrication Methods

Stone is obtained by quarrying. While the equipment used for quarrying has changed with time, the underlying process has changed little. Rock found on the Earth's surface may have numerous structural defects due to weathering, which causes fissures and clefts. When a quarry site for stone was identified, the overburden of weathered stone and soil was removed. The remaining exposed rock could then be cut into a series of terraces.

Although used to remove overburden, blasting was generally not used to cut stone since blasting created incipient cracks within the stone. These cracks later caused stone to deteriorate. Removing stone relies on the fact that stone has relatively low tensile strength. When pressure is applied, stone will separate along a series of channels or a row of holes drilled to a specific depth. The Romans used wooden wedges inserted into the channels or holes and then soaked them with water. The soaking caused the wood to swell, which split rock along the channel or row of holes. By the early eighteenth century, drills were used to cut a row of holes into the rock where a series of metal plugs and wedges, known as "feathers," were placed into each hole. The plug was placed between the wedges and the row of plugs in the seam was struck simultaneously to exert force on the wedges. When sufficient force was created by the wedges, the stone broke free. By 1880, the use of steam for channeling had sped up the process substantially (McKee 1973). As rock-cutting technologies improved in the late nineteenth and twentieth centuries, thinner sections of stone could be cut at the quarry itself or transported to saws specifically designed to cut the stone into veneer.

Rough-cut blocks of stone were dressed to the finished size and left to weather for as long as two years to assess their durability. Dressing finished the faces of the stone with a particular pattern (see Figure 5-4). This process could introduce small microcrystalline fractures at the surface of the dressed stone. Newly dressed stones that withstood the weathering process were used in aboveground exposed locations. Stones that did not weather well were used belowground or as rubble fill. Various dressing patterns evolved to suit the aesthetic qualities desired for the building. Dressing was and still can be done using a variety of mallets, chisels, picks, and hammers. After World War II, computer-numerical-control (CNC) milling machines were introduced into the construction industry. Today, this technology is used to create three-dimensional stone ornamental elements and relief features.

Quarries were created to serve local building needs, but as transportation networks emerged in the nineteenth century, stone could be transported throughout North America. The most prolific regions for marble and granite quarrying were New England (New Hampshire and Vermont) and the Southeast (Georgia). The Midwest (Indiana, Illinois, and Ohio) was noted for limestone. The eastern seaboard also had numerous slate quarries extending from Georgia to New England. The late-nineteenth-century shift away from load-bearing masonry systems forced many smaller quarries to close due to competition from the large regional quarries.

Structural Uses

Although smaller buildings were built with walls consisting of a single vertical layer of stone, most large buildings rarely relied on just a single thickness of cut stone. Two methods for stone construction were used. The first method consisted of

Figure 5-4 Common stone dressing patterns: (a) tooled, (b) picked, (c) punched, (d) rusticated, (e) rock-faced, and (f) vermiculated.

assembling two or more vertical layers of stone blocks with the dressed faces being exposed or a thinner layer of dressed stone secured to the underlying stone. The second method consisted of laying out parallel courses of stone and filling the space between them with rubble of lesser stone, gravel, or sand. Each vertical row of stone, known as a "course," formed the bonding pattern. A wall can be made from a series of uniformly dimensioned rows or random patterns (see Figure 5-5).

The development of the arch, vault, and dome allowed open spaces that otherwise would have been filled with masonry. The use of arches and vaults reduced

Figure 5-5 Stone can be laid in a variety of patterns called "courses." Shown here are uniform ashlar (left foreground) and random-coursed ashlar (background at right).

the overall weight of the structure so that taller buildings could be constructed. The need for larger mill buildings during the Industrial Revolution led to the use of flattened arches for more open spans, but even these were eventually succeeded by timber framing and subsequently by a combination of cast and wrought iron members that were the precursors of the modern skeletal steel-framed buildings.

Garden walls and foundation walls of small buildings could be laid up "dry" (without mortar). Most other stone building construction did use mortar to fill the voids surrounding the stone blocks as well as to consolidate any rubble fill between the exterior stone blocks. Mortar was used as a lubricant to help position stones. When the mortar cured, it helped stabilize the stone and seal joints from moisture penetration. Mortar selection was based on " . . . cohesiveness, adhesiveness, setting time, hardening time, handling ease, ability to set and harden underwater, . . . the degree of expansion and solubility . . . color and texture" (McKee 1973, 61). Traditionally, mortar was a mixture of sand, clay, and, when available, lime. Clay was commonly used for mortar if suitable lime resources from limestone or seashells were unavailable. Sometimes straw or horsehair was added to increase tensile strength. Fine-grained mortar using gypsum instead of lime has been traced back to ancient Egypt, Persia, and France. In the United States, gypsum mortar has been used to set marble trim.

Before the widespread inclusion of Portland cement in mortars, lime-sand mortars were used for applications above water. Underwater, mortar needed to be able to cure and not dissolve. The early Romans developed hydraulic mortars using volcanic pozzolana. The formulation for this hydraulic mortar was lost with the fall of the Roman Empire until the late eighteenth century, when experimentation in England led to its reintroduction as natural cement mortars. The mortars using Portland cement were considerably harder, stiffer, and less permeable, which caused compatibility problems when those mortars were used to make plastic repairs on earlier building constructions that used sand-lime mortar.

Much of the success in working stone was due to the recognition of its physical properties and how it may react to vertical, lateral, and horizontal forces. Stones can be shifted by lateral forces, settlement, construction defects, and poor strength characteristics of adjoining materials. As a result, several methods evolved to secure stone. One method was the use of buttresses or vertical piers that were placed perpendicularly along the face of a stone wall to counteract any potential bulging in a wall caused by thrust forces from arches, vaults, domes, and other lateral loads. Another method was the use of bonding stones that bridged the two vertical faces of the wall or projected into the rubble fill behind the face(s) of the wall. These bonding courses were held in place by friction created by the weight of the materials bearing upon them and were often an ornamental part of the decorative coursing or bond pattern of the exposed stone. A visible example of this is the use of quoins at the corners of stone buildings that act to stabilize the forces occurring there. Lastly, there was a method of incorporating metal ties holding stones together. The most common forms of these ties were bent rods known as "cramps," "clamps," or "dogs" that were cemented into the tops, sides, or backs of the stones to hold them together. The rods were inserted into drilled holes and secured using mortar or lead. The use of ferrous metals for these connections is often a problem when moisture-related corrosion creates stresses in the stone that cause the rod and stone to fail.

Ornamental Uses

Stone can meet its structural needs without having any ornamental qualities, but aesthetics led to its use as ornament. Ancient civilizations used ornament to convey a sense of identity, power, and religious beliefs. As construction evolved, opportunities emerged to express a sense of order that regulated form, massing, and proportion that included highlighting the building craft as well as the level of sophistication of the building owner and the nature of the community.

Stone has natural qualities that make it beautiful in its own right. The aesthetic qualities derived from mineral composition and grain size, overall color, and intrusions of other minerals can serve as ornament. Examples can be seen in the use of marble for wall surfaces that is dramatic in appearance even without being carved into three-dimensional ornamental elements. Ornamentation can be simple or it can be crafted into a composition where the individual elements are refined at a high level of detail and the entire ensemble works together to create a magnificent whole. The Greek classical orders strove to attain such compositions. The orders were expanded upon by the Romans and subsequently explored by architects and builders in the nineteenth and twentieth centuries. Along the way, other architectural trends of other eras influenced ornamentation. Until the modernist movement after World War I, a consistent pattern of exploration, reinvention, reinterpretation, and recollection of historical precedents and motifs that influenced ornamental stone details had been used. For example, gargoyles on cathedrals were based on the need for scuppers

Figure 5-6 Stone ornaments fill both a functional and an aesthetic need. This stone ornament (center) covers the joint at the edge of the roof and serves as a rain scupper.

to direct and control the flow of rainfall from the roof. The scuppers did not need to be highly ornamental, but many are considerably so (see Figure 5-6).

Despite the shift away from stone as a load-bearing system, the contribution that stone makes to the perception of a building is evident by the way so many materials have been introduced to simulate stone. Even the rejection of ornament that accompanied the modernist movement was a transient phase. The postmodernist movement of the late twentieth century revived the perception, if not the actual use, of stone as an ornamental medium.

BRICK

Brick is one of the most plentiful types of masonry and forms the most basic masonry construction systems. In modern times, although not as substantial as many stone buildings of the period, fired brick has been common wherever good sources of clay and lime could be found.

Types

Brick can be classified as unfired or fired. Its use has been documented back to 8000–10,000 BC. Brick made from mud that was left to dry is an example of unfired brick. The development of adobe grew from this approach, but adobe is "fired" by using the heat of the sun to cure the adobe bricks for several weeks or longer. True firing processes using heated kilns had come into use by 3500 BC. A kiln regulates temperature and oxygen usage to obtain a more predictable product that today is the modern brick (Campbell and Pryce 2003). The subsequent development of skeletal framing systems in the nineteenth century led to the use of terra-cotta as an ornamental non-load-bearing masonry unit and structural clay tiles for fire protection. These later masonry products are both fired in the kiln to achieve their final form.

Adobe and Sod

Adobe was composed of local soils, fibrous materials (e.g., straw or dried cactus), water, and/or cactus juice manually pressed into molds to create earthen blocks. Adobe was perhaps the earliest historic masonry material. After adobe blocks were formed, they were left in the sun to bake dry. The drying process took from several weeks to several months. The dried blocks were then stacked to form a wall. Since no kiln was involved, some consider adobe to be unfired brick or at best comparable to low-fired brick in terms of strength and performance.

Adobe usage was usually limited to regions with long periods of heat and sunshine, since freely moving water would erode adobe and melt it back into mud. Therefore, adobe had to be protected from moisture. The exterior exposed wall surfaces were typically coated with a layer of clay-sand that was patted and rubbed to a smooth finish. True adobe was most commonly found in the southern tier of North America that corresponds roughly to the extreme southeastern coastal region as well as the southwestern coastal and interior areas explored and colonized by the Spanish in the sixteenth and seventeenth centuries.

Adobe building construction blended load-bearing construction systems used by the Spanish and, particularly in the Southwest, the descendants of the pre-Columbian native tribes. Adobe dates back many centuries in this region. After the political upheavals of the eighteenth and early nineteenth centuries subsided, westward expansion from the newly formed United States brought homesteaders into the Southwest. The lack of other building materials, such as a plentiful wood supply for wood-framed construction, led to adoption of local practices that used adobe while adding such features as pitched roofs and wood millwork when available (see Figure 5-7). After the railroad came to the region, more contemporary building products became available and many adobe buildings constructed in the nineteenth-century settlements were covered with new exterior siding systems or buried within additions. When an entirely separate new house was built, these early buildings were used as outbuildings or abandoned.

Sod brick, a counterpart to adobe, was used in the Midwest in the nineteenth and early twentieth centuries. Instead of being formed from loose materials, sod was cut into individual blocks. The tightly bound mass of grass roots held the block together. Due to their physical similarity to adobe, sod brick buildings also needed to be protected from moisture-related problems. Many sod buildings were intended to be temporary quarters until a finer house could be built. Thus, they have frequently undergone changes similar to those of their adobe counterparts.

Brick

In contrast to adobe, brick was fired by mechanical means in a kiln. This type of brick making was used in the earliest Northern European settlements in North America in the first thirty years of the seventeenth century where fuel for firing kilns and raw materials for brick making were plentiful. While the composition of brick

Figure 5-7 Adobe must be protected from the weather or it will eventually disintegrate.

was largely based on local materials, the essential ingredients remained the same: clay and sand. Regional variations in brick color can be attributed to minerals found within local clays used to make the brick and how they reacted to the temperature and oxygen content of the kiln. Hence, colors varied from red to pink to buff to gray to blue and everything in between. By varying the temperature between firings or changing the atmospheric conditions to prevent oxygen from entering the kiln, variations in color could be obtained. Higher kiln temperatures produced darker brick, but iron in the clays of most bricks turned red in an oxidizing atmosphere and purple in a reducing atmosphere. In some locations, it was popular to apply a slip or glaze to the brick. When this glaze cooled, a shiny coating formed on the surface of the brick. These glazed bricks were used ornamentally or as face brick.

Early bricks were made in molds, hand-packed with clay and sand. The process combined water, sand, and clay in a pug mill that mixed them. This "soft mud" mixture was pressed into molds, removed from the molds, and left to dry. After several days of drying, bricks were moved to a kiln where they were fired. The firing process took up to two weeks.

Placement of unfired brick closer to the heat source caused variations in the physical properties of brick. Bricks farthest away usually did not reach a temperature sufficient to give them the desired strength characteristics of brick fired at higher temperatures. These bricks were referred to as "common bricks" and were used as interior infill rather than on the exposed exterior face of a wall. Bricks closer to the heat source were fired at an increasingly higher temperature that created the desired strength characteristics. These bricks were classified as "face bricks" and were used in exposed locations. Bricks closest to the heat source could actually begin to burn and deform into twisted shapes. These bricks, referred to as "clinker bricks," were used as rubble infill materials. The Arts and Crafts movement at the turn of the twentieth century incorporated the use of clinker bricks as an artistic statement along with the traditional face brick.

After the Revolutionary War, brick makers moved west. Since transportation was difficult and costly, bricks were often made on-site or in a local kiln. Most major towns and, subsequently, cities had one or more local brick-making operation. Even with railroad transportation, it was often more cost effective to use locally made bricks.

Until the introduction of a brick-making machine in 1792, brick making in the United States was done primarily by hand using the traditional soft mud process. Early-nineteenth-century advances transformed brick making into a "dry mud" process that used sand and clays with lower water content. This transformed brick making into a year-round activity as traditional drying processes were significantly reduced. Technological advances shifted brick manufacturing away from a human- and animal-powered operation to a water-powered and then steam-powered one. This period also saw the introduction of "stiff mud" processes that used extrusion-based manufacturing methods combined with bladed or wire-cutting methods. These advances led to the use of dry-pressed brick by the mid-nineteenth century. Dry-pressed or hydraulically pressed bricks, composed of finely pulverized shale, marl, or hard clay, were harder and more uniform than those produced by earlier brick-making processes. However, machine-made bricks did not comprise a significant proportion of the industry until the 1870s (Elliot 1993).

Methods used to fill or to remove bricks from a mold resulted in various textures. Hand-packed molds yielded a coarse surface, while machine-filled molds were consistently smooth. The earliest removal method used a water wash on the mold before it was filled. The brick, with a smooth, dense surface created by this process, is called "water-struck brick." The next method used sand instead of water and produced brick with a matte finish referred to as "sand-struck brick" (Allen 1999). The density and surface smoothness helped to determine how well brick weathered in an exposed location.

Terra-cotta

Terra-cotta literally means "baked earth." It is a mixture of fine-grained raw clay and grog (previously fired clay that has been ground into fine particles). Through weathering, milling, cleaning, drying, and aging processes, clay was transformed into a homogeneous consistency that made it pliable when placed in a mold. The shapes and decorative features were carved into a plaster mold that was slightly oversized to account for the shrinkage that occurred in firing and drying processes. The "blocks" formed by the mold were actually hollow and, when needed, internal ribs were crafted to provide sufficient structural support for the block to stand without collapsing under its own weight. Voids inside the block were later filled with mortar or cement when the unit was attached to the building.

Terra-cotta was originally developed by ancient civilizations, such as the Etruscans, Greeks, Romans, Persians, Indians, Chinese, and Central American Indians. Unfortunately, its formulation methods were lost in Europe after the fall of the Roman Empire. Terra-cotta use continued in central Asia, Persia, and Turkey. In thirteenth-century Europe, terra-cotta was revived and flourished down to the sixteenth century during the Renaissance, with a particularly high craft level in Italy. An example of this superior craft is "faience," a decorative flat panel of glazed terra-cotta; the term was coined to describe the material as used in Faenza, Italy. Terra-cotta use went through a second decline during the Reformation, which caused reduced trade with Italy (Jandl 1983; Weaver 1997).

Attempts to revive terra-cotta production in England began in the eighteenth century when manufacturers in England and France worked to develop materials comparable to the original terra-cotta. Coade stone, which began production in 1769, was one of the best-known terra-cotta products redeveloped in England in this period. It was a cast stone made from fine clay that was marketed as a low-cost alternative to natural stone and was imported into the United States in 1800 for use on the Octagon House in Washington, DC (McKee 1973).

Early terra-cotta resembled the typical reddish-brown brick in color since it was made with similar firing processes. However, the firing process for terra-cotta, done at higher temperatures, resulted in a harder and more durable product. As part of the manufacturing process, a liquid slip composed of diluted, fine-grained clay was added to surfaces that were expected to be exposed to public view and weathering effects of the atmosphere. Internal and otherwise concealed faces were left untreated. The glaze sealed exposed surfaces of the terra-cotta. This glazing made the exposed faces relatively moisture impermeable but also left the rest of the element vulnerable to moisture penetration. Later advances in the reformulation and combination of raw materials led to off-white, gray, and buff-colored terra-cotta. In 1894 the first true glazes for terra-cotta were introduced, and polychrome terra-cotta could be made to suit building styles of the early twentieth century (see Figure 5-8).

Figure 5-8 Terra-cotta was widely used in the late nineteenth and early twentieth centuries as a low-cost alternative to stone. The terra-cotta shown here, from the National Farmers Bank in Owatonna, Minnesota, was designed by the architect Louis Sullivan.

Throughout the first three decades of the twentieth century, terra-cotta gradually became a veneer. While any type of three-dimensional ornamentation could be created, many of the last buildings constructed of terra-cotta had only a nominal thickness of the decorative material cladding the substructure.

In the United States, the earliest attempts to make terra-cotta in the early nineteenth century were failures and terra-cotta, if used at all, was mostly imported from Europe. In 1840, terra-cotta production began on a limited basis in Worcester, Massachusetts, and other firms began operations in the 1850s. Several attempts to introduce terra-cotta in New York were made by such noted architects as James Renwick and Richard Upjohn; however, these efforts did not generate sufficient acclaim to create widespread interest.

Chicago was the first city to adopt the use of terra-cotta with the formation of the Chicago Terra-cotta Company in 1869 (Elliott 1993). Terra-cotta was even more widely used after the Chicago fire of 1871. After that time, terra-cotta was used nationwide in public, institutional, commercial, and some larger residential construction. For the next sixty years, terra-cotta experienced its highest level of popularity until rising labor costs and the Great Depression severely curtailed its use. Ceramic veneering systems developed in the 1930s remained for another twenty years until the rise of the metal curtain walls of the International Style

Today, historic terra-cotta is available through a limited number of manufacturers around the country. Those manufacturers have retained many of their original molds and are available for custom orders.

Structural Clay Tile

Structural clay tile was introduced in the 1870s. The tile was fabricated like terra-cotta in that clay was pressed into blocks and baked in a kiln. In the late nineteenth century, the growing need for fireproof construction led to products that could insulate structural components from the heat of a fire. This need was especially great for skeletal structural systems. If steel is left unprotected, the heat from a fire will compromise its physical strength, causing it to collapse. Structural clay tile systems were used to offset this problem. Structural clay tile was used in fire separation walls, floors (in a variety of arch configurations), and other assemblies that prevented the spread of fire and protected the enclosed structural elements. Structural clay tile was also used for load-bearing construction applications and competed with concrete block for that use.

In the 1920s, clay tile was modified to create a product line of glazed tiles that could be left exposed and provide a decorative finish. After World War II, however, the introduction of reinforced concrete flooring and metal decking systems began the decline of structural clay tile for these uses. Eventually, in the 1950s, the use of structural clay tiles for load-bearing walls and fire separation walls was replaced, respectively, by concrete blocks and a combination of wood framing and gypsum board (Paulson 1995).

Structural Uses

Typically, adobe was used in hot and arid climates. Consequently, adobe brick was used as a vertical load-bearing construction material in massive walls that were also used as thermal mass to absorb and reradiate solar energy. Window and door openings were spanned using wood lintels. Logs, timbers, or smaller wood members formed the roof. The limited strength of adobe prevented it from being used significantly in arches, vaults, and domes. As long as it remained protected from moisture, it was relatively durable.

Fired-brick construction methods roughly paralleled those of stone. A load-bearing brick wall was rarely constructed of a single vertical thickness, also known as a "wythe," of brick, as it was unstable. In one type of brick construction following that of stone construction, two wythes were laid parallel to one another and the gap between the two outer wythes was filled with rubble, broken brick, or other loose materials. In the other approach, the wall was built as a monolithic assembly using brick construction that incorporated face brick on the exposed surfaces, while interior wythes consisted of common brick and soft-fired clay brick. Monolithic walls consisting of two or three wythes were common. To stabilize the wythes, a bonding brick was placed across any two wythes. Various bond patterns developed based on local traditions (see Figure 5-9). Bricks on the interior side of the wall were usually of lesser quality since these would be covered by the interior wall finishes. By the early twentieth century, metal ties were used instead of a through-bond brick. The use of these ties continues to this day. Two wythes of brick together act as a cavity wall where moisture penetrating the face of the brick and the joints can be drained out of the wall assembly when it encounters the drainage plane created by the gap between the wythes.

Brick could be laid up to form arches, vaults, and domes. Spans across door and window openings could be completed with arch construction methods. Lintels were also used. Early lintels were crafted from a single block of natural stone or a wood timber that spanned the opening. The introduction of reinforced concrete also introduced a variety of reinforced concrete lintels. By the late nineteenth century, lintels were also made of a flat arch, a cast-reinforced concrete beam, a soldier or stretcher course of brick supported by a wooden board or steel shelf angle. This later construction was vulnerable to decay when prolonged moisture caused the wood to rot or the metal to corrode. This failure is often noticeable when the joints of the bricks above the window fail and admit additional moisture that accelerates the decay.

The post–World War II era saw the increased use of prefabrication technologies including the manufacture of brick panels that combined a brick veneer with a substrate of reinforced concrete that could be fabricated in a manufacturing plant and shipped to the job site. The late 1950s also saw the increased development of reinforced brick masonry systems in which reinforcement bars were placed between two wythes of brick and grouted in place.

While terra-cotta has been used as a structural material on small buildings, its predominant application has been for ornamental uses. In a structural sense, terra-cotta could be installed in a load-bearing condition using methods similar to those of fired brick, using mortar or concrete as a bonding mechanism in which the hollow cores were filled to attach the terra-cotta to the wall. The hollow cores were completely filled with mortar or concrete so that the otherwise hollow terra-cotta element would not be crushed. Once skeletal framing systems gained acceptance, terra-cotta was not typically used as a load-bearing material.

One particularly subtle but important detail of unit masonry construction was the combination of well-maintained pointing in joints and well-defined drip edges. The process of weathering could open the joints and admit water into the mortar bed and eventually into the interior of the wall. The recognition of this problem led to the inclusion of drip edges and tooling of the joint so that moisture would not rest on a horizontal surface or be drawn by capillary action into the joint. Horizontal linear mortar joint surfaces that stand proud of the adjoining vertical surface were

Figure 5-9 Bonds used to construct brick walls: (a) English bond, (b) Flemish bond, (c) American running bond, and (d) American common bond. These can be distinguished by the placement of header bricks that crossed the two (or more) wythes of brick to stabilize several courses of stretcher bricks. In the English bond, the header courses are in alternating courses with the stretcher course, while in the American common bond they appear with at least three courses of stretcher courses between them. The American running bond incorporates wall ties to stabilize the two wythes. Originally, these ties were bricks placed at an angle in notches cut into the backs of bricks in the exposed wythes. Eventually, these ties were made of metal.

used as a drip edge where moisture collects and falls from the surface by gravity. While this drip edge could occur within a joint or along the bottom edges of the masonry, the goal was to ensure that water flowed down the face of a building.

Ornamental Uses

Adobe, in its simplest form, was typically not considered ornamental. The composition of the construction materials used with adobe did make a unique visual statement. Corners tended to be rounded, whether by intent or as a result of weathering. Coloration was typically uniform and drawn from the local earth materials from which the adobe was made. The overall appearance tended to be simple, with most ornament coming from the windows, doors, and their associated trim. Some pueblos made of adobe featured multiple layers and terraces that reduced the overall sense of mass. Mission churches and other public buildings were large in comparison to other local buildings and houses.

Fired-brick was simple in shape and size. However, the placement of brick could form quite elaborate ornamental compositions. These arrangements (see Figure 5-10) added a finer-grained texture than stone. Like stone, brick could be ornamental in its color, shape, and texture. Examples included gauged bricks used as decorative elements in an arch or rubbed bricks that were crafted into various shapes using emery cloths or masonry shaping tools. The need for mortar to hold bricks in place provided greater opportunities for enhancing, complementing, or contrasting brick construction details. Flush joints or a color similar to that of the brick gave the appearance of larger monolithic materials when viewed at a distance. Conversely, stark contrasting colors made the brick appear as a composition of separate units.

With the advent of skeletal framing systems, terra-cotta was seen as a means to provide the ornamental aspects of stone at a fraction of the cost and weight. Terra-cotta was attached to the building by a variety of evolving methods. As terra-cotta became a lightweight ornamental enclosure system, mortars with lighter-weight aggregates, such as cinders or coke breeze, were used in combination with mechanical fastening systems. In this method, if the cavities were to be completely filled, a combination of a ferrous metal tie bar or anchor embedded in the mortar mixture and fastened to receiving fasteners on the substructure was used. In cavities that were not to be completely filled and were vulnerable to moisture penetration, bronze tie bars and anchors could be used to prevent corrosion.

IDENTIFYING DECAY MECHANISMS

Due to moisture absorption characteristics, chemical and physical properties, and their often similar or corresponding construction applications and locations within the construction, masonry products have many similar decay mechanisms. For brevity's sake, the decay mechanisms are generically described below. Decay mechanisms specific to a single masonry type will be described here individually wherever it is appropriate to do so.

While a simple visual inspection can be completed using binoculars as a preliminary assessment, masonry inspection should be done at close range using ladders, scaffolding, mechanical lifts, or other appropriate vertical access equipment. Assessment of decay may require the use of nondestructive testing techniques or the actual removal of surface materials to view the suspected damage directly. The important aspects of assess are (1) overall physical integrity; (2) surface conditions; (3) joint conditions; (4) conditions of the connections holding the masonry in place; (5) evidence of previous repairs or modifications; and (6) potential causes of any deterioration. Inspections should be performed by a qualified professional. This

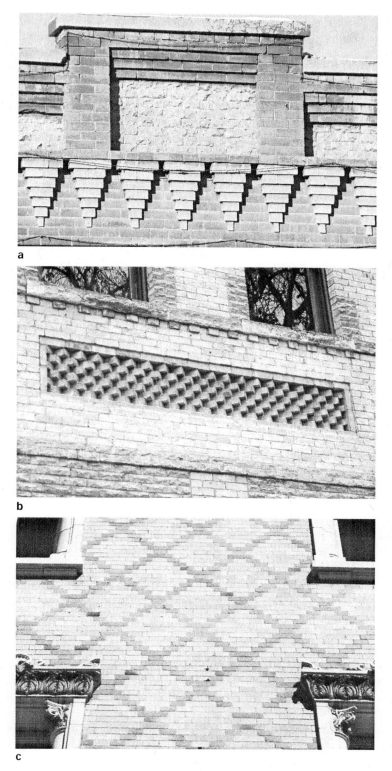

Figure 5-10   Brick can be used in a wide variety of ornamental ways, including (a) corbelling, (b) dogtoothing, and (c) diapering.

requirement is especially necessary for systems that include terra-cotta and veneer-oriented materials.

Construction Issues

For masonry, such as bricks and terra-cotta, the mixture of original materials may contribute to their failure. Although the masonry was sorted, certain defective units did make it into the final construction. For example, inadequately compressing materials into molds led to voids and delamination planes that eventually allowed moisture penetration. Also, the units could have been underfired but used anyway. This underfiring, especially in terra-cotta, decreases strength and increases moisture permeability.

Many unreinforced masonry buildings, built long before seismic codes were developed, lack safety features now integrated into modern masonry systems. Unreinforced masonry structures do not react well to changes in equilibrium. Equilibrium shifts are first revealed by cracking joints, spalling surface materials, or fracturing when cement mortars are stronger than the masonry. Next comes bulging and sagging within the construction. Finally, there is the collapse of the affected masonry, which may lead to a cascading collapse of adjoining masonry.

In ongoing decay, these defects occur slowly as the building settles differentially, groundwater levels fluctuate, or materials, such as wood or metal lintels, rot or corrode. Conversely, they may happen rapidly during or immediately following a seismic event. Modern reinforced masonry systems introduced in the twentieth century have been designed to withstand some shifts in their equilibrium and still maintain structural integrity.

Construction issues related to both the original construction and its subsequent maintenance are how watertight the constructed assembly was originally, how it has been allowed to deteriorate, and how water is permitted to collect or move along exposed surfaces. As originally constructed, joints should have been watertight and the tops of the walls capped and sealed at the cornices with suitable drip edges or flashing to prevent rainfall and melted snowfall from entering the assembly. Drip edges of even properly pointed joints and other construction details may have been compromised by lack of proper maintenance. Moisture entering a building has often been misidentified as coming through the masonry itself rather than correctly identified as entering through opened joints, failed flashing, and compromised drip edges that otherwise would have directed the water down the face of the assembly.

In some cases, the original construction itself may never have had the correct details to achieve this watertight assembly. For example, in the case of terra-cotta, which was advertised as a water-resistant material, a number of details may not have been included in the original design. The masonry assembly may also have been intended as a rain screen in which a certain amount of moisture would have been expected to be absorbed and released later when drier conditions prevailed. These two scenarios present a challenge to a restoration/rehabilitation team: how to integrate corrections without compromising the historic integrity of the building.

In sandstone and limestone, one construction problem is often prominently displayed. Weaknesses in adhesion between the sedimentary layers cause the stone to erode or spall in layers in a process known as "delamination." When the stone is placed into the building construction with the bedding planes in a nonhorizontal orientation, the interaction between gravity, freeze-thaw cycles, and moisture flow can cause delamination to occur (see Figure 5-11).

Figure 5-11 Failures in sedimentary rock cause delamination along the bedding planes and contoured edges of cut building stone, as shown in this deteriorated window sill.

One recurring and especially troublesome aspect of construction-related problems lies not in the construction itself but in how signs or other features were attached to and later removed from a building. While terra-cotta is most vulnerable to this problem, all masonry can suffer from it. Numerous holes made to attach these items may penetrate into hollow cavities within the wall. These holes can weaken the material and allow moisture penetration if not properly sealed. Likewise, the removal of these items can leave a hole where moisture can penetrate. Caution must be used when adding or removing fastening devices so that the process does not further damage the material and all penetrations are made watertight.

Moisture

Masonry is a porous material that will absorb water. Once introduced to the material, moisture can migrate through the masonry and into the living space, potentially causing untold damage in the process. Water from leaks, surface runoff, condensation, and other internal sources must be accounted for in identifying sources of moisture damage. Moisture penetration can be predicted by porosity (the percentage of void spaces in a material). Highly porous materials absorb moisture more readily. Unless a mechanism is in place to shield the masonry from moisture, masonry will absorb it.

Face bricks with hard continuous surfaces were more moisture resistant. Common bricks were more porous and were not used in exposed locations. The surface of slightly overfired bricks would melt slightly and form a glazed surface that would seal the pores. The exposed surfaces of terra-cotta were treated with a glaze of slip that sealed the pores. Stone could have naturally tight or open pores, and this factor was part of the basis for selection of quarrying sites. Highly porous stone typically did not make good structural material.

Water Runoff

The lack of appropriate roof and site drainage contributes significantly to the decay of masonry products. Moisture can migrate through joints between the masonry or through the masonry itself. Older traditional constructions of brick and adobe, as well as any sand-lime mortars that bind them together, are vulnerable to erosion caused by water from an improperly drained roof splashing back onto the masonry when it hits the ground. The scouring action of water moving across the surface or dripping is compounded by freeze-thaw cycles and the continued declining structural integrity of the materials. Eventually, sufficient material is leached out of adobe, low-fired brick, or mortar joints to cause the assembly to fail (see Figure 5-12).

The use of ferrous metal ties and anchors to hold masonry together can cause problems where poor moisture control within the assembly accelerates corrosion. Masonry ties and anchors, used from the late nineteenth century through the 1960s, were often made from ferrous-based materials, such as iron and steel. Since ferrous metals expanded as they corroded, not only could tie rods and connectors corrode and fail but also, when embedded within the masonry element, the expansion could cause the material to crack and/or spall. One historic practice was to use bronze ties or anchors where moisture problems were anticipated.

Efflorescence

Mineral salts within masonry dissolve in moisture as it migrates through the element. At the exterior surface, moisture evaporates and crystals form efflorescence (see Figure 5-13). Efflorescence is an indicator of moisture movement through a masonry unit. The leaching out of mineral salts, if continued for a long period of time, can have structural implications for the element itself. Salt crystals can also form within the material in a process known as "subflorescence" and create internal stresses that cause the surface material to spall or fail.

Freeze-Thaw Cycles

Continuous, repeated cycles of freezing and thawing cause stress on materials that absorb moisture. Masonry is no exception. In adobe and permeable unfired or

Figure 5-12 Inadequate protection from water runoff has led to back-splashing of water that has seriously threatened the integrity of this adobe house.

Figure 5-13 Efflorescence is the crystallization of dissolved salts from within the brick that form on the exterior surface of the brick when moisture migrating through the brick evaporates.

low-fired common brick, the expansion and contraction that occur degrade structural bonds. In more durable brick, early mortar systems used a sand-lime mortar mixture that readily allowed expansion and contraction of the brick itself. However, when later portland cement–based mortar mixes were inappropriately used to repoint these older construction systems, no allowance for expansion was made. Consequently, the resulting internal pressure caused the surface to spall off in reaction to the stress (see Figure 5-14). Left uncorrected, spalling can affect a wider area and destroy the ability of the brick to restrict moisture flow further into the unprotected core of the brick. As freeze-thaw cycles continue, the brick is slowly pulverized.

Stones with lower tensile strength, such as sandstone, may suffer considerably as freeze-thaw cycles disrupt bonds in bedding planes and cause the stone to delaminate (see Figure 5-15).

Figure 5-14 A combination of soft brick repointed with portland cement–based mortar has caused spalling of the bricks' exposed outer surfaces and severe erosion of the bricks themselves.

Figure 5-15 The combination of freeze-thaw cycles and de-icing salts has damaged the base of this sandstone retaining wall along the end of these stairs.

Rising Damp

This phenomenon occurs when high groundwater or moisture-laden soils come into contact with porous masonry. Water is drawn by capillary action into the masonry. Layers of masonry above the dampened material draw water upward. In areas of poor soil drainage or prolonged periods of high humidity, this water is less likely to evaporate and continues rising until an equilibrium point is reached. This moisture penetration may be located in the basement alone or may extend to the first floor (and higher). As the moisture migrates through masonry, it may also find its way into the interior finishes of the living space and damage them as well.

A similar phenomenon that mimics the appearance of rising damp occurs when landscape sprinkler systems spray masonry surfaces continuously without ample time for drying. Instead, the moisture is absorbed into the masonry, where it acts like rising damp. In either case, the source of moisture must be confirmed before taking corrective action. Failure to do so may do more harm than good if the problem is misdiagnosed and the true water source remains.

Soil Instability

When unstable soil conditions occur, buildings can begin to sink into the ground or be lifted out of the ground. This movement causes considerable stresses in building materials and can create deformation, displacement, or fracturing. Soil instability has several causes. First, the soil itself compresses if it does not have sufficient bearing capacity to support the weight placed on it. In some instances, this compression may be a naturally occurring problem or may be related to the placement of a heavy structure on poorly compacted fill. Second, fluctuation of the groundwater creates variations in moisture content that make expansive materials, like clay, expand or contract. Recent heavy rains, removal of large trees, or changes in the slope or elevation of soil can all contribute to this process. The third phenomenon is differential settling that occurs when a building rests upon materials of two different bearing capacities, such as bedrock, ledge, or well-compacted soils, next to expansive clays or poorly draining soils. Excessive subsurface water flow will remove the finer-grained particles of soil, causing the remaining soil to collapse. This collapse can create a sinkhole that, depending on geologic conditions, can expand to form an extremely large crater.

Soiling

While soiling is a major aesthetic concern in the appearance of historic buildings, the problems it can create can go much deeper. Soiling indicates opportunities for loss of historic fabric due to surface crust formations and biological growths that hasten moisture penetration and accelerate decay. Air pollution, dust, bird droppings, and acid rain can react with the masonry to stain the building or form gypsum or sulfate crusts on surfaces, along detail profiles, and within crevices of ornamental details. These crusts are composed in part of minerals from the masonry element that have reacted chemically with the soiling elements. Plants (e.g., moss, lichens, and other botanicals) can take root in the crusts, crevices, or joints and can introduce new penetrations that permit precipitation to enter the interior of the overall masonry assembly, where it can cause damage hidden from direct view. As these penetrations grow or weather, they open further and admit even more moisture. When these crusts and biological growths are removed with aggressive cleaning methods, parts of the historic fabric are often removed as well.

Abrasive Cleaning

Early misguided efforts that considered cleaning a primary step, if not the only step, in exterior restoration aggressively removed soiling as quickly and to the greatest extent possible. These attempts to clean masonry surfaces soiled by decades of pollution or paint used cleaning and stripping methods that are now considered too abrasive. These processes included mechanical grinding, scraping, or sandblasting, high-pressure washing, or extremely caustic and reactive chemicals. While these methods did immediately appear to improve the building's appearance in the short term, they often removed historic material along with the soil, stains, and paint coatings and accelerated long-term decay.

Such aggressive cleaning practices often led to unnecessary removal of portions of the finish. For brick or terra-cotta, if sufficient surface material was lost, the exposed softer interior core decayed more rapidly (see Figure 5-16). For stone, especially sandstone and limestone, ornamental details lost their original appearance as finer details were eradicated. Likewise, weathered or loose mortar was removed and all too commonly repointed using inappropriate portland cement–based mortars or other elastomeric materials.

A second scenario for losses occurred where paint had been applied to soiled buildings with slightly weathered masonry or had a particularly strong bond between the paint and the original surface. In a sense, the paint acted like a consolidant; when it was removed, it took surface materials with it.

Compounding these inappropriate approaches, after the cleaning was completed, sealant was commonly applied to prevent moisture penetration. Many sealants actually trapped moisture inside the masonry. This moisture further damaged the masonry as it repeatedly expanded and contracted during freeze-thaw cycles. Unfortunately, many buildings were treated this way, and the effects of the cleaning and sealing practices went unnoticed for years.

Figure 5-16 Sandblasting destroys the hard outer shell of a brick and exposes the softer inner core. The accelerated decay of sandblasted brick, as shown here, can be noted by the rounded edges and surface erosion.

REMEDIATION METHODS

As the effects of decay mechanisms become known, preserving the remaining historic fabric with perhaps only nominal repairs may be possible. The primary precaution mentioned in regard to wood stabilization and conservation is applicable here— remove the source of the decay even when no other work is deemed necessary.

Stabilization and Conservation

Compared to wood conservation, stone conservation is more problematic. The stripping, cleaning, and coating systems used early in the preservation movement of the mid-twentieth century demonstrated how complex and fragile masonry systems were and that these early methods were detrimental in the long run. The physical and chemical properties of materials used to make a repair need to be carefully considered. The mixed results of seemingly simple procedures, like repointing mortar, cleaning, and repairing either surface or structural defects, continue to demonstrate the need for an acute understanding of potential short- and long-term impacts. Compressive strength, thermal expansion, chemical interactions, visual integrity, and a host of other factors must be weighed when selecting a treatment. These characteristics must be assessed in terms of how they relate to the corresponding properties of the masonry they are intended to conserve. An extremely common example is the devastating impact of one "simple" mistake: replacing lime-sand mortars with portland cement mortars.

"Stabilization" refers to protecting the physical integrity of the item and involves remediation by using temporary measures that may later be removed as part of the final treatment. Temporary measures may be prudent while evaluating long-term treatments. These measures include protecting decayed masonry from further damage by covering it with tarpaulins, installing lateral supports and bracing, removing unnecessary loads from within the building itself, and making nominal repairs to flashing and openings in masonry joints.

As in any cleaning project, it is necessary to identify the goals. A thorough investigation must answer the following questions:

Cleaning and Stripping

- Is cleaning needed for aesthetic or remedial purposes?
- What caused the soiling or paint conditions that currently exist?
- What previous cleaning efforts have been made?
- What previous coatings and other treatments have been applied?
- How is the masonry reacting to the soiling or previous treatments?
- How should built-up crusts, if present, be treated?
- What should the cleaning process remove?
- What are the appropriate cleaning processes for this building?
- What, if any, are the historically appropriate paint coatings?

Based on lessons learned from the use of earlier, more destructive methods, the object is to remove soiling by the gentlest means possible. Processes are now available to conduct cleaning projects that satisfy this requirement. In certain instances where crusts have formed, the best treatment may be to leave them in place so as not to destroy details of the element being cleaned.

Practices for cleaning soiled surfaces include soaking, low-pressure washing systems that may include using nonionic detergents, steam/hot water cleaning methods, and, in certain cases, chemical washes. The critical factors in using these approaches include the physical stability of the existing material, the presence or absence of crusts, and the chemical composition of cleaning agents and how they react with the masonry and any adjacent mortar being cleaned. For example, an acidic cleaning product should not be used on limestone or a lime-sand mortar since these materials are sensitive to acid, while granite, sandstone, and portland cement–based mortars are not. Correspondingly, some masonry products may be reactive to alkaline-based cleansers.

Practices and products for removing paint include a range of chemical cleaners and strippers. Two types are most commonly available. The first type includes alkaline strippers that use potassium, ammonium hydroxide, or trisodium phosphates as a basic ingredient. The second type includes organic solvents that use methylene chloride, methanol, acetone, xylene, and toluene. There are also a number of other products specifically formulated to remove such things as graffiti and stains of industrial, metallic, or biological origin.

Good practice dictates using small test areas to determine the effectiveness of products before initiating the overall project. In any cleaning or stripping project, the product is applied using an applicator spray, a natural-bristle brush, or a poultice. The product is left in place for a period of time, referred to as the "dwell time," and then removed. Some processes use plastic- or chemical-resistant sheets to cover the exposed product and make penetration of the chemical into soiled or painted areas more effective. When the recommended dwell time is reached, the product is removed and the surface is washed with water or a neutralizing agent. If need be, a stiff natural-bristle brush may be used. Metal bristles are to be avoided, as they may scratch terra-cotta glazes or break off and become embedded in the masonry, causing later discoloration as they corrode. Likewise, nylon bristles may react with the chemicals involved and leave a residue that will need to be removed. Once the test area dries, the effectiveness is assessed and the process is repeated until satisfactory results are obtained. A negative aspect of the application process and the rinsing/neutralization process is that they can introduce water and sometimes certain salts into the masonry. Care must be taken to note open joints and observe the effects of the process to ensure that efflorescent salts or chemical films do not result.

Overly aggressive cleaning methods have generally been eliminated as inappropriate for historic buildings. While the prohibition largely involves sandblasting, grinding, and other mechanical removal methods, some leeway has been accorded to recent methods that use microabrasion techniques first developed in museum curatorial processes. These techniques make use of micro-balloons, finely ground walnut shells, glass powder, sponges, rubber pellets, and a wide variety of other materials. Although these techniques have been used with success on materials such as metals, they are still not considered appropriate for masonry.

Concerns about the environmental impact of phosphates discharged into aquatic habitats and off-gassing of volatile organic compounds (VOCs) in part fostered by the LEED program have led to the reformulation and introduction of a number of cleaning and stripping products that reduce the use of these compounds. In using any cleaner or stripper, local health and life safety regulations must be confirmed and observed. All products and processes have specific safety precautions regarding respiratory, eye, and skin protection while they are being used or stored on-site. Likewise, environmental safety laws must be followed for both on-site collection and off-site disposal of effluents generated by cleaning and stripping.

Water-Repellant and Waterproof Coatings

The long-term damage from inappropriate coatings on historic masonry has led to the development of better coating products. There has been an ongoing debate on whether applying any coating to historic masonry is appropriate. Waterproof coatings prevent any form of moisture, either vapor or liquid, from entering the coated material, while water-repellant coatings allow water vapor to migrate into and out of the material. The National Park Service (Mack and Grimmer 2000) has two observations regarding these coatings: (1) water-repellant coatings are not necessary on watertight buildings and (2) waterproof coatings should not be applied to historic masonry.

At the center of that debate is the recognition of differences between water-repellant and waterproof coatings. The critical factor is that the coating must be vapor permeable to allow moisture vapor within the masonry to escape. While this characteristic may seem to run counter to the intent of keeping moisture out of the wall assembly, vapor-permeable coatings do prevent liquid moisture from entering the material. The key difference is that vapor-permeable or "breathable" coatings do allow moisture inside the material to escape. If water vapor cannot escape, the material containing it may succumb to the ravages of freeze-thaw cycles or the moisture may migrate to the interior of the building and damage interior finishes.

Breathable coatings have been developed that are vapor permeable and allow migration of water vapor. These coatings have been formulated using a variety of water-based solutions containing modified siloxanes, silanes, alkoxysilanes, or modified metallic stearates, and have largely replaced the acrylic and silicone resins used in the organic solvent-based formulations of earlier generations. These coatings must be periodically reapplied.

Waterproof coatings used in modern construction do not allow any water to pass through them. For new construction, measures can be incorporated into the design to eliminate moisture migration from the ground and interior spaces, thus making waterproof coatings viable. For historic construction where moisture control measures may have been compromised or may never have existed, these coatings neither solve the problem nor allow moisture to exit the assembly in a predictable manner unless all other moisture sources are controlled as well (see Figure 5-17). For example, in the case of rising damp, applying waterproof coatings without controlling all paths of moisture migration may force the damp to rise higher than before coatings were applied. Waterproofing below-grade surfaces, such as basement walls and foundation assemblies, should only be done as a last resort when other drainage and dampproofing measures are unavailable, unfeasible, or ineffective.

To offset problems in selecting coatings, it is necessary to assess the potential effects of the appearance and performance of the coating itself. While a series of test sections can be used to assess how the coating penetrates the surface and how shiny the resulting surface appears, these sections are largely ineffective in determining moisture control performance. Unless the entire assembly that the test sections are

a   b

Figure 5-17 Masonry coatings must allow the moisture trapped within the masonry to escape. The detail in (b) shows the impermeable shell formed by the coating.

on has been made watertight (e.g., repointed mortar, repaired flashing), the true moisture behavior will still be affected by uncontrolled moisture entering through adjacent nonwatertight openings.

Consolidants and Epoxies

Consolidant and epoxy use follows application processes similar to those for treating deteriorated wood. Consolidants penetrate into small voids in the material being treated. They then cure to stabilize the weakened material. Unfortunately, many consolidants have proven ineffective in many masonry products like brick, adobe, terra-cotta, and cast stone, as they seal in moisture, may not bond well, and may discolor the original material. Consolidants can sometimes be used in stone repairs before an epoxy treatment that is used to fill the larger voids created by the decay mechanism. These treatments bind the existing decayed material into a more stable condition and therefore are not reversible.

Infill Repairs

When damage is extensive, it may be necessary to repair or replace damaged materials. In repairing a structural member, it is crucial to consider the visual and structural aspects of the replacement materials. Certain methods conceal repairs from public view, while others are more obvious. Selecting the appropriate methods involves balancing the acceptable level of visual intrusion with the need for health and life safety. The selection must also be tempered by the economic feasibility of completing repairs.

Plastic Repairs

In earlier repair attempts, one strategy used various mortar mixes applied to damaged masonry. These are called "plastic repairs" since mortar was applied while in its pliable or plastic state and then fashioned to the contour of the original surface. When mortar was compatible with the existing historic materials, this was a plausible repair. Unfortunately, many early repairs did not account for strength, flexibility, and permeability characteristics. This incompatibility was common in adobe repairs and in repointing sand-lime mortar joints. While in the short term these repairs seemed appropriate, long-term stresses and ensuing damage to the historic materials revealed that they did more harm than good. Because of incompatibility between the repair materials and original materials, original materials began to deteriorate.

The use of plastic repairs should be sensitive to the physical characteristics mentioned above and to other factors, such as color, texture, reversibility, and long-term durability. The material being repaired is an important consideration as well. Broken terra-cotta, for example, may prove difficult to match color and appearance in piecemeal repairs. Instead, it may be feasible to replace broken elements with recast replacements.

The most misunderstood plastic repair is repointing. To the inexperienced, repointing appears to be a simple insertion of mortar into a degraded joint, with little regard for how compatible the color, strength, or permeability of the new mortar is with the existing mortar and masonry. Until the 1871 introduction of portland cement–based mortars, previous mortars used a sand-lime mixture that allowed masonry to expand and contract with thermal and moisture conditions. Incompatible replacement mortars cause stresses within masonry that eventually hasten decay. In addition to physical properties, visual properties need to be considered. Laboratory analyses can determine the color, size, and proportion of the aggregate (i.e., sand) used in a historic mortar. Colorants can match the color if the original or comparable

Figure 5-18 This is an example of poor repointing. When repointing masonry, the mortar should be made visually and physically compatible with the original mortar by matching the color, profile, strength, and expansion characteristics of the original mortar.

source of the aggregate cannot be located. Lastly is the need to understand the process of repointing. Due to weathering, both the masonry and existing mortar may have deteriorated. Loose mortar must be removed and replacement mortar inserted into the open joint. Two precautions exist. First, the original profile of the mortar joint must be identified and repointed as exactly as possible. The profile should include a drip edge or means to prevent water from resting along the joint. Second, the edges of the masonry may have eroded to form a wider joint. Repointing must take this into account and provide visual continuity based on the width of other adjacent original joints. The repointing mortar should not be simply slathered onto the opening to create a flush profile; this is a common mistake of the inexperienced worker (see Figure 5-18).

Dutchman

A dutchman can provide a continuous surface on the exposed face of masonry. It may be possible to remove the damaged portions of masonry and infill the cleared void with a patch made of compatible material. Compatibility refers to strength, size, permeability, and appearance. When a compatible unit is not available, other masonry units may be removed from the existing construction, cut parallel to the exposed face, and positioned in the void spaces. The remaining void behind the now thinner unit can be filled with mortar.

A dutchman can also conceal an epoxy repair or any other internal structural reinforcement treatment. In the case of terra-cotta, consideration should be given to replacing the entire piece rather than compromise its visual appearance and moisture resistance. A dutchman can be dressed and worked to provide a continuous surface on an ornamental surface as well. A masonry dutchman is usually attached by either bedding it in compatible mortar, using an epoxy or adhesive resin, or drilling into the material beneath the damaged portion and inserting a combination of rods and epoxy filler to secure it.

Structural Reinforcement

Using structural reinforcement processes similar to those used for wood structures, limited opportunities may arise to repair and reinforce existing defective construction in place. However, structural reinforcement generally means dismantling the masonry assembly to access internal damage to the assembly, tie rods, and anchors. In a load-bearing system, loads from above must be isolated from the area under repair using temporary support systems (e.g., needle beams or temporary bracing) to relieve weight from the bearing surfaces. Once loads are stabilized, damaged areas beneath the supports can be repaired or rebuilt.

A common problem that can be corrected this way is the repair of lintels over openings. Many lintels installed prior to the use of stainless steel, aluminum, or other nonferrous metals were constructed from wood or ferrous metal shelf angles that have since rotted or corroded away. These deteriorated elements can cause masonry above the lintel to fail. Prior repairs for this condition can be seen where mortar in joints opened up by the settling has been repointed one or more times without correcting the failed lintel.

Replication/ Replacement

Masonry systems are not made to be easily dismantled, moved, rebuilt, or readily replicated. This fact raises the question of how to find compatible materials. Research may identify original manufacturers of brick and terra-cotta or local stone quarries, but that does not mean that these products are still available (unless through salvage). Lack of availability can force the use of available materials; replication of missing or deteriorated features in the original historic materials; or the use of appropriate alternative modern materials.

Since finding suitable replacements has proven difficult, replacement products have been developed. Some can be used structurally or ornamentally, while others are for ornament only (see Table 5-1).

Although each product is well suited for specific applications, local, state, and federal guidelines and ordinances governing historic properties may preclude their use. Therefore, consult with the relevant review agencies and manufacturers to verify appropriateness. In making a final determination on appropriateness, the compatibility criteria should be met. Other criteria to consider include weather resistance, color fading, ultraviolet resistance, modeling ability, paint/coating adherence, rigidity, expansion coefficients, fire rating, specific connection requirements, and reversibility (Park 1988).

TABLE 5-1: Structural vs. Ornamental Products

PRODUCT	STRUCTURAL	ORNAMENTAL
Cast aluminum	X	X
Cast stone (dry-tamped)	X	X
Glass fiber reinforced concrete (GFRC)		X
Precast concrete	X	X
Fiber-reinforced polymers (fiberglass)		X
Epoxies	X	X

Source: Park, 1988, 10–13.

With any replacement product, it is critical to investigate these issues before proceeding, since it may be difficult and expensive to switch if the product is later deemed inappropriate. Products introduced each year for new construction raise the possibility that they may be used on older buildings. While this is possible, these crossover products should be researched and approved before the project design is considered complete.

References and Suggested Readings

Allen, Edward. 1999. *Fundamentals of building construction: Materials and methods.* 3rd ed. New York: John Wiley & Sons.

Ashurst, John and Nicola Ashurst. 1988. *Practical building conservation, Vol. 1: Stone masonry.* New York: Halsted Press.

———. 1988. *Practical building conservation, Vol. 2: Brick, terra-cotta and earth.* New York: Halsted Press.

———. 1988. *Practical building conservation, Vol. 3: Mortars, plasters, and renders.* New York: Halsted Press.

Ashurst, Nicola. 1994. *Cleaning historic buildings, Vol. 1: Substrates, soiling, and investigations.* London: Donhead Publishing.

———. 1994. *Cleaning historic buildings, Vol. 2: Cleaning materials and processes.* London: Donhead Publishing.

Barnes, Mark R. 1975. Adobe bibliography. *Bulletin of the Association for Preservation Technology,* 7(1): 89–101.

Beckman, Poul and Robert Bowles. 2004. *Structural aspects of building conservation.* 2nd ed. Burlington, MA: Elsevier Butterworth Heineman.

Belle, John, John Ray Hoke, Jr., and Stephen A. Kliment, eds. 1994. *Traditional details for building restoration, renovation and rehabilitation from the 1932–1951 editions of Architectural Graphic Standards.* New York: John Wiley & Sons.

Borgal, Christopher. 2001. Thin-stone systems: Conflicts between reality and expectations. *APT Bulletin,* 32(1): 19–25.

Brown, Paul Wencil and James R. Clifton. 1978. Adobe. I: The properties of adobe. *Studies in Conservation,* 23(4): 139–146.

———, Carl R. Robbins, and James R. Clifton. 1979. Adobe. II: Factors affecting the durability of adobe structures. *Studies in Conservation,* 24(1): 23–39.

Campbell, James W. and Will Pryce. 2003. *Brick: A world history.* New York: Thames & Hudson.

Cliver, E. Blaine. 1974. Tests for analysis of mortar samples. *Bulletin of the Association for Preservation Technology,* 6(1): 68–73.

De Mare, Eric, ed. 1951. *New ways of building.* London: Architectural Press.

Edison, Michael P. 2002. Color and long-term color retention in composite patching systems for stone and masonry. *APT Bulletin,* 33(2/3): 23–32.

Elliott, Cecil. 1993. *Technics and architecture.* Cambridge, MA: MIT Press.

Fidler, John. 2002. Plastic dreams: Weathering of glass-reinforced plastic facsimiles. *APT Bulletin,* 33(2/3): 5–12.

George, Eugene. 1973. Adobe bibliography. *Bulletin of the Association for Preservation Technology,* 5(4): 97–103.

Grimmer, Anne. 1984 (reprinted 1997). *A glossary of historic masonry deterioration problems and preservation treatments.* Washington, DC: United States Department of the Interior.

———. 2000. *Dangers of abrasive cleaning to historic buildings.* Preservation Brief No. 6. Washington, DC: United States Department of the Interior.

Harris, Samuel Y. 2001. *Building pathology: Deterioration, diagnostics, and intervention.* New York: John Wiley & Sons.

Henry, Allison, ed. 2006. *Stone conservation: Principles and practice.* Dorset, UK: Donhead Publishing.

Hornbostel, Caleb. 1991. *Construction materials: Types, uses, and applications.* 2nd ed. New York: John Wiley & Sons.

Iowa, Jerome. 1985. *Ageless adobe: History and preservation in Southwestern architecture.* Santa Fe, NM: Sunstone Press.

Jandl, H. Ward. 1983. *The technology of historic American buildings: Studies of the materials, craft processes, and the mechanization of building construction.* Washington, DC: Association for Preservation Technology International.

Jester, Thomas C., ed. 1995. *Twentieth century building materials: History and conservation.* New York: McGraw Hill Book Company.

Labine, Clem and Carolyn Flaherty, eds. 1983. *The Old-House Journal compendium.* Woodstock, NY: Overlook Press

London, Mark. 1988. *Masonry: How to care for old and historic brick and stone.* Washington, DC: National Trust for Historic Preservation.

Loth, Calder. 1974. Notes on the evolution of Virginia brickwork from the seventeenth century to the late nineteenth century. *Bulletin of the Association for Preservation Technology,* 6(2): 82–120.

Lukas, Robert and Jerry Stockbridge. 1988. Soil-related distress in older structures. *APT Bulletin,* 20(1): 4–7.

Lynch, Gerard C. J. 2006. *Gauged brickwork: A technical handbook.* 2nd ed. Dorset: Donhead Publishing.

Mack, Robert C. and Anne Grimmer. 2000. *Assessing cleaning and water-repellant treatments for historic masonry buildings.* Preservation Brief No. 1. Washington, DC: United States Department of the Interior.

—— and John P. Spiewak. 1998. *Repointing mortar joints in historic masonry buildings.* Preservation Brief No. 2. Washington, DC: United States Department of the Interior.

Mark, Robert, ed. 1995. *Architectural technology up to the scientific revolution: The art of structure of large-scale buildings.* Cambridge, MA: MIT Press.

McGettigan, Edward. 1995. Factors affecting the selection of water-repellent treatments. *APT Bulletin,* 26(4): 22–26.

McGrath, Thomas L. 1979. Notes on the manufacture of hand-made bricks. *Bulletin of the Association for Preservation Technology,* 11(3): 88–95.

McKee, Harley J. 1973. *Introduction to early American masonry: Stone, brick, mortar and plaster.* Washington, DC: Preservation Press

Merritt, Frederick S., ed. 1958. *Building construction handbook.* New York: McGraw-Hill Book Company.

National Park Service. 1978. *Preservation of historic adobe buildings.* Preservation Brief No. 5. Washington, DC: United States Department of the Interior.

New York Landmarks Conservancy. 1997. *Historic building facades: The manual for maintenance and rehabilitation.* New York: John Wiley & Sons.

Newman, Alexander. 2001. *Structural renovation of buildings: Methods, details, and design examples.* New York: McGraw-Hill Book Company.

O'Connor, Thomas. 2001. Guide to the use of new ASTM standards for sealants. APT Bulletin, 32(1): 51–57.

Park, Sharon.1988. *The use of substitute siding on historic building exteriors.* Preservation Brief No. 16. Washington, DC: United States Department of the Interior.

—— 1996. *Holding the line controlling unwanted moisture in historic buildings.* Preservation Brief No. 39. Washington, DC: United States Department of the Interior.

Paulson, Conrad. 1995. Structural clay tile. In Jester 1995, 150–155.

Peters, Tom F. 1993. *Building the nineteenth century.* Cambridge, MA: MIT Press.

Peterson, Charles, ed. 1976. *Building early America: Contributions toward the history of a great industry.* Radnor, PA: Chilton Book Company.

Pieper, Richard. 2004. *The maintenance, repair, and replacement of cast stone.* Preservation Brief No. 42. Washington, DC: United States Department of the Interior.

Prudon, Theodore H. M. 1989. Simulating stone, 1860–1940: Artificial marble, artificial stone, and cast stone. *APT Bulletin,* 21(3/4): 79–91.

Prueher, Brooks. 1995. Consolidants, coatings, and water repellent treatments for historic masonry: A selected, annotated bibliography. *APT Bulletin*, 26(4): 58–64.

Rabun, J. Stanley. 2000. *Structural analysis of historic buildings: Restoration, preservation and adaptive reuse for architects and engineers.* New York: John Wiley & Sons.

Ritchie, T. 1973. Notes on the history of hollow masonry walls. *Bulletin of the Association for Preservation Technology*, 5(4): 40–49.

Rose, William. 2005. *Water in buildings: An architect's guide to moisture and mold.* New York: John Wiley & Sons.

Scheffler, Michael J. 2001. Thin-stone veneer building facades: Evolution and preservation. *APT Bulletin*, 32(1): 27–34

Scheffler, Michael J. and Edward A. Gerns. 1995. Thin stone veneer. In Jester 1995, 168–173.

Schuller, M. P., R. H. Atkinson, and J. L. Noland. 1995. Structural evaluation of historic masonry buildings. *APT Bulletin*, 26(2/3): 51–61.

Searls, Carolyn L. and Cece Louie. 2001. The good, the bad, and the ugly: Twenty years of terra-cotta repairs reexamined. *APT Bulletin*, 32(4): 29–36.

Searls, Carolyn L. and David P. Wessel. 1995. Guidelines for consolidants. *APT Bulletin*, 26(4): 41–44.

Selwitz, Charles. 1995. The use of epoxy resins for the stabilization of deteriorated masonry. *APT Bulletin*, 26(4): 27–34.

Slaton, Deborah E. and Harry J. Hunderman. 1995. Terra cotta. In Jester 1995, 156–161.

Smith, Major Percy Smith. 1875 (reprinted 2004). *Rivington's building construction (Vols. I–III).* Dorset, UK: Donhead Publishing.

Speweik, John P. 1995. *The history of masonry mortar in America, 1720–1995.* Arlington, VA: National Lime Association.

Stockbridge, Jerry G. 1986. Evaluating the strength of existing masonry walls. *APT Bulletin*, 18(4): 6–7.

———. 1989. Repointing masonry walls. *APT Bulletin*, 21(1): 10–12.

Tabasso, Marisa Laurenzi. 1995. Acrylic polymers for the conservation of stone: Advantages and drawbacks. *APT Bulletin*, 26(4): 17–21.

Tiller, de Teel Patterson. 1979. *The preservation of historic glazed architectural terra-cotta.* Preservation Brief No. 7. Washington, DC: United States Department of the Interior.

Tunick, Susan. 1998. The reign of terra cotta in the United States: Enduring in an inhospitable environment, 1930–1968. *APT Bulletin*, 29(1): 43–48.

———. 2001. The evolution of terra cotta: "Glazing new trails." *APT Bulletin*, 32(4): 3–8.

Volz, John R. 1975. Brick bibliography. *Bulletin of the Association for Preservation Technology*, 7(4): 38–49.

Warland, Edmund George. 1929 (reprinted 2006). *Modern practical masonry.* Dorset, UK: Donhead Publishing.

Weaver, Martin E. 1995. *Removing graffiti from historic masonry.* Preservation Brief No. 38. Washington, DC: United States Department of the Interior.

———. 1997. *Conserving buildings: A guide to techniques and materials.* New York: John Wiley & Sons.

Weiss, Norman R. 1995. Chemical treatments for masonry: An American history. *APT Bulletin*, 26(4): 9–16.

Winkler, Erhard M. 1977. The decay of building stones: A literature review. *Bulletin of the Association for Preservation Technology*, 9(4): 52–61.

CHAPTER 6 Concrete

Concrete is a material that can be cast into blocks, structural members, and various simulated stone products and panels. While many modern buildings have been constructed using concrete, the use of concrete dates back to ancient times when constructions were made from rammed earth and eventually Roman cement. Concrete has since become one of the most versatile building products of the past two centuries. While slow to gain acceptance in the early nineteenth century, it had become a major construction material by the mid-twentieth century and has been used in a wide variety of structural and ornamental applications.

MATERIAL OVERVIEW

Concrete differs from other forms of masonry by how it is formed into a building material. While stone can be cut and brick and terra-cotta can be created through a firing process in a kiln, the firing methods used in brick and terra-cotta making do not apply to cast concrete. Instead, casting is a chemical reaction process that occurs as the moisture in the concrete cures or dries out. Unlike stone, concrete has a plastic state when it is first mixed that allows it to be molded into any number of shapes. The mixture of cement, aggregate, and sand then bonds in the curing or hydration process that results in a material with qualities similar to those of stone.

Fabrication Methods

The essential process involves mixing these materials together and applying them over a substrate or placing them in molds or formwork where they harden and cure. Early unreinforced concrete constructions used a stiff mixture that was tamped into place by hand. When placed in a form, the original plastic or flowable nature of the mixture took on the shape of the mold in which it was placed. The adoption of reinforced concrete resulted in the use of thinner mixes to allow concrete to fill in around the reinforcement materials. Individual blocks and decorative panels were created this way using individual castings that were removed from the mold and used in the construction.

Physical Properties

Concrete is valued for its compressive strength but suffers from low tensile strength. When well made, concrete provides a strong, durable surface that can withstand a variety of climates and uses. In its simplest form, concrete consists of sand, aggregate, cement, and water. These ingredients are mixed together to form a wet mixture that can be forced into forms while in its plastic state and then left to cure. The curing process has two phases. First is the hydration process, in which the concrete hardens and becomes structurally stable but shrinks as it dries out. Second is the carbonation process, in which the concrete reacts with carbon dioxide. In this process, the alkalinity of the concrete is reduced as the concrete gets stronger.

An important physical property of concrete is alkalinity. This property is particularly strong in reinforced concrete, where it protects the metal reinforcement materials from corrosion. Other important properties are salt content and sulfate content, which are introduced when sands and aggregates containing them are used in the mixture. These "dirty" sands and aggregates react with other materials, and can weaken the strength of concrete and accelerate deterioration.

Romans used pozzolana derived from volcanic ash for their concrete mixture; this was natural hydraulic concrete that cured underwater, a significant trait of Roman construction. The fall of the Roman Empire led to the loss of the recipe for this mixture for many centuries. Although concrete was used in the Middle Ages by the Moors and the Spanish and minimally in Europe, it was considered inferior to stone in Northern Europe. Although a variety of mixtures were formulated, it was not until the eighteenth century that a method using portland cement was devised that emulated the hydraulic properties of Roman concrete.

In modern concrete, additives and alternative aggregates are added or substituted to provide lighter weight, faster curing, coloration, or other desired properties. Portland cement, originally an admixture used to decrease the curing time in colonial lime-sand mortars, has replaced the lime component altogether. This modern cement process has been refined throughout the past two centuries and forms the basis for nearly all cement products made today.

CONSTRUCTION METHODS

Concrete can be used as a load-bearing construction material or as a skeletal frame using castings created in two ways. In the first approach, the structure is "cast in place" by hand or using forms based on a variety of vernacular and local building traditions. This method is both old and new; it was used as a rammed earth system in ancient times and is now used in reinforced concrete framing and panelized construction systems. Examples of these processes include unreinforced constructions based on rammed earth construction techniques (e.g., tabby, pise, cob) that may or may not have included forms of natural cement (see Figure 6-1) as well as modern reinforced concrete framing systems and concrete shells. In the second process, individual units can be precast in molds and cured. Precast concrete blocks or panels can be created on-site or brought to the site from a manufacturer's factory. The individual concrete units can then be employed in constructions using traditional masonry methods.

For modern buildings, skeletal framing systems made from reinforced concrete or steel use a curtain wall to enclose the structure. Concrete products can therefore become simply an enclosure system where individual castings (e.g., cast stone, precast panels, and other concrete assemblies) are connected to the structural frame.

Figure 6-1 Tabby, as shown here, was used in the southeastern United States as a masonry system for walls and foundations. It consisted of equal portions of sand, lime, oyster shells, and water.

Load-Bearing Construction

Even the earliest constructions made from concrete, which are relatively primitive and weak by today's standards, could be used in load-bearing applications, such as foundations and load-bearing walls. The concrete could be used by itself or as a binding material in masonry joints. Due to its hydraulic nature, concrete could also be used in wet locations and even underwater in foundation walls and footings for bridges and piers. This form of construction persisted for thousands of years and has been revived as part of the sustainability movement that emerged in the late twentieth century that has brought back the use of rammed earth techniques (sometimes introducing a cement component) in the construction workplace.

Until the rediscovery of portland cement, a lime-sand mortar mixture had been used that had fairly low compressive strength. This mixture was not hydraulic and disintegrated with prolonged exposure to liquid water or freeze-thaw cycles. For this reason, large constructions composed primarily of lime-sand mortar were considered to be infeasible, thus relegating it to use as the binder between masonry units. However, when portland cement was introduced in 1871, higher compressive strength and lower moisture permeability were two of its most valued properties. When used in constructions of that period that included higher-strength bricks, stone, and other masonry and concrete products, portland cement–based mortars performed admirably. However, as mentioned in the previous chapter, the unfortunate outcome from the attempt to use the earlier and somewhat weaker bricks with the new portland cement led to failure of the brick. This incompatibility still haunts preservationists when inexperienced property owners or contractors repoint older masonry walls with the later, and almost always incompatible, portland cement–based mortar.

Construction Types

Concrete construction can be classified as one of two types: unreinforced and reinforced. Unreinforced concrete does not include reinforcing materials, such as individual rods or woven mesh, and has low tensile strength. Reinforced concrete includes a network of metal rods or a woven metal mesh.

Unreinforced Concrete

Although the Romans used bronze rods to strengthen some of their concrete constructions, their concrete is largely categorized as unreinforced. Unreinforced concrete is generally a composition of sand, aggregate, water, and cement (either portland cement or a material that acts as natural cement). The addition of the cement provides a bond between the separate materials that is stronger and more durable than that of the earlier simple rammed earth wall. In unreinforced wall systems, sections of wall are built in "lifts" as tall as the form allows. The form is filled with the mixture, which is then tamped or packed until all voids within the form are filled. The process is repeated until the desired height is reached.

The use of unreinforced concrete systems often reflects the availability of local materials. While some early buildings were constructed of unreinforced concrete, today unreinforced concrete constructions are typically non-load-bearing and ornamental in nature. For instance, in the sixteenth century, the Spanish in the southeastern United States introduced the use of tabby, a mixture of lime, sand, oyster shells, and water. The formulation was based on the local abundance of oyster shells that could be used as an aggregate, as well as crushed and burned for lime. Tabby could be used for walls and foundations. Unless there is a significant proportion of lime or natural cement to act as a binder, these systems are, like adobe, vulnerable to moisture problems. Exposure to water will cause tabby to disintegrate.

In the United States, a number of notable buildings were constructed throughout the middle to late nineteenth century using unreinforced concrete as the primary construction material (see Figure 6-2). The introduction of cast stone in 1868 (Cowden and Wessel 1995), concrete block in 1900 (Simpson, Hunderman, and Slaton 1995), and architectural precast concrete in 1920 (Freedman 1995) were notable advances in the use of concrete as both structural and ornamental features of a building, but the widespread introduction of reinforced concrete in 1885 (Slaton et al. 1995) signaled a significant shift away from simple load-bearing constructions

Figure 6-2 The Ponce de Leon Hotel (now Flagler College) in Saint Augustine, Florida, was constructed in 1885–1887 using unreinforced concrete construction methods.

for walls, footings, and other foundations to the use of the skeletal structural frame. While precast concrete products are used in load-bearing applications, the vast majority of concrete used in buildings from the early twentieth century on is for reinforced skeletal frames, foundations, and slabs.

Reinforced Concrete

Reinforced concrete was developed in France and England in the second quarter of the nineteenth century before it was introduced in the United States. In 1860, S. T. Fowler received a patent in the United States for a reinforced concrete wall, but it was not until the last quarter of the century that research allowed this concrete to be more commonly used. The work of Thaddeus Hyatt, Ernest Ransome, and William Ward fostered the acceptance of reinforced concrete. As an example, William Ward completed the first reinforced concrete building in the United States in 1875 in Port Chester, New York (Elliott 1993).

Reinforced concrete was used in place of unreinforced concrete constructions as well as an alternative to steel in a skeletal framing system. The key to the success of reinforced concrete lies in the integration of reinforcement bars, or "rebars," developed in the late nineteenth and early twentieth centuries. Early experimentation revealed that smooth steel reinforcement bars permitted creep (linear displacement) to occur as the concrete shrank during curing. This creep allowed cracks to open within the concrete that, in turn, weakened the concrete, as well as allowing water to penetrate and corrode the rebar. A variety of "deformed" rebars were investigated that led to the configurations used to this day.

Reinforced concrete was cast in place using formwork that includes an integrated assemblage of rebars. The concrete mixture was placed in the formwork and hardened before the formwork was removed. To ensure that all voids were filled, the mixture was tamped or vibrated in place. Tamping was done by hand until the late 1940s, when pneumatic vibrators were introduced (Beckman and Bowles 2004). A pneumatic application system known as "Shotcrete" that sprayed wet concrete onto the rebars was introduced in 1910 (Sullivan 1995).

Throughout the nineteenth century, experimental applications were tested to determine how best to use reinforced concrete. By the early twentieth century, adequate progress had been made to permit the construction of the Ingalls Building in Cincinnati in 1903, the first "skyscraper" using a concrete structural frame (Elliott 1993). Having passed that milestone, designers were ready to push concrete to new limits in design.

An increasing number of buildings for academic, institutional, residential, commercial, and industrial uses were built of reinforced concrete. Using structural systems modeled on those of mill buildings, the construction of new manufacturing plants changed how construction was done and how loads were supported by the structural system. In this period, the work of Ransome was augmented by innovations in manufacturing processes, construction techniques, and reinforcing bar designs by the inventor Thomas Edison and by such notable designers and architects as Julius and Albert Kahn, as well as continuing work by English, French, and German researchers (Coney 1987; Elliott 1993; Slaton et al. 1995).

The dimensional changes and creep that occur when concrete shrinks as it cures or is exposed to long-term compressive loads became a problem when concrete was used in conjunction with reinforcement materials. The concept of "prestressing" reinforced concrete to anticipate expected dimensional changes was explored in the

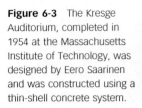

Figure 6-3 The Kresge Auditorium, completed in 1954 at the Massachusetts Institute of Technology, was designed by Eero Saarinen and was constructed using a thin-shell concrete system.

first half of the twentieth century, with several investigations occurring in the 1920s and 1930s. In 1949 prestressed systems gained acceptance with the design of the Walnut Lane Bridge in Philadelphia (Newlon 1995). While originally used in civic works, such as bridges, these systems have also been used in building construction.

Architects like Eero Saarinen and Frank Lloyd Wright experimented with flowing forms and cantilevered construction. The mid-twentieth century saw the ever-expanding use of reinforced concrete in new construction techniques, such as precast panels, thin shells, folded plates, and corrugated slabs (Michaels 1950; Raafat 1958). Reinforced concrete eventually came to be used in numerous civic buildings where large free span areas were desired (see Figure 6-3).

Other Concrete Applications

Cast concrete has been used in the United States for more than 150 years. By varying the formulation, concrete products have been produced with finishes ranging from coarsely pitted to extremely smooth while providing excellent compressive strength. Concrete castings were intended to become a less expensive alternative to natural stone while simulating the appearance of stone.

Cast Stone

Basically made from portland cement, clay, very fine aggregates, and water, cast stone was introduced to simulate fine-grained, evenly colored natural stone but was produced less expensively. While the use of cast stone can be traced back to the Middle Ages, it was not until the late eighteenth century that a cast stone product based on a combination of the terra-cotta formulation of Coade stone mixed with portland cement was used in the United States for architectural ornament and interior decoration. Cast stone saw only limited usage prior to the production of natural cement or hydraulic lime around 1820 (Pieper 2004). The cast stone industry evolved slowly until 1868, when a patent was awarded for Frear stone. Cast stone can be used in limited structural applications and in a number of ornamental applications, such as window moldings, door surrounds, and small columns. In recent years, cast stone has also been used as a substitute for deteriorated natural stone on rehabilitation and restoration projects when the use of actual stone was not feasible.

Concrete Block

The desire to produce a simulated stone product from concrete to reduce construction and labor costs led to the introduction of the concrete block, also known as the "concrete masonry unit" (CMU) in 1900 (Simpson 1999). The block consisted of a combination of portland cement, water, sand, and stone aggregates. To reduce the weight of the block, the casting process included two or more hollow cores within each block. The machine used to make the blocks could be configured to simulate a variety of decorative treatments commonly used on dressed stone. The two most common patterns were the "simple plain faced" and the "rock face," and examples of both can be found throughout the country (see Figure 6-4). The rapid, widespread adoption of this product was due to its availability through mail order distribution. By 1907, over 100 companies were selling the block-making machines through catalog distributors, such as Sears and Roebuck, which also promoted their use in the construction of their Sears and Roebuck kit homes. The machines were sold through a distribution system that took advantage of the Rural Free Delivery system used to deliver mail to rural areas and thus were rapidly dispersed across the country.

The popularity of the handmade block-making machine increased dramatically in the following two decades and led to the eventual automation of the process. By 1910, more than 1000 firms were producing concrete block across the country. Initially, the new industry was plagued by lack of standardization in sizes, so in 1924 the standard size was set at 8 inches × 8 inches × 16 inches, and by the end of the decade, 90 percent of the manufacturers were producing the standard-sized block. In 1917, the industry introduced lightweight aggregates such as coal cinders that made the blocks easier to handle in construction. The 1930s and 1940s saw the introduction of other lightweight aggregates, such as pumice, expanded shale, slag, cinders, and slate, which were marketed under a variety of licensed product names (Simpson 1999). In 1934, a machine was introduced that could make nine blocks

Figure 6-4 Cast concrete blocks were introduced at the turn of the twentieth century and were used as foundations and walls for houses, commercial buildings, light manufacturing buildings, garages, and outbuildings. Rock-faced concrete block, shown here, was a popular choice.

Figure 6-5 Simulated masonry has been insensitively added to these row houses in the Pullman landmark district in Chicago. Compare the transom window location above the doors for each house.

per minute, and by 1940 most other aspects of the block-making process had been automated as well. As time went on, the plain-faced block became increasingly popular as they were used as the basis of modern cavity walls (Simpson, Hunderman, and Slaton 1995).

Simulated Brick and Stone

During the period when masonry increasingly became used as a veneer on otherwise wood frame buildings, the demand for cheaper, quicker, and easier construction systems made its mark on the residential sector with products that imitated brick and stone. One of the more common ones was a simulated masonry system (e.g., permastone) that used a series of molds to imprint a masonry pattern on a veneer of wet concrete to give the appearance of separate masonry units (see Figure 6-5) (McKee 1995).

IDENTIFYING DECAY MECHANISMS

Due to the moisture absorption characteristics, chemical and physical properties, and their often similar or corresponding construction applications and locations within the construction, concrete has many decay mechanisms similar to those of masonry construction. Decay mechanisms specific to concrete will be presented here individually wherever it is appropriate to do so, but the reader is urged to review the mechanisms described previously for masonry to gain a complete sense of the factors affecting concrete construction.

As with masonry construction, assessment of decay in concrete construction may require the use of nondestructive testing techniques to obtain a true indication of existing conditions and internal decay. The important aspects to assess are (1) overall integrity of each concrete element, (2) surface conditions, (3) condition of the connections holding the concrete element in place, (4) evidence of poor original construction practices, (5) evidence of previous repairs or modifications, and (6) the potential causes of any deterioration found during the inspection process. The inspection should be performed by a qualified structural engineering consultant with a specialization in concrete structures. This consultant should also be able to determine what seismic upgrades are needed or how the structure can be modified to change the floor slab within the existing shell of the building if that is being considered in the rehabilitation or reuse of the building.

Early efforts in designing and building concrete structures were conservative by today's standards. Beyond any potential design flaws introduced into the original design, however, the fabrication processes for the in-place casting of these elements also introduced numerous construction flaws and material deficiencies that would not meet today's construction standards. These flaws include the incorrect proportioning of sand, aggregate, water, and cement; lack of curing protection during cold weather construction; inadequate or no allowance for thermal expansion and contraction; and the use of salt-laden or inadequately cleaned sand and reactive aggregates in the concrete mix itself. These factors can cause significant reductions in expected load capacity. Even well-proportioned mixes, if not properly mixed and/or tamped when they were placed, can have voids and honeycombs that allow moisture to collect and accelerate deterioration. These factors can lead to erosion, cracking, or failure of the construction and are frequently recognized by staining from moisture-related phenomena. However, since many of these unreinforced constructions are typically concealed beneath or behind other materials, the degree of their inadequacy may be concealed and will need to be exposed before a correct determination of conditions can be made. Another construction issue is whether the individual lifts or pours were allowed to cure before the next lift was placed on top of them. This delay could have caused the formation of a "cold joint" that may not have bonded fully to the previous lift. This joint could create a weak zone in the construction that could fail and admit water, cause the construction to collapse, or both.

Construction Issues

Reinforced masonry can also display such notable problems as surface scaling, surface cracks, and abnormalities due to creep and shrinkage from the original time of construction. One factor that has led to problems in meeting modern performance requirements is the improper placement of rebar. Rebar, especially if it is made of ferrous metals, must have a consistent cover of concrete (typically 2 inches in modern construction practice) to protect it from carbonation activity, moisture penetration, and subsequent corrosion. The combination of carbonation and long-term moisture penetration reduces the alkalinity of the concrete and promotes rebar corrosion. When ferrous materials corrode, they expand and can cause the concrete cover to delaminate and further expose the rebar to the elements. These poor construction practices, together or separately, can all act to reduce the structural capacity of concrete assemblies.

During inspections, it is critically important to assess the cause and severity of any cracks or deflections in structural components. An assessment must be made to determine if the cracks and deflections are actively growing or have reached equilibrium. Unless the potential long-term stability of a concrete construction is evaluated initially, any proposed changes to the structural loading of the building may accelerate the destruction of these members. Assessing the hardness and strength of the concrete in light of the proposed use is also an important part of this evaluation. What may have been adequate for the prior use may not be so in the new one.

When many early but now historic reinforced concrete buildings were built, they may have been overdesigned by modern standards. In many instances, this overdesign means that there is sufficient additional bearing capacity to allow increased loads within the building, such as adding intermediate floors between existing floors in an adaptive use project. On the other hand, the construction may have serious design defects with regard to modern codes. Both unreinforced and reinforced concrete were in common use well before modern seismic codes were developed and therefore may lack many of the common safety features of modern building codes. The structural consultant should be able to advise the design team on what options are available in this regard. In either case, sufficient analysis must be completed to meet the current seismic code requirements for the expected use of the finished project. Methods for seismic retrofitting will be discussed in further detail later in this book.

Moisture

Water is perhaps the primary enemy of concrete construction due to the underlying chemical nature of concrete and its manufacturing process. Concrete is a porous material that will absorb water readily unless protected by a coating or moisture barrier. Once introduced to the material, moisture can react with residual minerals and salts within the concrete and accelerate any decay mechanisms already in place. Therefore, water must be controlled to account adequately for surface runoff, eliminate efflorescence and reactions with sulfates and other salts in the concrete, and eliminate moisture penetration into cracks and voids.

Water Runoff

As with masonry, the flow of water on concrete must be controlled. If this is not done properly, staining, erosion, and decay can occur. The use of drip edges and guttering systems on exterior surfaces can keep water from forming a path across the concrete surface, where it may find its way into any crack or void. Given a sufficiently long period of time, the erosion of finer aggregates can severely weaken a wall (see Figure 6-6).

Figure 6-6 The repeated back splashing from rainfall has eroded the base of this wall (lower left center) and, if left uncorrected, could result in the failure of the wall altogether.

Likewise, as water finds its way into cracks and disruptions in the surface, the combined effects of freeze-thaw cycles can work to degrade the concrete further. Lack of original water controls, or later modification to the construction that may have compromised the original controls built into the design, should be corrected so that continued staining, erosion, and decay can be mitigated. Maintenance practices should also be evaluated to determine if deferred or poor maintenance has contributed to inadequate runoff control that has resulted in the retention of water at drain inlets, roof scuppers, and gutters.

Chemical Reactions

In the 1950s, a number of buildings were constructed of exposed concrete structures and cladding. Because of the exposed nature of the construction, protection of rebar was unsuitable to prevent carbonation attacks. Carbonation occurs when atmospheric carbon dioxide mixes with water and subsequently forms carbonic acid. The corroded rebar and adjoining surfaces required substantial repair in less than twenty years. The decay was unsightly but, if repaired quickly, was rarely problematic from a structural safety perspective. Beyond the mechanisms that produce efflorescence in other masonry products, concrete is also vulnerable to a number of chemical reactions due to materials used within the mixture itself and local environmental conditions (see Figure 6-7).

While carbonation is considered a natural process in the curing of concrete, other chemical reactions occur as a result of local conditions. One of these processes occurs when chlorides within the concrete itself reduce the alkalinity of the concrete that normally protects the reinforcement from corrosion, and these chlorides then act on the reinforcement itself. Similarly, sulfates in the soil can attack the concrete composition and reduce the structural integrity of the concrete. If sufficiently severe, sulfate attacks can destroy a concrete structure in as little as twenty years. These chlorides and sulfates may stem from the use of unwashed sands, especially

Figure 6-7 The staining on the underside of this concrete canopy slab is the result of the crystallization of dissolved salts that leached from within the concrete.

in marine environments, where salts and sulfate compounds are readily found in the soil. Likewise, certain areas have acidic soils that affect groundwater or experience acid rain conditions that can act to neutralize the alkalinity of the concrete. In some instances, these compounds come in contact with the concrete via groundwater contaminated with these compounds or overspray from local bodies of seawater. In any event, these salts show up as efflorescence on the surface of the concrete. In some instances, the decay appears as fine sand particles on the surface of the concrete. These particles are actually decomposing fine aggregates from the concrete that often can be removed with minimal brushing force. Other reactions occur when reactive aggregates containing these salts and sulfates are used in the original mixture to create the concrete.

An alkali-silica reaction can occur when alkalis in portland cement react with silica in the sand and aggregates. This reaction occurs when reactive mineral particles in the concrete encounter a sufficient alkali content and moisture level. The reaction forms an expansive gel that, if located close to the surface, can cause cracks in the surface to appear. These unsightly cracks can disrupt the surface of the concrete and subsequently allow varying-sized pieces of the construction to break away (Beckman and Bowles 2004).

Freeze-Thaw Cycle

Like masonry, concrete is vulnerable to freeze-thaw cycle damage. Left uncorrected, this deterioration can continue to affect an ever-increasing area and allow deeper penetration of moisture into the concrete. In reinforced systems, this deterioration can accelerate the corrosion mechanisms acting on the rebar as more concrete spalls from the covering concrete and the rebar is exposed to increased effects of carbonation and direct contact with moisture itself.

Soiling and Abrasive Cleaning

The lime-sand concretes used in the first half of the nineteenth century do not have the durability and strength characteristics of modern portland cement–based concretes. Early lime-sand-based concrete ornaments are susceptible to soiling mechanisms similar to those that attack lime-sand mortar and soft fired bricks. Combined air pollution, water penetration, and chemical reactions can cause the formation of crusts that, when removed, take historic building fabric along with them. Similarly, as noted in the previous section, chemical reactions can create salt particles and crystals to form on and within concrete using portland cement.

Many of the precautions for abrasive cleaning of masonry also apply to the cleaning of concrete. While the problems are more common in repointing joints, concrete assemblies may not have adequate stability to withstand any abrasive cleaning techniques. Abrasive methods can remove the outermost layer of unconsolidated materials and cause pitting, etching, and erosion.

REMEDIATION METHODS

As with decay mechanisms, several remediation approaches similar to those used in masonry can be applied to concrete construction. Remediation methods specific to concrete applications will be described individually below wherever it is appropriate to do so.

Stabilization and Conservation

As the inspection process proceeds, stabilizing the concrete structural elements that have become deteriorated or are in danger of failing may be necessary. When the deterioration poses a threat to safety, immediate stabilization must be started.

Figure 6-8 Temporary secondary supports have been erected in this shell to keep this building on Alcatraz Island in San Francisco from collapsing.

Stabilization ranges from erecting netting to prevent loose or broken pieces from falling to the ground and striking passersby to covering the exposed surfaces with tarpaulins to protect the concrete from the elements and then erecting a secondary support system to prevent the structure from collapsing (see Figure 6-8). Secondary support systems can be constructed from structural steel framing that may include lateral bracing along walls and various configurations of beams, columns, and plates that are temporarily secured to the structure before and during the construction phase of the project. As the defects are corrected, these stabilization systems may be removed.

Due to its chemical reactivity, concrete conservation is problematic. The physical and chemical properties of the materials used to complete the repair need to be carefully considered. Compressive strength, thermal expansion, chemical interactions, and visual integrity must be assessed in terms of how they relate to the corresponding properties of the concrete they are intended to stabilize or conserve. Lastly, the reversibility of the proposed measures must be evaluated so that they may be readily removed if they are later found to be inappropriate.

As with masonry, the selection of coatings for concrete must be done with care to prevent water from being trapped within the construction material, particularly in areas subjected to freeze-thaw conditions. For concrete construction, in addition to keeping out moisture, there is also the goal of repelling acids, salts, and sulfates that may

Coatings

be contained in the groundwater as well as in airborne pollution. When below-grade repairs are considered, caution should be used in applying coatings. Unless all surfaces of the structure can be protected, the treatment may be compromised by exposure from otherwise inaccessible surfaces (e.g., the underside of a basement floor slab).

Many exposed concrete structures built throughout the twentieth century have suffered decay due to local environmental conditions. While the use of coatings is feasible aboveground to mitigate staining and airborne decay mechanisms, most coatings available to treat exposed concrete impart a glossy finish that would seriously alter the visual integrity and appearance of the building. These products therefore have seen extremely limited use in concrete remediation.

Infill Repairs

Infill repairs can be completed using new concrete, grout, or epoxy mortars. A compatible mixture of concrete can be applied in a simple patching project. Grout, which is a liquefied mortar, and epoxy mortar, which is a combination of epoxy mixed with sand and cement, can be used to fill open voids and cracks in an unreinforced concrete construction using application processes similar to those used in treating deteriorated stone. These treatments bind the existing decayed material into a more stable condition. Depending on the material used and the location where it is applied, these repairs may or may not be reversible.

In repairing ornamental concrete elements, a replicated ornamental element or a true dutchman could be crafted from the variety of substitute materials described in the previous chapter. However, the use of plastic repairs is common in both structural and ornamental concrete elements. Many of these repairs are done using new concrete, grout, or epoxy mortar mixes. It is advisable to investigate carefully compatibility and reversibility. Color, color fastness, texture, tooling, weathering characteristics, and durability are other important criteria to consider for areas seen by the public or building occupants.

In reinforced concrete systems, damage often occurs at the surface where structural rebars are corroding. In many instances, the concrete covering the rebar has cracked, scaled, spalled, or delaminated, leaving the rebar exposed to more moisture and carbonation damage. The deteriorated concrete can be cleared and tooled to provide a stable, firm surface for a replacement concrete or epoxy mortar patch. The corrosion in the rebar must be halted before applying the patch to the concrete member. Tooling the concrete to provide a sufficient thickness of replacement concrete is recommended since simply applying a concrete patch will result in a vulnerable bond in areas where the replacement patch is thinner.

Structural Reinforcement

Beyond the temporary stabilization provided at the start of the project, it may be necessary to either permanently reinforce structural components or provide additional reinforcement for the expected use of the building after completion. Seismic upgrading may need to be addressed at this time. A thorough analysis must be completed by a structural consultant before proceeding. In completing the infill repair work noted above, concerns related to overall load capacity should be addressed. If the existing system, with its repairs in place, cannot support the expected loads, then additional structural capacity must be added.

While the exact nature of these reinforcement methods could and does fill several volumes, an overview will be provided here. The additional capacity can be added in several forms. First, the concrete cover on the rebar can be removed and additional rebar installed. This new rebar is then covered with new concrete applied to the original structural element. This process may increase the dimension and cross-sectional area of the element being upgraded. This additional size must be accounted for in terms of its impact on clearances and the tolerances needed to install any applicable finishes. Second, instead of installing new rebar and replacing the concrete cover, the existing construction can be stabilized and repaired as needed and external reinforcement applied. These stabilizers take on numerous configurations but largely consist of structural steel reinforcements, such as those described in the section on stabilization and conservation.

These reinforcement strategies must be carefully considered in terms of how they will be viewed by the public. Aesthetic appearance is important in maintaining the visual integrity of historic interior spaces and exterior forms. These proposed adjuncts must be concealed from view where possible or they may be rejected by oversight design review agencies.

In certain instances, deterioration or inadequate capacity may require the structural element to be completely removed and replaced with appropriate new construction. The old construction is removed and the new construction is installed. This strategy is dangerous in that the structural stability of the remaining construction must be securely maintained so as not to cause nearby structural elements or the building as a whole to collapse when the old construction is removed. This strategy is not impossible but, depending on the scope of proposed removal or structural system modification, it can become quite expensive. As adaptive use projects continue, the tendency to modify the structural system is a constant issue. However, one of the primary tenets of the *Standards* regarding reuse is that the new use must be compatible with the existing construction. Therefore, wholesale replacement of structural elements may negate any opportunity to meet the standard for rehabilitation.

While substitute materials, such as glass fiber reinforced concrete (GFRC), cast stone, and precast concrete replacement components, may be used to replicate ornamental aspects of concrete construction, their structural properties must be reviewed to determine if they can meet the needs of the finished project. In any event, as with all of the remediation strategies described above, new components must visually match the original elements and be physically compatible with adjoining concrete or other materials remaining in place after the damaged element is removed or a missing element is replaced.

Replication/ Replacement

References and Suggested Readings

Allen, Edward. 1999. *Fundamentals of building construction: Materials and methods.* 3rd ed. New York: John Wiley & Sons.

Association for Preservation Technology International. 1997. *Historic concrete investigation and repair.* Chicago: Association for Preservation Technology International.

Beckman, Poul and Robert Bowles. 2004. *Structural aspects of building conservation.* 2nd ed. Burlington, MA: Elsevier Butterworth Heineman.

Belle, John, John Ray Hoke, Jr., and Stephen A. Kliment, eds. 1994. *Traditional details for building restoration, renovation and rehabilitation from the 1932–1951 editions of Architectural Graphic Standards.* New York: John Wiley & Sons.

Boothby, Thomas E., M Kevin Parfitt, and Charlene K. Roise. 2005. Case studies in diagnosis and repair of historic thin-shell concrete structures. *APT Bulletin,* 36(2/3): 3–11.

Coney, William B. 1987. *Preservation of historic concrete: Problems and general approaches.* Preservation Brief No. 15. Washington, DC: United States Department of the Interior.

Cowden, Adrienne B. and David P. Wessel. 1995. Cast stone. In Jester 1995, 86–93.

de Jonge, Wessel. 1997. Concrete repair and material authenticity: Electrochemical preservation techniques. *APT Bulletin*, 28(4): 51–57.

De Mare, Eric, ed.1951. *New ways of building.* London: Architectural Press.

Elizabeth, Lynne and Cassandra Adams. 2000. *Alternative construction: Contemporary natural building methods.* New York: John Wiley & Sons.

Elliott, Cecil. 1993. *Technics and architecture.* Cambridge, MA: MIT Press.

Freedman, Sidney. 1995. Architectural precast concrete. In Jester 1995, 108–113.

Frens, Dale H. 2002. Restoration of the concrete roof of the Mercer Museum in Doylestown, Pennsylvania. *APT Bulletin*, 33(1): 13–19.

Harris, Samuel Y. 2001. *Building pathology: Deterioration, diagnostics, and intervention.* New York: John Wiley & Sons.

Hornbostel, Caleb. 1991. *Construction materials: Types, uses, and applications.* 2nd ed. New York: John Wiley & Sons.

Jester, Thomas C., ed. 1995. *Twentieth century building materials: History and conservation.* New York: McGraw Hill Book Company.

McKee, Ann Milkovich. 1995. Simulated masonry. In Jester 1995, 174–179.

Michaels, Leonard. 1950. *Contemporary structure in architecture.* New York: Reinhold Publishing Corporation.

Newlon, Anthony. 1995. Prestressed concrete. In Jester 1995, 114–119.

Newman, Alexander. 2001. *Structural renovation of buildings: Methods, details, and design examples.* New York: McGraw-Hill Book Company.

Park, Sharon.1988. *The use of substitute siding on historic building exteriors.* Preservation Brief No. 16. Washington, DC: United States Department of the Interior.

—— 1996. *Holding the line controlling unwanted moisture in historic buildings.* Preservation Brief No. 39. Washington, DC: United States Department of the Interior.

Peters, Tom F. 1993. *Building the nineteenth century.* Cambridge, MA: MIT Press.

Peterson, Charles E., ed. 1976. *Building early America: Contributions toward the history of a great industry.* Radnor, PA: Chilton Book Company.

Pieper, Richard. 2004. *The maintenance, repair, and replacement of cast stone.* Preservation Brief No. 42. Washington, DC: United States Department of the Interior.

Prudon, Theodore H. M. 1989. Simulating stone, 1860–1940: Artificial marble, artificial stone, and cast stone. *APT Bulletin*, 21(3/4): 79–91.

Raafat, Aly Ahmed. 1958. *Reinforced concrete in architecture.* New York: Reinhold Publishing Corporation.

Rabun, J. Stanley. 2000. *Structural analysis of historic buildings: Restoration, preservation and adaptive reuse for architects and engineers.* New York: John Wiley & Sons.

Ritchie, T. 1978. Roman stone and other decorative artificial stones. *Bulletin of the Association for Preservation Technology*, 10(1): 20–34.

Sickels-Taves, Lauren B. 1997. Understanding historic tabby structures: Their history, preservation, and repair. *APT Bulletin*, 28(2/3): 22–29.

Simpson, Pamela H. 1999. *Quick, cheap and easy: Imitative architectural materials, 1870-1930.* Knoxville: University of Tennessee Press.

Simpson, Pamela H., Harry J. Hunderman, and Deborah Slaton. 1995. Concrete block. In Jester 1995, 80–85.

——. 1999. *Quick, cheap and easy: Imitative architectural materials, 1870–1930.* Knoxville: University of Tennessee Press.

Slaton, Amy E., Paul E. Gaudette, William G. Hime, and James D. Connolly. 1995. Reinforced concrete. In Jester 1995, 94–101.

Slaton, Deborah E. and Harry J. Hunderman. 1995. Terra cotta. In Jester 1995, 156–161.

Speweik, John P. 1995. *The history of masonry mortar in America, 1720–1995.* Arlington, VA: National Lime Association.

Sullivan, Anne. 1995. Shotcrete. In Jester 1995, 102–107.

Weaver, Martin E. 1995. *Removing graffiti from historic masonry.* Preservation Brief No. 38. Washington, DC: United States Department of the Interior.

—— 1997. *Conserving buildings: A guide to techniques and materials.* New York: John Wiley & Sons.

CHAPTER 7 Architectural Metals

MATERIAL OVERVIEW

Numerous building products have been composed of architectural metals. Metal objects were crafted by the blacksmith, who created many things associated with daily life as well as those specifically related to construction—nails, hinges, door latches, and so forth. Over time the use of hand-wrought metals faded with the rise of cast iron and the eventual introduction of steel and other alloys. By the twentieth century, a wide range of architectural metals had been developed. This chapter traces the development of architectural metals used throughout the past four centuries in the United States and discusses their fabrication, decay, and remediation.

Historic Origins and Sources

The use of metal in civilized cultures can be traced to those cultures' emergence from what archaeologists refer to as the Stone Age, when they shifted from using stone to using metals for tools, jewelry, and other items. The first step was using copper, but it was the later Bronze Age, when copper and tin were used to create bronze, that indicated a growing maturity in using metal. Archaeological evidence indicates that the first use of bronze in Asia occurred as early as the fourth millennium BC. Bronze making spread across Asia to Europe and northern Africa throughout the early part of the first millennium BC and developed in what is now South America in the early first millennium AD (Hornbostel 1991). The next period of development was the Iron Age, in which iron smelting emerged. Methods for increasing the temperature in smelting processes allowed iron, which has a higher melting point than earlier smelted metals, to be melted from raw ore.

131

Pure metal can occur naturally, but normally metal must be extracted from ore containing it. This separation occurred in a smelting process in which a high-temperature oven (and later the open hearth blast furnace) melted the various metals into a liquid state. These metals could be removed and allowed to cool in a purer form than the one in which they existed before smelting. Smelting systems gradually increased in sophistication, efficiency, and temperature control to allow a great variety of metals to be obtained. In several instances, early metal usage was limited due to high costs. Both steel and aluminum were initially considered too costly for architectural applications until production improvements made them significantly more available and less expensive.

Certain metals, such as gold and copper, in their purest form are relatively soft and malleable. Mixing two or more pure metals together, however, created an alloy that was stronger and still presented a highly valued color and workability. As different metals became identified in their pure forms, they could be mixed to make various alloys, such as combining tin and copper to make bronze. These features were important, as many of these metals were used in making jewelry and ornaments for the rulers and the richest members of the local society.

Several methods were used historically to create or shape metal forms. Metal can be cast, extruded, rolled, stamped, and hand wrought. Molten metal can be cast into a shape by "puddling" or pouring it into a mold, usually made from densely compressed sand or ceramic materials. Extrusion involves pushing molten metal through an opening shaped in a desired cross section. Rolling can be done using either hot or cold methods. For hot-rolled steel, a large roller presses metal into a desired profile or a flat plate. Cold rolling performs similar shaping but in much thinner thicknesses. Stamping is done on cold metal to impress figured shapes into the metal that is used for siding, roofing, ceilings, and wall panels. Reheating cold metal and shaping it with a hammer or other devices, which has been the process used by blacksmiths for several centuries, creates hand-wrought or machine-forged products.

In the earliest American colonial settlements, wrought iron was the primary architectural metal in use. Wrought iron objects ranging from hardware to household goods were created by blacksmiths using traditional metal fabrication processes brought from Europe. Some settlements, such as the English colonies, featured ironwork that was simpler and a bit less decorative than the French and Spanish varieties (University Museums and Art Galleries 1976). Eighteenth-century casting processes developed in England spread to North America in the early nineteenth century. With the Industrial Revolution came many improvements in metal manufacturing and the introduction of numerous alloys that combined two or more metals. The success of steel and other metals introduced into widespread commercial production led to the eventual reduction of wrought iron and cast iron use; today, they are used largely for ornamental objects.

Types of Metals

Like wood and stone, metals can be classified on the basis of their material characteristics, such as strength, density, color, and other physical properties. Architectural metals can be made of one pure elemental metal (e.g., lead, iron, copper, tin, or zinc), but more commonly they are alloys. For architectural purposes, metals can be classified as ferrous (containing iron) and nonferrous (not containing iron). Ferrous metals are vulnerable to oxidation or rusting when exposed to chemical compounds in air or water (see Figure 7-1).

Figure 7-1 Ferrous metals contain iron that corrodes when exposed to chemical compounds in air and water. As shown here, the corroding steel has cracked the base of this pylon.

Many metals used in the past 150 years in North America have been alloys. While pure metal was used in jewelry making, alloys provided a stronger and usually a more durable product. Purer forms of metal have also been used as coatings, such as zinc, lead, or tin coatings, applied to ferrous alloys to prevent corrosion. The properties of the more common alloys and elemental metals used for architectural purposes will be discussed below.

Copper, Bronze, and Brass

Although originally used in jewelry, copper, which is a nonferrous metal, was later combined with tin or zinc to form, respectively, the alloys bronze and brass. Bronze is typically composed of 60 percent copper and 40 percent tin. Brass is 50 percent or more copper and between 5 and 20 percent zinc. The use of these metals flourished until the fall of the Roman Empire and then severely declined in Europe until the Middle Ages. Because of their lower strength, these metals were used in a variety of nonstructural applications, including castings and extrusions for doors, window framing, and other ornamental or utilitarian objects.

These metals were also hammered or rolled into sheets and used in a variety of cladding applications. By the late nineteenth century, copper and bronze sheeting were being used in roofing as stamped shingles, standing metal seamed roofing, and flashing and guttering systems.

In the nineteenth century, improvements in manufacturing led to the use of these metals for piping, door and window hardware, plumbing fixtures, banisters, newel posts, and a variety of nonstructural ornamental uses. The use of copper for electrical wiring has long been standard.

Cast Iron

The earliest of the ferrous-based architectural metals, cast iron contains the highest amount of carbon, typically 2.2–5 percent (Campbell 1997). The carbon is a residual element that remains in the cast iron after the raw iron ore has been heated in an oven. The iron melts and is drawn off into a mold, usually made of packed sand, where it is allowed to cool. This iron, referred to as "pig iron," can then be melted and cast into the mold of the object needed. Cast iron can withstand compression and was often used in columns in the nineteenth century.

Wrought Iron

Wrought iron is formed when an iron bar is worked or "wrought" using a hammer and an anvil. The reheating to a nearly molten state makes the iron more workable, and the repeated cycles of hammering and reheating until the iron is in the desired shape drives out a certain amount of the carbon and aligns the crystalline iron within the object. As a result, the carbon content is reduced to less than 0.1 percent of the overall content. The reduction in carbon and the forging sequence give wrought iron a higher tensile strength than cast iron. Wrought iron came to be used for objects that took advantage of this property. These uses included hinges, door latches, and structural beams. While wrought iron is not necessarily ornamental in intent, certain regions have more highly ornamental and decorative ironwork based on local vernacular traditions.

Steel

With the transition from coke to coal as a fuel source came the opportunity for higher temperatures inside ovens and, later, blast furnaces. In the early nineteenth century, steel was largely unavailable due to the inability to produce sufficient quantities for useful construction applications. The earliest industrialization processes of the nineteenth century were an extension of hand working of iron. Large rollers and other mechanized processes were introduced with the intention of extending and replacing hand-wrought iron in building construction. These early efforts further reduced the carbon content of the final product, and the use of steel in architectural applications slowly began. The introduction of the Bessemer process, a refinement that improved the way steel was processed in the blast furnace, significantly changed that situation. Developed by Henry Bessemer in England in 1855 and further refined in 1879 by Sidney Gilchrist Thomas and Percy Gilchrist, the process substantially increased the opportunity to use low-grade ores in the production of steel. Along with the regenerative open-hearth furnace introduced in the 1860s in England by William Siemens, this process is credited with enabling the expanded steel production capacity needed in the United States to make steel readily available for structural and architectural applications (Gayle and Gayle 1998). By the 1880s, skeletal steel framing systems for architectural applications were being constructed throughout the country. The Home Insurance Building completed in Chicago in 1885 was considered to be the first modern skyscraper using a skeletal structural steel framing system (Elliott 1993).

Steel is an alloy formed when carbon is removed from iron by a smelting and refining process. Steel can be classified as shown in Table 7-1.

Table 7-1: Carbon Content of Steel	
Low-carbon steel	<0.02%
Mild steel	0.25%
Medium-carbon steel	0.25–0.45%
High-carbon steel	0.45–2.0%

Source: Weaver (1997), 182.

Throughout the late nineteenth and early twentieth centuries, various steel alloys were crafted with differing amounts of nickel, manganese, tungsten, and silicon. Molybdenum, vanadium, and copper alloy steels were introduced prior to World War I. The mid-twentieth century saw the development of alloy steels using "formerly rare and entirely new metals" that met a variety of engineering requirements (Hornbostel 1991, 752) ranging from tools to machinery components to building products. The ingredients in these alloys inhibited corrosion, improved strength characteristics, and enhanced durability.

One steel product introduced in 1933 attempted to take advantage of the expansion qualities of the iron oxides that formed as the steel corroded. The material, called "weathering steel," was used in cladding and structural applications. The theory behind its development was that the low-carbon steel was to be left untreated and exposed to the natural elements. As the expected corrosion progressed, a patina of self-sealing rust would inhibit further penetration of moisture and corrosive chemical compounds (Scott and Searls 1995). Unfortunately, the rate of corrosion was sometimes misunderstood, and in those instances the metal rusted completely through. Another disadvantage of this material was the staining of adjacent surfaces as the iron oxides migrated to adjoining surfaces during rainstorms.

Aluminum

In 1821, the aluminum-bearing ore bauxite was discovered in France. At first, aluminum was considered a precious metal due to the difficulties of processing it. Production methods devised in 1886 greatly improved its availability and, like steel, it became readily available for architectural applications (Swanke, Hayden Connell Architects 2000). Although aluminum is not as strong as steel, it has found its way into a number of architectural products, particularly ornamental work and cladding systems. During World War II, it became a major structural component in lightweight structural systems for aircraft. The advances in fabrication, assembly, and connection methods during that war led directly to its use as a major component in curtain walls and storefront systems introduced in the 1950s and beyond.

Although aluminum is now given many different surface treatments, historically it was left with a smooth polished surface or a variety of machined finishes. Innovations in coating aluminum led to the introduction of (1) porcelain enamel panels (see Figure 7-2) in 1924, on which ceramic coatings of assorted colors were applied to an otherwise flat panel and (2) anodized paint was applied on the surface of aluminum in the late 1920s (Kelley 1995). Anodizing uses an electrical current to attract paint particles to the surface, where they bond more securely with the aluminum than in a simple sprayed or brushed-on application. These products and processes became widely popular in the 1950s.

Lead

The use of lead dates back several millennia BC and has found its way into many products. Perhaps the earliest architectural use was for lead piping in ancient Rome. Lead does not corrode as rapidly as ferrous metals, and the lead oxides formed when it does corrode act to seal the coated metal from moisture penetration. Lead also has a low melting point and is quite malleable. These characteristics encouraged its use in pipes, coatings, solder, window glass and cames and as an additive in various

Figure 7-2 Porcelain enamel panel systems were used for new and (shown here along the first-floor facade) renovated buildings.

alloys. Lead was long used as an additive in paints because of its superior adhesion, drying, and covering characteristics (Park 1995).

The use of lead in architectural building products was common until well into the twentieth century, when growing concerns in the 1960s and 1970s about lead poisoning resulted in its removal from a number of products, particularly painting products. While new non-lead-based products have been introduced, many lead-based building products still exist in historic buildings. Further details of lead-related health issues and lead abatement practices will be presented later in this book.

Tin

Tin has been used for many millennia. Pure tin is relatively soft and malleable and, like lead, is not used for structural purposes. The dominant uses for tin were as an ingredient in corrosion-resistant coatings, metal alloys (e.g., terneplate, bronze, and brass), gilding, and solder (Hornbostel 1991). Tin coatings on stamped metal panels have often led to the mistaken use of the term "tin ceilings" when in fact the substrate material was steel.

Other Alloys

Various metal alloys have been developed for use in architectural products. Three of them are nickel silver, Monel, and stainless steel.

Nickel silver, an alloy composed primarily of copper, zinc, and nickel, was developed in Europe and came into commercial use in the United States in 1835. Originally imported from China as "paktong," it was produced in England and Germany and was known as "German silver." Throughout the nineteenth century, nickel silver was used as a less expensive alternative to silver and silver-plated tableware. Only in the 1920s did it come into significant architectural use with the emergence of the Art Deco movement. Nickel silver then was used for plumbing fixtures, decorative panels, grilles, railings, doors, and as dividers between sections of terrazzo flooring. The use of German silver, by then known as "nickel silver," continued into the 1940s but was curtailed by nickel shortages during World War II. By the 1950s, nickel silver had been largely replaced by stainless steel and aluminum (Cowden 1995).

Monel, an alloy composed primarily of copper and nickel, was introduced for experimental architectural use in 1907. However, it was not fully accepted until the 1920s. Prized for its resistance to corrosion, the alloy was used in harsh environments or in locations where regular access was difficult. These uses included roofing and skylights, heating and ventilation ductwork, gutters and flashing, light fixtures, countertops, and sinks. Products made from Monel were fabricated by casting or assorted rolling methods. By the 1950s, stainless steel and aluminum began to replace Monel in architectural applications.

Stainless steel is an alloy composed of steel, chromium, and nickel. Simultaneous experiments in Europe and the United States began at the turn of the twentieth century, but stainless steel did not become commercially available in the United States until 1927. The use of stainless steel was driven by its corrosion-resistant properties that came from the use of chromium. The Empire State Building and the Chrysler Building in New York City were among the first high-profile buildings to use stainless steel. The steel content made the material stronger than Monel and aluminum and allowed stainless steel to be used in many of the same applications as those two alloys. In addition, stainless steel was used in window and door framing, curtain walls, storefronts, and other assorted exterior cladding applications. Unlike regular steel, stainless steel has been produced using the electric furnace that had been introduced in 1906 specifically for making high-grade alloy steel. The use of stainless steel and aluminum for curtain wall assemblies increased substantially in the 1950s. This occurred both in new construction and in updating the facades of existing buildings (Hornbostel 1991; Score and Cohen 1995).

CONSTRUCTION METHODS

While metals had been used in various aspects of construction for window cames, a variety of hand-wrought constructions, and simple castings, the process for fabricating architectural metal elements on a large production scale was limited due to the manual methods used and the resources needed to smelt the metals. In the Industrial Revolution, a variety of methods were devised to increase the production and quality of the available metals used for creating architectural building components.

Once individual pieces are formed, the two primary forms of connection in metal fabrication are welding and mechanical fastening. Welding is a process in which a compatible metal is melted and bonded to the pieces being joined using a torch. As the weld cools, it forms a permanent bond between the pieces. Mechanical fastening systems include rivets or bolts placed in aligned holes in the pieces to be joined. In some cases, the holes can be aligned and the fastener inserted to make the attachment. In others, a piece of plate steel or angle steel may overlap both pieces, and the rivets are inserted and secured through the plate and the joined pieces.

Casting

Iron, bronze, and brass objects made from castings were typically assembled as a sequence of plates with an ornamental finish on one side and a plain finish on the other. A negative mold was created by pressing a positive model into a bed of densely packed moist sand. The model, usually a plaster sculpture or a wood carving, was then removed and the metal was cast into the resulting void created by the model. Three-dimensional objects could be built up from a series of sectioned castings or cast using a lost wax process. On the sides that were not exposed to view, an assortment

Figure 7-3 Cast iron facades were popular throughout the third quarter of the nineteenth century because they provided a relative quick and inexpensive way to develop a substantial presence in both the population centers of the East and the growing cities and towns of the West.

of connection elements were cast or attached. The final object being created could then be assembled by connecting separate components together, leaving a hollow interior. In this manner, a larger element (e.g., a newel post or an entire building facade) could be assembled from smaller pieces. Casting such objects or assemblies as one solid, continuous piece was impractical for size and weight reasons, not only during the manufacturing process but for shipping and installation as well.

Historically, because of its carbon content, cast iron was brittle and low in tensile strength, but it did have relatively high compressive strength. As such, cast iron came to be used in objects that took advantage of its compressive strength, such as structural columns, as well as for a variety of ornamental uses. Cast iron columns were used throughout much of the mid-nineteenth century for building construction. James Bogardus introduced the first facade in the United States made completely from cast iron in 1849 (Elliott 1993). With this came the opportunity for facades to be created almost directly from a builder's guide that included well-articulated details, such as pilasters, pediments, and other ornamentation, according to the classical orders of architecture that were popular at the time, but without the expense and time needed to carve the multitude of subcomponents from stone. Once the casting molds were created, they could be reused a number of times to accommodate facades of varying lengths and buildings with different numbers of stories. For the next three decades, buildings around the country were constructed using traditional framing methods but with a cast iron facade installed that gave the building the immediate presence and respectability of stone without its attendant expense (see Figure 7-3). With the introduction of advanced steel production processes in the late nineteenth century, the use of cast iron for structural construction diminished significantly. Although these facades were an early forerunner of modern curtain wall systems, the brittleness of the material and the relative ease of fabricating curtain walls using steel and, later, aluminum and other architectural metals and plastics largely eliminated the use of cast iron for architectural applications by the mid-twentieth century.

Hand Wrought

The local blacksmith traditionally was the source for many household and utilitarian items found in the early colonial settlements. The presence of a blacksmith gave a settlement credibility as an established and growing community. Beyond household goods, the blacksmith was responsible for fabricating anything that was made from iron: nails, hinges, door latches, horseshoes, rifle barrels, tools, wheel rims, and so forth. As the industrial era of the early nineteenth century evolved, many of these products were siphoned off to other manufacturers. Among the first such shift was the production of cut nails. Eventually, the fabrication processes for many of these products were industrialized. While the slowly ebbing scope of the blacksmith trade continued into the twentieth century, the trade still remained useful in more ornamental crafts, such as the ornamental work found on entrance gates, balconies, window guards, and other such constructions (University Museums and Art Galleries 1976). The largest concentrations of this ornamental work (as well as cast iron) can be found in many of the former Spanish and French colonial settlements, such as St. Augustine, New Orleans, and San Diego, or in the former English settlements, such as Charleston and Savannah (see Figure 7-4).

Rolling and Extruding

With the technological improvements of the nineteenth century came the ability to press molten metals into specific shapes and thick slabs known as "plates" using large rollers to impress a profile into the metal. Similarly, by forcing the metal through a profile, an extruded shape could be made.

Figure 7-4 Wrought iron ornamentation is a character-defining feature of buildings in areas originally settled by the Spanish. Shown here is a balcony on this building in San Diego that celebrates that traditional form of metal work.

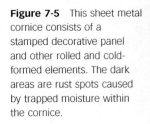

Figure 7-5 This sheet metal cornice consists of a stamped decorative panel and other rolled and cold-formed elements. The dark areas are rust spots caused by trapped moisture within the cornice.

Sheet and Strips

In the hot-rolled process that occurred when the metal was still in a molten state (hence the name), metal sheets and strips were made that served a variety of architectural and structural purposes. In 1867, a cold rolling process was developed to accomplish the same thing. These metal sheets are rated based on a standardized gauging system that denotes thickness. The gauge number actually runs inversely to the thickness (i.e., a higher gauge number represents a thinner sheet). The modern gauge designation runs from 1 (0.2813 inch) to 30 (0.0125 inch) (Hornbostel 1991).

Sheets could be made of steel, copper, bronze, tin, or other metals. The thinness of the sheet relative to its length and width resulted in a significant amount of flexure, which in turn reduced the sheet's stiffness. The thinness of the sheets, however, permitted them to be stamped in decorative patterns for shingles, roofing, and ceilings.

The flexibility of sheet metal also made it suitable for cold forming (e.g., bending and rolling) sheets into a variety of ornamental elements consisting of an external skin supported and attached to the building by a lightweight metal frame. A number of cornices and other types of building ornamentation were fabricated in this way in the late nineteenth century to replicate the more expensive stone and cast iron elements used earlier in the century (see Figure 7-5).

The ability to create bends, ridges, and corrugations enabled the sheet to be stiffened without adding weight. In certain roofing and flooring systems, this stiffening process enabled the use of corrugated metal sheets in place of thicker, heavier, and more expensive deck plating.

Bars and Plates

Metal plates ranging from $1^1/_2$ to 4 inches thick were made for various structural uses. Bars were no greater than 4 inches wide. Plates were elements that were wider than 4 inches. Ships, large tanks, and pressure vessels could be constructed from bars and plates welded together along their seams and attached (using rivets, bolts, or welds) to skeletal framing.

Many standardized sections used after the turn of the twentieth century came about through refinement of the shapes used in earlier cast iron columns and wrought iron beams and the practice of building structural sections from steel plates and bars. Bars and plates could be further worked by cold rolling them into angles, channels, and other folded shapes. The cold rolling actually strengthened the steel through the physical realignment at the crystalline level of the material (Baker and Langland 1947). These elements could be used by themselves or built up into a variety of truss configurations both large and small. These trusses were used in building roof and floor framing and other structural applications. Modern glued-laminated beams and engineered lumber have their roots, in a sense, in the innovative practices used for assembling small subcomponents of steel plating and bars for structural purposes.

Structural Framing Sections

By the late nineteenth century, standardized shapes were being made to keep pricing competitive. The structural steel frame emerged as an alternative to the then traditional load-bearing masonry construction. Combining the concept of timber framing with the material advantages of cast iron and wrought iron, steel framing finally overtook load-bearing masonry as a way to achieve taller buildings in the late nineteenth century without sacrificing significant floor space on the lower floors of the building (see Figure 7-6).

Figure 7-6 Standardized structural sections were developed to keep pricing competitive. Shown here is a column in an early-twentieth-century building that is undergoing renovation in Atlanta, Georgia.

IDENTIFYING DECAY MECHANISMS

As part of an inspection, the condition of structural materials and their structural capacity must be evaluated. While there may be external indications of structural problems, such as sagging floors or bulging walls in load-bearing masonry or wood-framed buildings, skeletal steel-framed buildings can conceal defects behind curtain walls and interior room finishes. In certain instances, this concealment may seem to warrant the removal of interior finishes and exterior curtain walls or cladding to enable a direct visible inspection. However, given the numerous nondestructive testing methods available, a qualified structural engineering consultant should be hired to make the assessment.

Construction Issues

Many codes and mathematical formulations for structural loading design have evolved, and the earlier construction may or may not be suitable for the intended use of the building after the rehabilitation, restoration, or renovation has been completed (Friedman 1995; Newman 2001). The *Secretary of the Interior's Standards* emphasize that a major factor to consider is the compatibility of the intended reuse of the building with the existing structural capacity as defined by contemporary codes and design standards.

One dominant construction issue in structural upgrading is the insertion of seismic upgrades and improving the building's ability to resist forces from winds, such as those generated by hurricanes and tornados. Careful consideration must be given to how upgrading will impact the visual and historic integrity of the building.

Moisture and Corrosion

Surface corrosion on metal objects can initially be of some benefit since the oxidation process may form a patina that seals underlying metal from further exposure. In other instances, if the material is thin or does not have substantial mass, the object will continue to corrode until it has completely disintegrated. Corrosion is first indicated by a discoloration of the surface finish, either as a dulling of the luster or as an actual color change. Left unabated, the corrosion can proceed to flake off as scale or fine powder.

Ferrous metals do not react well to chemical compounds in air or moisture. As a result, the iron oxidizes and expands in volume. When free to do so, the oxidation or rust can scale off from the object and expose underlying material to further corrosive action. Left untreated, the material eventually disintegrates into loose iron oxide pieces. Nonferrous metals, like zinc, lead, bronze, and copper, can form a patina that seals the surface from further corrosion. However, repeated cleaning of these objects may remove enough patina to enable corrosion to continue. Unfortunately, for objects located in wet areas, a portion of this oxidized surface can be washed off from the object and stain adjoining surfaces as the oxidation flows to the ground (see Figure 7-7). Likewise, repeated and intermittent moisture contact can accelerate corrosion, particularly when combined with acid rains or reactive compounds in groundwater.

For ferrous metal objects, such as reinforcement materials within a concrete element or a cast iron facade assembly, corroding metal expands and causes an increase in stresses in other portions of the adjoining assembly. Cracks and rust stains appear on the surface of concrete, rust penetrates coatings from the interior, and joints and seams can open between components of a larger assembly. In all cases, moisture is typically the culprit and finding the source of the water penetration is critical in correcting the problem.

Figure 7-7 This street light in Chicago has been exposed to corrosive atmospheric conditions, and the resultant oxidized material has stained the adjoining surfaces.

Galvanic corrosion occurs when two dissimilar metals come together in the presence of an electrolyte. The electrolyte, commonly a solution of water and salts, creates a path for ions to leach out from the positively charged anode metal and migrate to the negatively charged cathode metal. Left unchecked, this process can cause the source metal to disintegrate until it collapses and fails. For example, cast iron will corrode in the presence of lead or copper.

Metal Fatigue and Failure

Repeated cycles of thermal expansion and contraction, loading and unloading, and restrictions due to lack of proper expansion control can cause internal stresses that weaken metal. Metal fatigue also is a problem for architectural metals. Continuous flexure and stresses can force a metal into a plastic state where it loses strength even after the load is removed. Subsequently, the compromised material fails when additional loads are placed upon it. Depending on the metal's ductility properties, it will fail in an attempt to relieve the internal stress. Failure is indicated by cracking, rupturing, or jagged tears in the metal when in tension. In compression, stiff, brittle metals, like cast iron, crack or shatter. Ductile metals, like brass and steel, stretch until they rupture. Other fatigue and failure conditions may be present when the building has undergone alterations or changes in use that did not take into account the structural capacity of the materials involved.

In structural applications, the member becomes deformed and twists in reaction to torsion and bending stresses. Metal structural systems of the early nineteenth century displayed these tendencies particularly in reaction to the heat of a fire. As structural steel was introduced and despite its higher strength, it too succumbed to the ravages of metal fatigue and plastic failure when exposed to the high heat associated with fire. As a result, hollow clay tile was introduced to act as an insulator to protect the steel member from the heat. This approach was used well into the twentieth century until pneumatically applied fire insulation systems and other fire separation enclosures were developed.

For nonstructural applications, repeated thermal expansion cycles may disrupt the material's integrity or the connections holding the metals together, particularly if the expansion has not been accounted for in the existing construction. Stresses can open joints at the seams and actually crush or tear the metals. These opened joints and torn materials can admit water that, in turn, can corrode and further weaken the internal workings of the assembly. Evidence of this failure can be seen in the presence of open joints, corrosion growing from the inside of the assembly, or even a layer of multiple earlier repairs that have continued to fail.

Impact loads can fracture brittle materials, like cast iron, and bend or deform ductile materials. However, long-term failures may not be immediately obvious to the casual observer except when the accumulation of fractures and deformations is viewed in a comprehensive fashion. In investigating these problems, it is necessary to check for the presence and performance of expansion joints, unusual loading conditions, and any recent changes in the building that may account for new load conditions.

Connections

Connection systems have evolved through time and include mechanical fasteners, such as bolts, rivets, and screws, and physical processes, such as welding and soldering. In most instances, these connections are concealed from public view or access. Inspection of the physical connections may be done by removing protective coverings for a direct view or by nondestructive investigation techniques using fiber optic probes.

In architectural metals, connection design is a critical factor in the continued integrity of the overall construction. The design relates not only to the physical connections themselves but also to protection from destructive forces due to heat, moisture, and corrosion. This design issue is especially relevant for otherwise unprotected ferrous metals. When protective coatings or enclosures have become compromised or were improperly designed, constructed, or applied, the potential for accelerated decay is inevitable. For structural materials located in concealed areas, this decay may become extensive before being recognized through failed, sagging, or bulging structural components, as well as binding doors and windows whose frames have been distorted through stresses created by failing structural components. Connections and joints in architectural metals used in window and door frames, cladding, and other ornamental applications may also suffer from metal fatigue and vandalism.

REMEDIATION METHODS

As with all treatment strategies, the best solution is to correct the source of the decay even when no other work is deemed necessary. Treating the symptom without addressing the source itself will simply mean that problems will recur over time. Four basic strategies are available for the remediation of architectural metals. The historic value of the component will often dictate strategy selection. If the item in question is in good condition and has significant historic value, then the best course of action may be to leave it as is and protect it during the remainder of the project. The other strategies, in increasing order of intervention, are stabilization and conservation, structural reinforcement and repair, and replication and replacement.

Stabilization addresses physical integrity, and generally involves remediating corrosion and compromised connections that can destabilize structural elements of the building or the surface integrity of nonstructural or ornamental metals. Conservation involves working with the existing materials and protecting them during later phases of construction.

Stabilization and Conservation

For structural elements, stability is influenced by the change in loads on the structural system. For all architectural metals in a building, the corrosive effect of prolonged exposure to moisture and chemical compounds on both ferrous and nonferrous metals may also need to be stabilized. As the sources of decay are identified, the damaged metals may possibly be stabilized with only nominal repairs while effecting repairs to eliminate the source of the decay. In either case, remediation of these problems must be accounted for in the final design treatments so that the performance and visual integrity of the metals can be maintained appropriately.

Temporary Reinforcement

Over time, damage or alterations to the building, changes in the building's use or loading conditions, and deterioration of the skeletal framing members themselves can all act to create unusual loading or overloading conditions in the existing structural members. During the investigation and inspection phase, the structural analyses should reveal any unstable or potential life safety problems that need to be addressed prior to the construction phase. Accordingly, it is important that these conditions be addressed and stabilized before applying any preservation treatments so that the conditions can be corrected to meet life safety codes. Temporary reinforcement must be installed not only to protect the historic materials but also to safeguard the workers present during the construction phase of the project. The temporary reinforcement should be installed to minimize damage to the historic building materials adjoining the reinforced structural members.

At times, the safest course may be to add reinforcement elements directly to the structural framing or support it using scaffolding, temporary supports, and bracing. However, this support should take into account the historic significance of the material to be supported and should therefore be selected to minimize permanent changes to the visual appearance and historic integrity of the building where appropriate. In other situations, conditions may require disassembling the compromised structure and reassembling it at a later time, reusing as much of the original materials as possible. The treatment decision must be based on the historic significance of the structural materials being stabilized. Once the structure has been stabilized to meet the expected loading and safety needs, the temporary support system should be protected during the construction phase so that the structure remains stable until permanent stabilization treatments are completed.

Moisture Control

Exposure to moisture and the chemical compounds in moisture is a primary cause of deterioration in architectural metals. Leaking roofs, open joints, failed sealants, gaps caused by metal fatigue, failed flashing and gutters, and inadequate maintenance of protective coatings create conditions in which the architectural metal is allowed to deteriorate more rapidly.

The control of moisture migration is often the single most important remediation strategy in conserving architectural metals. While the damaged material or faulty construction that admits the moisture can often be repaired or replaced, moisture damage within a wall cavity is often much more expensive to correct. However, correcting the problem becomes more difficult as the historic significance of the building increases. Repairing problems within the wall will require the removal of either exterior or interior enclosing materials or both. To obtain a visual assessment, the scope of this removal ranges from small sections to enable insertion of fiber optic inspection devices to larger sections for the direct physical inspection of certain areas. The repair treatments themselves may then require large sections of historic fabric to be removed.

Temporary stabilization measures may include nominal repairs to the existing building enclosure (e.g., roofing, flashing, gutters, and cladding) or simply sealing off the moisture entry points using tarpaulins or sheathing. Here again, these temporary measures need to be protected during the subsequent construction phase until the permanent treatments can be installed.

Corrosion Converters, Corrosion Inhibitors, and Epoxies

When the problem is corrosion or surface damage, it is possible to correct the damage once the source has been eliminated. The corrosion can be removed or a corrosion converter can be applied. A corrosion converter works chemically to transform corrosion into an inert material. Next, a corrosion inhibitor can be applied to stop corrosion from recurring. Last, an epoxy material can be worked into the damaged areas to build up a continuous surface. This surface is then painted or coated to match the finish of the adjoining undamaged metal.

For the most part, these treatments are not reversible, but they are used when the historic significance of the material warrants maximizing its retention or when replacement of the material or replication of the element is not possible or desired. Superficial surface damage and limited internal damage to a structural element can be repaired in this manner without destroying the remaining intact metal. Epoxies are appropriate where exposed surfaces will be repainted or concealed by subsequent treatment processes.

Conservation

Ornamental metal objects, decorative panels, grilles, or lighting fixtures may need to be removed to an off-site location for repair and conservation as well as protection. Surface corrosion and deterioration assessments may need the special attention of a trained conservator, as the objects themselves may have a historic value unrelated to the significance of the building. Any ornamental architectural metal objects remaining on-site need to be covered and protected from damage that might occur during the construction phase of the project.

Structural Reinforcement and Repair

In rehabilitating a building, installing structural reinforcement of structural framing and completing infill repairs are often necessary. Careful attention to the structural analyses should provide the needed information to design an appropriate solution that can meet future needs while not compromising the visual and historical integrity of the original structural system and the spaces that system creates. Similarly, infill repairs may be needed for nonstructural and ornamental metals that include the exterior cladding and the surfaces of the enclosed spaces.

Strengthening and Repair Strategies

In addition to temporarily stabilizing structural components, a permanent reinforcement system must be designed. This system is especially important where the expected loading for the future use of the building cannot be supported by the existing framing. The *Standards* expressly state that the planned building use should be compatible with the original existing structural framing rather than imposing additional loads that force major upgrades. Insensitivity to load requirements can result in significant alterations to the existing structural system and thus substantially damage the historic integrity of the building.

That being said, there may be additional capacity in the existing structural framing. Due to the experimental nature of iron and steel frame construction of the nineteenth century, conditions may be present that vary significantly from the practices of today. Connection practices and sizing formulas have varied throughout the past 150 years. Earlier codes and design approaches may have introduced a conservative approximation of the loading capacity. The actual capacity of the existing system should be verified before introducing structural reinforcement treatments that may or may not be needed.

Even when the expected new uses are compatible with the existing structural systems, reinforcement may be needed when the decay or damage is severe or when corrosion treatments and surface repairs using epoxies alone may not be sufficient. Damage to the structural member may have created more stresses that cause connections to fail or the member to sag or move, vertically, laterally, or in both directions. In this situation, where the damage has not totally destroyed the member, reinforcing the member at the location of the damage rather than replacing the member may be possible.

There are several strategies to achieve the needed structural reinforcement:

- Adding supplemental framing systems
- Attaching strengthening systems
- Adding external strengthening systems
- Changing the structural action of the framing

The most invasive approach to adding new structural members is to install a supplemental framing system to relieve the loads from the original structure. This strategy is often employed when doing seismic upgrading of historic buildings. It is problematic from the very beginning. In many instances, the structural framing system in part defines the visual and physical sense of the historic spaces that it encloses. The addition of new framing to meet future load conditions must be weighed against the impact the addition has on the visual and historic integrity of the affected spaces. In some cases, it may be possible to find appropriate locations to insert the new framing by investigating secondary spaces, such as vertical chases, stairwells, restrooms, and ancillary spaces, within the portion of the building needing the additional framing. The use of concealment and the minimization of disturbances to historic materials are critical to the success of this strategy.

The next method is to increase the structural capacity by adding supplemental strengthening elements to the structure. In the nineteenth century, many of the early framing members were built up from separate plates, angles, and channels, and fastening systems such as rivets, bolts, and welding were devised. Following this historical precedent, weakened structural members can be upgraded to meet expected conditions by attaching any number of these structural elements. In addition to the traditional materials and fastening methods, adhesively bonded fiber composites can be used. However, upgrading undamaged members, especially those in public view, raises the question of how well it maintains historic integrity and should be carefully considered before proceeding.

A third approach to enhancing the structural capacity is to add external structural elements. For example, if there is sufficient space between the beams and the ceiling, it may be possible to install a truss system along the bottom of the beam to upgrade the bearing capacity of the original beam. Horizontal structural assemblies can be installed as well when the expected lateral load conditions require them.

Lastly, the structural action of the framing can be modified. In this approach, materials between structural framing members can be replaced with continuous concrete slabs to create a diaphragm floor. This method introduces concerns about the shearing forces at the interface between the new slab and the original structural frame (Beckman and Bowles 2004).

Patching/Splicing

Infill repairs incorporate some form of new material into the historic fabric that goes beyond simple modifications to the surface damage. Beyond structural reinforcement, many repairs involve replacing damaged portions of ornamental exterior cladding, roofing, and guttering systems as well as interior stamped metal ceilings. Often, less ornately detailed but equally important are the flashing, sheet goods, and other cladding that are neither stamped nor highly decorative.

Infill repairs in ornamental objects and stamped metal cladding systems consist of removing the damaged metal and splicing in a matching replacement piece of similar metal or physically compatible material. The connection is made secure by soldering, welding, screwing, or using adhesives to bind the new metal with the old. This strategy is to be used with caution. The replication and repair of the damaged metals on an ornamental object should be done by a skilled artisan or craftsperson. Replacement of damaged sections of decoratively stamped exterior cladding and roofing or interior ceilings may be accomplished using salvaged materials of similar design or a contemporary reproduction of historic systems.

For unadorned metal surfaces that are flat or constructed with simple bends, curves, and angles, removing only the most severely damaged portions and then filling the hole with a patch cut to match the opening is a possible repair method. The dutchman patching process can be used to provide a continuous surface on the exterior exposed metal. The patch may be attached by soldering, nailing, or screwing it into place. It may also be fastened to the material by using adhesives, such as bituminous asphalt or other contemporary adhesives, when the repair is not in public view. In locations exposed to moisture, all penetrations and seams should be made weathertight using an appropriate sealant.

Patching and splicing can be done in a variety of locations, particularly if the surface is to be painted. These repairs are generally done where the work will not be in public view, as they alter the visual appearance of the surface. Whether the repairs are in a public location or not, they should be discreetly marked to indicate their date of installation so that future consultants will understand that they were not part of the original historic fabric of the building.

When the damage is extensive, replacement of damaged materials may be needed. In replacing a structural member, consider both the visual and structural aspects of the replacement materials. Certain methods can be used that conceal the replacement from public view, while others are visually intrusive. In selecting the method, it is necessary to balance the level of acceptable visual intrusion with health and life safety issues. The selection must also consider the economic feasibility of completing a possible replacement.

When a structural member or ornamental object is beyond repair, replacement or replication may be the only appropriate course of action (see Figure 7-8). In a rehabilitation project, replication and replacement of the failed or missing item is done when the cost of physically repairing the object far exceeds its historic value. In saving the building, replacing inadequately performing architectural metal features is sometimes necessary. A key concept in all strategies is to minimize the visual intrusion of the repairs. The *Secretary of the Interior's Standards* allow replication of missing elements based on physical evidence and also permit replacement construction so long as the new construction is differentiated from the original one. Therefore, missing features can be replicated. However, the building must be safe to occupy and the structural repair methods must meet modern code requirements and be approved by an accredited structural consultant. A key precaution is that when any damaged structural member is removed, the remaining structure must be stabilized to eliminate movement or deflections in the remainder of the building and the removed member must be replaced using visual and physical evidence to create the replacement member.

For nonstructural components, replication and replacement may be problematic. Ornamental cladding materials may no longer be made or may be available

Replication/ Replacement

Figure 7-8 Shown here is a cast iron storefront that was restored as part of a late-twentieth-century urban revitalization project. Missing or damaged portions of the facade were replicated to maintain its overall visual integrity.

only from salvage operations. Likewise, ornamental castings may be impossible to get without a custom fabrication process. Castings can be made from existing elements to replace missing or unrepairable parts and pieces. However, these castings may need to be done using contemporary materials that simulate the appearance of the original assembly but at a fraction of the cost of replication. These materials are similar to those used to replicate masonry, as described previously, and include cast aluminum and fiberglass-reinforced polymer-based products.

Cleaning and Coatings

Air pollution, contact with occupants, dust, and soiling from birds all degrade the visual appearance of structural and ornamental metal objects. Left unattended, the combination of these factors can accelerate corrosion and decay. Starting with a dulled finish, the soiling continues to build up and conceal more advanced problems, such as pitting and etching. Conversely, frequent and overzealous cleaning or inappropriate cleaning products can degrade the finish, leaving etching and pitting as well as disturbing or even removing the protective patina created by oxidized metals or any protective coatings applied to the surface of the metal.

There are several cleaning methods that can be used to remove failing paint and corrosion from a metal surface. These include hand scraping, wire brushing, low-pressure grit blasting, wet sandblasting, the use of chemical strippers and rust removers, and flame cleaning. The key criterion is to select the gentlest method possible that is appropriate to the metal being cleaned. For example, relatively soft metals, such as brass, bronze, or copper, do not react well to aggressive abrasive methods, like grit blasting, but may respond well to chemical cleaners formulated specifically for these metals. Conversely, stronger metals, like steel and cast iron, may be well suited for grit blasting techniques. The best approach to use, when in doubt, is to conduct sample tests on less prominent locations to determine the specific strategy to use on the particular building, material, and conditions.

For cleaning soiled surfaces, it is necessary to determine how the soiling agents have interacted with the metals. As corrosion proceeds, there may be a simultaneous disintegration of original material along with an amalgamation of the soiling agents and the corrosion products themselves. This disintegration is particularly characteristic of ferrous metals. The corrosion can be treated using the methods previously described.

As with corrosion and paint removal, acceptable cleaning agents and processes vary with the metal and the local conditions. In general, cleaning processes can be categorized as follows:

- Low-pressure washes with ionized detergents
- Chemical washes formulated specifically for a particular type of metal
- Buffing agents and polishes with low-abrasion properties

Again, the key factor is that these processes must act as gently as possible in removing the soiling agents. Once a surface has been cleaned, apply a protective coating (e.g., wax or lacquer) to ornamental metals to slow the effects of future soiling agents and apply corrosion-inhibiting coatings to structural elements.

References and Suggested Readings

Allen, Edward. 1999. *Fundamentals of building construction: Materials and method.*, 3rd ed. New York: John Wiley & Sons.

Ashurst, John and Nicola Ashurst. 1988. *Practical building conservation, Vol. 4: Metals*. New York: Halsted Press.

Association for Preservation Technology International. 1997. *Metals in historic buildings* (workshop held September 28–30. 1997) Chicago: Association for Preservation Technology International.

Baker, Earl P. and Harold S. Langland. 1947. *Architectural metal handbook*. Washington, DC: National Association of Ornamental Metal Manufacturers.

Beckman, Paul, and Robert Bowles. 2004. *Structural aspects of building conservation*. 2nd ed. Burlington, MA: Elsevier Butterworth Heineman.

Belle, John, John Ray Hoke, Jr., and Stephen A. Kliment, eds. 1994. *Traditional details for building restoration, renovation and rehabilitation from the 1932–1951 editions of Architectural Graphic Standards*. New York: John Wiley & Sons.

Campbell, Marian. 1997. *Decorative ironwork*. New York: Henry N. Abrams.

Cowden, Adrienne B. 1995. Nickel silver. In Jester 1995, 58–63.

De Mare, Eric, ed. 1951. *New ways of building*. London: Architectural Press.

DeSilets, Robert and Stuart MacDonald. 1973. To strengthen the girder. *Bulletin of the Association for Preservation Technology*, 5(1): 50–57.

Elliott, Cecil. 1993. *Technics and architecture*. Cambridge, MA: MIT Press.

Freitag, Joseph K. 1895. *Architectural engineering with special reference to high building construction including many examples of Chicago office buildings*. New York: John Wiley & Sons.

Friedman, Donald. 1995. *Historic building construction: Design, materials, and technology*. New York: W. W. Norton & Company.

Gayle, Margot and Carol Gayle. 1998. *Cast-iron architecture in America: The significance of James Bogardus*. New York: W. W. Norton & Company.

Gayle, Margot, David W. Look, and John G. Waite. 1992. *Metals in America's historic buildings: Uses and preservation treatments*. Washington, DC: United States Department of the Interior.

Harris, Samuel Y. 2001. *Building pathology: Deterioration, diagnostics, and intervention*. New York: John Wiley & Sons.

Hornbostel, Caleb. 1991. *Construction materials: Types, uses, and applications*. 2nd ed. New York: John Wiley & Sons.

Jackson, Neil. 1989. Metal-frame houses of the Modern Movement in Los Angeles: Part 1: Developing a regional tradition. *Architectural History*, 32: 152–172.

———. 1990. Metal-frame houses of the Modern Movement in Los Angeles: Part 2: The style that nearly. . . . *Architectural History*, 33: 167–187.

Jester, Thomas C., ed. 1995. *Twentieth century building materials: History and conservation*. New York: McGraw Hill Book Company.

Kelley, Stephen J. 1995. Aluminum. In Jester 1995, 46–51.

Michaels, Leonard. 1950. *Contemporary structure in architecture*. New York: Reinhold Publishing Corporation.

Newman, Alexander. 2001. *Structural renovation of buildings: Methods, details, and design examples*. New York: McGraw-Hill Book Company.

Park, Sharon. 1988. *The use of substitute siding on historic building exteriors*. Preservation Brief No. 16. Washington, DC: United States Department of the Interior.

——— 1995. *Appropriate methods for reducing lead-paint hazards in historic housing*. Preservation Brief No. 37. Washington, DC: United States Department of the Interior.

Peters, Tom F. 1993. *Building the nineteenth century*. Cambridge, MA: MIT Press.

Peterson, Charles, ed. 1976. *Building early America: Contributions toward the history of a great industry*. Radnor, PA: Chilton Book Company.

———. 1980. Inventing the I-beam: Richard Turner, Cooper & Hewitt and others. *Bulletin of the Association for Preservation Technology* 12(4): 3–28.

———. 1993. Inventing the I-beam, part II: William Borrow at Trenton and John Griffen of Phoenixville. *APT Bulletin*, 25(3/4): 17–25.

Prudon, Theodore H. M. 1979. Installing new non-corrosive anchors in old masonry: Some examples. *Bulletin of the Association for Preservation Technology*, 11(3): 61–76.

Rabun, J. Stanley. 2000. *Structural analysis of historic buildings: Restoration, preservation and adaptive reuse for architects and engineers.* New York: John Wiley & Sons.

Score, Robert and Irene J. Cohen. 1995. Stainless steel. In Jester 1995, 64–71.

Scott, John C. and Carolyn L. Searls. 1995 Weathering steel. In Jester 1995, 72–79.

Streeter, Donald. 1973. Early American wrought iron hardware: H and HL hinges, together with mention of dovetails and cast iron butt hinges. *Bulletin of the Association for Preservation Technology,* 5(1): 22–49.

Swanke, Hayden Connell Architects. 2000. *Historic preservation: Project planning and estimating.* Kingston, MA: R. S. Means.

Trelstad, Derek H. 1995. Monel. In Jester 1995, 52–57.

University Museums and Art Galleries. 1976. *Iron: Solid wrought/USA.* Carbondale: Southern Illinois University at Carbondale.

Waite, John G. 1991. *The maintenance and repair of architectural cast iron.* Preservation Brief No. 27. Washington, DC: United States Department of the Interior.

Weaver, Martin E. 1997. *Conserving buildings: A guide to techniques and materials.* New York: John Wiley & Sons.

Weiss, Norman, Pamela Jerome, and Stephen Gottlieb. 2001. Fallingwater part 1: Materials-conservation efforts at Frank Lloyd Wright's masterpiece. *APT Bulletin,* 32(4): 44–55.

Weygers, Alexander G. 1997. *The complete modern blacksmith.* Berkeley, CA: Ten Speed Press.

PART III

Building Fabric

CHAPTER 8 Roofing

The roof is the most critical building assembly in maintaining the integrity of the building. Throughout time, builders have tried to construct durable roofs that provide protection from moisture from rainfall and melted snow. Failure of the roof to resist moisture penetration rapidly leads to the decay of the materials within the building. As moisture penetrates and migrates by gravity through the building, the affected materials are vulnerable to rot, corrosion, staining, warping, and other decay mechanisms, such as mold and freeze-thaw problems. The structural decay caused by these mechanisms can cause further failure of the roof as connections fail and joints open to admit additional moisture.

OVERVIEW

While roofing systems originally were constructed from locally available materials based on traditional building practices, modern roofing practices have evolved based on the type of building being constructed. Many early roofing systems featured a steep-sloped construction that allowed water to drain directly from the surface. As time passed, commercial, retail, industrial, and residential buildings incorporated low-sloped roofs (often now commonly and inaccurately referred to as "flat roofs") into their designs. In modern building practice, a low-sloped roof is one where the pitch is less than 3 inches of vertical rise per 12 inches of horizontal run (Allen 1999). Since one of the main purposes of the roof is to shed water and snow, a flat, level roof would let rainwater and snowmelt remain on the roof to eventually penetrate into the building as the roofing material ages and decays. In this regard, even what appears to be a flat roof is actually constructed with a pitch to accelerate drainage from the roof through scuppers along the parapet at the edge of the roof or into sumps that collect and remove the water through drains. For consistency in this book, roofs having a pitch that exceeds a 3:12 ratio will be referred to as "steep-sloped" roofs.

ROOF TYPE

155

Figure 8-1   Steep–pitched roofs are a character-defining feature of a building: (a) gabled roof, (b) gambrel roof, (c) hip roof, and (d) mansard roof.

Throughout the past four centuries, a variety of roof shapes evolved. The roof became a character-defining feature of the building on which it was located (see Figure 8-1). When altering or expanding a building, it is necessary to respect the historic roof shape and materials to ensure the retention of its historic integrity.

ROOF CONSTRUCTION

Roofs consist of five major components: framing, sheathing, cladding, flashing, and drainage systems. Although construction methods varied based on local traditions and available materials, these components can be found in some form on every roof. Durability of the roof depended on four factors: physical properties of the materials, the way in which those materials were fabricated, the installation techniques, and regular and timely maintenance. Poor-quality materials, poor fabrication and installation methods, and lack of maintenance all significantly reduced the useful life of a roof.

Framing

The earlier steep-sloped roofs used in timber framing systems consisted of pairs of rafters placed at a pitch greater than 3:12. These rafters were classified as principal rafters (the larger timbers that framed each bay of the building) and common rafters (the smaller members that were evenly spaced between the principal rafters). Each rafter pair met at the roof peak or was attached to a continuous ridge beam. A series of structural members known as "purlins" were attached at right angles to the principal rafters and provided intermediate span support for the common rafters. The sheathing system was then attached to the rafters or the purlins themselves, depending on which construction tradition was used.

As balloon framing and the subsequent platform framing came into use in the early nineteenth and twentieth centuries, respectively, construction techniques were adapted to meet the stylistic needs of the succession of building styles popularized by the design tastes of the period. While the distinction between principal and common rafters receded, the stick framing methods still used the basic principles of load transfer found in timber framing techniques. The major difference was that the loads were carried by uniformly dimensioned and evenly spaced, lighter framing members.

In larger buildings that required spans longer than those that could be provided with a single length of rafter, built-up truss systems were used. Trusses were originally crafted from wood timbers using mortise-and-tenon connections. However, wrought iron reinforcements were in use by the mid-eighteenth century (Rifkind 1980). Throughout the nineteenth century, truss construction evolved to include steel gusset plates and rods that relied on threaded nuts and bolts for connections. As steel production technology improved in the late nineteenth century, trusses were constructed completely of steel using threaded, welded, and riveted connections. By the mid-twentieth century, trusses made from metal alloys were also being used to reduce the weight of the roofing system. The mid-twentieth century also saw the introduction of high-tension steel cable roof structures, such as the North Carolina State Fair Arena in Raleigh (Buchholdt 1999), which spawned a number of variants well into the late twentieth century.

Except for Spanish-influenced colonies, low-sloped roofs were uncommon in early colonial buildings. The low-sloped roof gained wider acceptance during the Industrial Revolution of the early nineteenth century and by the end of the twentieth century was standard practice for industrial and commercial buildings. The early use of wood timber framing for low-sloped roofs was replaced in the mid- to late nineteenth century by steel beams and trusses as well as reinforced concrete versions of the same members. In the early twentieth century, architectural revival styles (e.g., Spanish Colonial, Mission, and Mediterranean) and modern styles (e.g., Art Deco, Art Moderne, and International) were introduced that included the use of low-sloped roofs as a character-defining feature.

Sheathing

Once the framing was completed, the sheathing substrate for the cladding was added. Two techniques were used on steep-sloped roofs. The first was open sheathing, which consisted of an assembly of boards or battens that did not form a continuous enclosure over the roof framing. Overlapping layers of cladding were then attached to this assembly. The second was closed sheathing, which consisted of boards or planks secured to the roof framing with no significant gaps between them. The introduction of metal roof cladding dictated the use of the closed sheathing method. By the mid-twentieth century, plywood-based products had largely replaced the use of open sheathing methods in steep-sloped roofs.

Allowing the underside of the sheathing (and cladding) to dry is important in extending the useful life of the roof. The open sheathing method enabled air to flow through the roof assembly to accelerate drying. To accommodate the ventilation needs of the cladding when closed sheathing was used, the application method could be adapted to include roofing felts or battens and cleats between the roof substrate and the cladding materials.

Low-sloped roofs were sheathed with a decking assembly consisting of planks, boards, and, when they became available in the nineteenth century, corrugated metal or concrete (poured or precast panels). After World War II, the use of metal on low-sloped roofs broadened to include extruded alloys as part of the cladding/sheathing assembly.

Cladding

In primitive societies, thatch, bark, and animal hides were used as cladding. Advanced societies used materials that reflected more advanced technology. The earliest of these cladding materials was clay tile. Later materials included slate, wood, metal, asphalt-based products, and rubberized membranes (see Figure 8-2). Thatch was used in the early North American colonies, but practices quickly moved to wood shingles. Documented evidence shows that clay tile and slate were also used in the early colonies. Tile was known to have been made in the Dutch colonies by 1650 but was largely imported from Europe until the eighteenth century (Park 1989). Similarly, slate roofing was in use by the mid-seventeenth century, but it was not until 1785 that a commercial slate quarry opened in the United States (Levine 1992). Concerns for fire safety by the mid- to late seventeenth century led to wider adoption of clay tile and slate in more congested settlements, such as Boston and New York (Sweetser 1978).

For high-sloped roofs, tile, slate, and shingles represent the oldest forms of roof systems that were laid up in successive offset layers (or courses) where the lower course was overlapped by several inches by the upper course. The remaining exposed portion or "reveal" of any given lower course ranged from a few inches to several feet. The overlapping course protected the joint below it based on the principle that water will flow downhill by gravity without being drawn into the joint.

By the late nineteenth and early twentieth centuries, manufacturing processes had evolved to allow the use of stamped and some rolled roofing products. The earliest forms of these were metal panels and strips that interlocked or were mechanically fastened together, as well as bitumen- and asphalt-saturated felts and other textiles. Low-sloped roofs, previously used only in dry, hot climates, were constructed for many commercial and mill manufactories along the East Coast due to the demand created by the Industrial Revolution. The continued development of the steel industry throughout the nineteenth century increased the ability to produce large quantities of rolled and stamped steel and other metal roofing products for low-sloped and steep-sloped roofs. By the early twentieth century, standardized built-up roofing systems had been introduced for the low-sloped roof market that remained the dominant form of low-sloped roof construction until the introduction of single-ply membrane roofs in the late twentieth century (National Park Service 1999).

Clay Tiles

The use of clay tiles for roofing has been documented as far back as 10,000 years ago in China. Clay tile can be used to form specific shapes for various common and ornamental roof elements. While it is possible to make a plain tile similar to a flat shingle, these roofs commonly consisted of tiles formed as convex and concave surfaces that were overlapped so that water could not enter the joints between them. The most common form was the pantile tile, which had a horizontally oriented

Figure 8-2 Roof cladding systems have been composed of a variety of materials, including (a) tile, (b) slate, (c) wood, and (d) metal.

S-shaped profile that allowed the convex portion to overlap the concave portion of the adjoining tile. While tile roofs were used in early colonial buildings, their popularity waned in the nineteenth century with the introduction of metal roofing products. Revival styles of the early twentieth century brought them back into popular usage.

Tile is vulnerable to wind loads and mechanical failure when they are struck by tree limbs or walked on. If the fastening system corrodes or fails, tiles may slide out of position and allow moisture penetration. Clay tile roofs can last for more than 100 years with good maintenance, but more commonly they last for only 50 years or so.

In the early twentieth century, a number of tile substitutes were developed, including concrete and an asbestos-cement combination (National Park Service 1999; Tobin 2000), that were intended to make the roof fireproof while making the installation faster, lighter, and less expensive. What appears from ground level to be clay tile may in fact be one of these substitute products.

Asbestos Cement

Concerns for fire safety and durability led to the introduction of asbestos-cement products for both roofing and siding in the early 1900s. Early roofing versions of these products were created using a wet mix process that combined asbestos, cement, and water in a slurry that was fed into a laminating machine to form interlaced layers. These layers were then pressed into flat or corrugated sheets. Later, a dry mix approach was introduced, in which the asbestos and cement were combined on a conveyor belt. Water was added, and the mixture was then rolled under a hydraulically weighted cylinder. An embossing cylinder was then used to create the desired texture. When the product had cured for twenty-four hours, the sheets were cut, punched, or steam-cured into the finished shape. Color could be added using pigment mixtures or by pressing ceramic granules into the panel while it dried (Arkansas Historic Preservation Program 2004).

Slate

Slate was imported from Europe (primarily Wales) until a commercial slate quarry opened in Pennsylvania in 1785. Local slate deposits were known from Maine to Georgia, but lack of economical transportation methods limited their use. As railroad networks expanded in the nineteenth century and Welsh miners migrated to this country, American slate became more readily available. As the country expanded westward, other slate deposits were found but the most desired slate still came from the East Coast. Slate production flourished throughout the nineteenth century. However, due to market competition from other roofing materials manufacturers, by the end of the twentieth century active quarries remained only in three concentrated areas in Vermont-New York, eastern Pennsylvania, and Virginia.

Slate is a metamorphic stone that was split and cut into a variety of shapes of fairly uniform thickness. Roof slate was made by cutting the raw slate blocks vertically and horizontally along the grain and cleavage planes, respectively, into a variety of standardized sizes. Slate varied in hardness and color. Most hard slates could last for more than a century, while softer slates might last for only forty to seventy-five years. Colors included blue, black, green, gray, red, maroon, and purple. After a slate was cut and installed, weathering caused it to lighten or darken to varying degrees, as well as change color somewhat as minerals in the slate oxidized (Jenkins 1997; Levine 1992).

Slate roofs were constructed either using open sheathing with a network of battens or laths to which the slate was attached or using closed sheathing. The individual slate shingles were laid in overlapping offset courses and secured in place using a wood peg, a wire, or a nail (now commonly made of copper). The exposed surface was rectangular in shape but could also have been "knapped" or trimmed into hexagonal, triangular, or semicircular forms.

The major source of decay in slate was from weathering. Chemical components within the slate reacted with water and temperature to form gypsum crystals that could break the slate apart along the horizontal cleavage plane. The result was seen as flaking or delamination of the surfaces. As the slate decayed, moisture penetrated it and accelerated the process through repeated freeze-thaw and thermal expansion-contraction cycles. Once moisture began to degrade the slate, the underlying wood sheathing and structural framing became vulnerable to damage as well. Installation problems could occur when the nails used were driven tightly into the slate (causing stresses in the slate) or were prone to corrosion (corroding ferrous metals expand, and therefore copper nails are recommended). A second installation problem occurred when the slates overlapped by less than 2 inches and therefore allowed water to penetrate past the slate and migrate into the building by capillary action. Lastly, the mechanical force of walking on or something hitting the slate could cause slate to crack or become dislodged (Levine 1992).

Wood

Wood roofing is typically found in two forms: a board and batten system and a shingle system. In a board and batten roof, the sheathing consisted of boards running lengthwise down the slope of the roof with the joints between them covered by a batten, a piece of wood also running lengthwise down the slope of the roof. Shingles were thinner pieces of wood that were typically laid in offset overlapping courses between the eaves and ridge.

Shingles have been the more common wood roofing product on high-sloped roofs for many years. Shingles of the early colonial period were shaped to a uniform thickness to present a consistent appearance. A variation on traditional shingles was a shake, which actually was a shingle that was much thicker and less uniform in appearance. Historically, both types were nailed to open sheathing where the multiple overlapping courses formed a weathertight assembly, and the gaps in open sheathing permitted air flow that accelerated the drying of the shingles after a storm. The introduction of rolled asphalt sheet goods, shingles, and plywood sheathing at the turn of the twentieth century began the transition to a method for completely enclosing the roof that eventually reduced concerns over moisture penetration and increased fire resistance. Wood shingles have been used for both roofing and siding on a number of historic buildings throughout the past four centuries, perhaps reaching the height of their popularity with the Carpenter Gothic and Queen Anne styles of the nineteenth century and seeing renewed acceptance with the revival styles of the early twentieth century.

Wood roofs are susceptible to moisture absorption through capillary action at exposed end grains. Untreated or unpainted wood decays more quickly on northern exposures and suffers ultraviolet decay on sunny southern and western exposures. The most direct evidence of decay includes curling and splitting, moss growth, dark water stains, and rot. Wood shingle roofs that are properly treated and maintained can last for twenty-five to thirty years.

Metal

Initially made from wood, shingles were later fabricated from stamped or rolled metal pieces that imitated wood shingles or clay tiles in appearance. For irregularly shaped roof elements, such as domes, cones, and bell-shaped roof sections, panels could be cast, stamped, or rolled and then fastened together. This approach allowed for elaborately ornamental roofing elements that simulated the appearance of more expensive materials, like stone or tile, at a fraction of their cost and weight. Efforts to make a lightweight metal roofing system that could be quickly installed led to the manufacture of a wide variety of metal roofing products by the early twentieth century.

Rolled sheet metal roofing products were composed of various nonferrous metals, such as copper, aluminum, and zinc, as well as coated ferrous metals, such as tin-plated steel. The construction assembly was completed in three basic forms. First, the edges of the individual sheets of metal were mechanically fastened or folded together and soldered to seal the seam. Second, standing seam roofs were composed of metal panels with edges that were bent upward at a right angle to the roof and secured together by folding the edges together or crimping a U-shaped strip of metal over the common joint between them. Third, sheets of corrugated metals were applied like large shingles and secured either to a substrate or directly to the structural framing.

Metal roofs could typically last for thirty years or more with good maintenance. Two of the most common problems included corrosion and metal fatigue. Corrosion was caused by exposure to moisture and atmospheric pollution that resulted in the short-term staining, streaking, or pitting of the surface or in the long-term decomposition of the metal itself. Sources included rain and snowmelt, atmospheric pollution, bird droppings, the presence of accumulated dirt and plants (e.g., moss, grass, small plants) that retained moisture, and standing water around clogged drains and gutters. One specific type of corrosion, galvanic action, occurred when dissimilar metals were in contact with each other. For example, direct contact between a copper object and an iron object caused the galvanic action that corroded metal. The long-term thermal expansion cycles common on metal roofs could cause metal fatigue. Over time, as the metal expanded and contracted, stress within the metal and at the connections reduced the integrity of the material. Repeated flexures and buckling caused creases that eventually began to tear the metal apart. Metal fatigue could be accelerated by walking on the roof, unusually high winds, and, in some cases hailstorms.

Asphalt

The continued search for lightweight and easily installed roofing materials led to the development in 1903 of asphalt shingles that were made from roofing felt treated with bituminous asphalt and covered with small stone or colored ceramic granules (Jackson 1995). Asphalt shingles had largely replaced the use of other materials for shingles by the mid-twentieth century. The color of the shingle is dictated by the granules, which also block the ultraviolet energy from the sun and enhance fire resistance.

Asphalt shingles have several different configurations. The most common type is the three-tab shingle, where the tabs strive to emulate the previous standard appearance of wood shingles or slate, although they are not as thick as those earlier materials. A variety of shingles were introduced in the early twentieth century, including hexagonal, diamond, and other interlocking shapes that may no longer be available locally. One recent shingle that has been permitted in historic districts nationwide is the "architectural" shingle, which is thicker and more durable than the standard three-tab shingle and has an irregular tab sequence that is believed to better emulate the appearance of historic roofing.

Rolled asphalt roof products come in two forms. First, for flat roofs, a built-up roofing (BUR) system was introduced in the 1850s and was in wide use by the 1880s across the country. This system consisted of alternating layers of bitumen-saturated felt and a bituminous binder that were installed directly on the roof decking (see Figure 8-3). When sufficient thickness was achieved, the exposed surface was covered with a ballast (e.g., crushed gravel, river stone, or other stone aggregate) to offset uplift forces from the wind and to block the ultraviolet radiation from the sun and other weathering and decay mechanisms.

The BUR system continues to be used on many historic buildings to this day, although the energy crises of the 1970s led to the development in the late twentieth century of alternative roofing products that were less reliant on petroleum-based materials. One of these is the modified bitumen roofing system, which can be laid in multiple plies and is secured in place by applying heat (e.g., torching) so that the membrane bonds to the substrate as it cools.

The second form of rolled roofing products consists of rolls of "tar paper" or building felt that were directly attached to the substrate. Tar paper has been used as a moisture and wind barrier, but more frequently it is used in modern construction as underlayment for other roofing products or as part of a BUR system. The product can be secured using nails, staples, or bituminous binders. Some of these products have been known to contain asbestos.

Inappropriate mechanical attachment methods, wind uplift, and loss of surface continuity due to wind scouring and ultraviolet radiation exposure are the major weaknesses of asphalt roofing products. Seasonal exposure to sunlight and temperature

Figure 8-3 BUR consists of layers of bitumen-saturated felt that were applied to achieve the desired thickness. In many cases, gravel was then placed on top of the roofing to provide protection from wind and sunlight.

extremes leads to expansion and contraction of the material, causing it to dry out, break apart, and reduce the overall integrity of the material. Evidence of these problems includes brittleness, cracking, and cupping. Poor flashing practices permit moisture to penetrate the surface and migrate between the layers of material. In BUR systems, "alligatored" or undulating surfaces containing puddles of water are indicators of declining integrity. Another problem is clogged roof drains and scuppers. Longevity varies widely due to the range of thicknesses and application methods. Residential-grade shingles typically can last for fifteen to twenty years, while BUR systems may last for twenty to twenty-five years.

Single-Ply Membrane Systems

Single-ply membrane systems were introduced in the 1950s in the United States. This system was only slowly accepted until the 1970s, when it was viewed as an economical alternative to the petroleum-based BUR systems. Membrane systems available today compete with the installation and replacement of BUR systems. A common example is EPDM (ethylene propylene diene monomer rubber) roofing, which is essentially a synthetic rubber membrane covering rigid insulation placed directly on the roof decking. Other single-ply membrane roofs have been made from polyvinyl chloride (PVC), thermoplastic polyolefin (TPO), and numerous other polymers and elastomeric materials in reaction to the failures of earlier products. Like BUR roof systems, single-ply membrane roofs are secured to the roof and covered with gravel or ballast stone to prevent wind and ultraviolet damage.

Poor installation practices (e.g., improper folds, seam adhesives, flashing, and sealants), as well as inadequate maintenance inspection and repair, hasten the decay of these materials. Exposure to ultraviolet light, wind scouring, and inappropriate alterations should also be avoided. This family of products should last for twenty to thirty years if well maintained.

Flashing and Sealants

Flashing consists of a system of metal sheets installed at joints to protect the joints from moisture penetration. Flashing has been made from lead, terne metal, copper, hot-dipped galvanized metals, and various other metals. Flashing is used where vertical elements, such as chimneys, skylights, and dormers, as well as penetrations from vent piping and drains project from the roof. Replacement flashing can use the traditional materials or employ new products like terne-coated stainless steel. Flashing was placed on the sheathing and attached to the vertical face of the projecting element. In certain instances, a piece of counter-flashing was installed to cover the joint between the flashing and the vertical face and secured into the mortar joint of a chimney or under the coping at the top of the vertical projection (e.g., a roof parapet or skylight). Otherwise, the cladding on the vertical facing was placed over the flashing. Similarly, the roof cladding was placed over the flashing and secured to the roof sheathing. Flashing typically should extend at least 2 inches under the cladding to eliminate space where water can be drawn in by capillary action (see Figure 8-4).

All joints in the roof that result from discontinuous materials must be protected from leaks. The cladding material itself could be installed in an overlapping succession of offset layers to protect the joint between courses. However, joints also occurred at the intersection of the roof planes at ridges, hips, and valleys. To protect

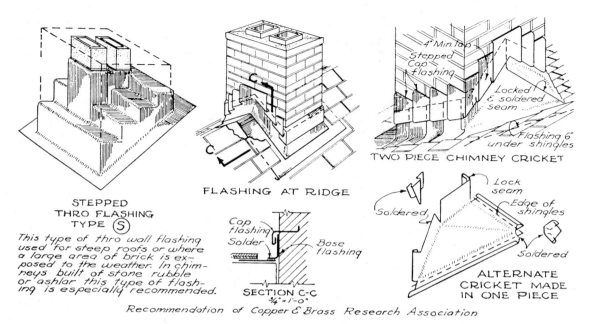

STEPPED
THRO FLASHING
TYPE (S)

This type of thro wall flashing
used for steep roofs or where
a large area of brick is exposed to the weather. In chimneys built of stone rubble
or ashlar this type of flashing is especially recommended.

FLASHING AT RIDGE

Cap
flashing
Solder

Base
flashing

SECTION C-C
¾"=1'-0"

4" Min.lap
Stepped
Cap
flashing

Locked
& soldered
seam

Flashing 6"
under shingles

TWO PIECE CHIMNEY CRICKET

Lock
seam

Edge of
shingles

Soldered

Soldered

ALTERNATE
CRICKET MADE
IN ONE PIECE

Recommendation of Copper & Brass Research Association

Figure 8-4 Flashing protects joints by diverting water past the joint. Flashing (and other roofing products) usually overlaps the joint by 2 or more inches to eliminate the effects of capillary action (from Ramsey/Sleeper, *Traditional Details for Building Restoration, Renovation, and Rehabilitation,* ©1991 by John Wiley & Sons, Inc. Reprinted with permission of John Wiley & Sons, Inc.).

joints along the ridge or hip intersections, the cladding on the windward side of the joint was extended to protect the joint from wind-driven moisture or the joints were capped with either a separate piece of cladding or ornamental cresting. This cap could be of the same material as the other cladding used or could be made from metal or wood. In the valleys, cladding could be installed continuously across the joint or the flashing could be installed over the joint with the cladding overlapping the flashing at the end of each course. Leaving a portion of the flashing exposed was known as an "open valley." Completely covering the flashing with cladding was known as a "closed valley." Other flashing elements, such as chimney crickets (an assembly of flashing or cladding on the uphill side of a chimney that diverted water around the chimney) and drip edges (protruding strips that ensured that the water dripped from the edge of the roof rather than flowed down the face of the wall below the roof) along the eaves, also acted to enhance water resistance.

Sealants for joints are defined by four characteristics: the ability to stick to material (adhesion), the ability to stay together (cohesion), the ability to accommodate movement (elasticity), and the ability to withstand weathering (Scheffler and Connelly 1995). Sealants have varied over time, including such materials as lead, oil or resin-based caulks, and various forms of pine pitch or coal sands tar. Research on elastomeric sealants began in the late 1920s and by the 1950s and early 1960s led to the availability of butyl (synthetic rubber), polysulfide, acrylic, and silicon-based sealants. Caution should be used in removing older sealants, as some polysulfide-based sealants contain PCBs and certain oil-based sealants contain asbestos or lead.

Roof Drainage Systems

By the early eighteenth century, gutters were used to direct water from the roof to a specific location (e.g., cistern, underground storm water system, or simply away from entrances). For steep-sloped roofs this was done using either an internal or an external guttering system. An internal gutter, also known as a "box gutter," was typically a trough constructed within the roof plane at the eaves. Usually lined with metal, these gutters were less visually obtrusive, although some were constructed on the roof itself. An external or "hanging gutter" was attached directly to the base of the eaves. Unlike the external gutters of today, many of these were integrated into the eaves and are not as readily apparent. Early external gutters were made from hollowed logs, curved lead sheets, or wood boards. Metal gutters became popular during the Industrial Revolution and remain so today.

The drainage system for low-sloped roofs used troughs, scuppers, and sumps to direct water from the roof. Troughs, typically made of metal, would collect and direct water at the perimeter of the roof to downspouts or scuppers. In some buildings, these scuppers were extremely ornamental and incorporated figural forms (e.g., gargoyles, animals, humans). Sumps were common on large industrial buildings, where the water drained down through pipes inside the building to storm water systems underground.

The storm water collection system at the perimeter of the roof drained through a leader (downspout) to the storm drain or an on-site drainage field. On larger buildings, an ornament leader head was used as a transition between the collection system and the downspout. Like gutters, downspouts have varied over time. Originally, they were made from wood. By the mid-twentieth century, construction practices had generally shifted to the use of metal downspouts.

TYPICAL PROBLEMS

Conducted after each major weather event or at least annually, inspections should check for loose, broken, damaged, or missing roof elements. Tree limbs and leaves should be removed, and guttering and flashing should be checked for splits, tears, and insecure and discontinuous connections. The underside of the roof should be checked for water stains and decay. When identified, these problems should be corrected as soon as possible. If immediate correction is not possible, then the roof should be adequately stabilized and covered with tarpaulins, plywood, roofing felt, or roofing paper to minimize further water penetration.

While the many different roofing materials and systems each have specific problems, several generic problems are common to all of them. These problems include moisture penetration, mechanical abrasion, mechanical failure, and ice dams.

Moisture Penetration

When the roofing systems begin to fail, two events may typically occur. First, the material itself begins to break apart and decay. As the decay continues, moisture can migrate into the roofing material, which can accelerate damage due to freeze-thaw cycles as production of decay, either rot or corrosion, continues to degrade the stability of the material. Second, the bonds at the joints connecting two portions of the roofing materials weaken and lose their moisture resistance. This creates direct paths for water to enter through the roof and migrate to the interior of the building (see Figure 8-5).

An important determinant of the longevity of the material is its surface condition. Spalling, chips, cracks, and flaking in clay tile, slate, and cementitious roofing products increases their vulnerability to moisture penetration. Likewise, wood, metal, and bituminous roofing products that lose their protective coatings (e.g., preservatives

Figure 8-5 Water penetration is clearly evident on the underside of this open sheathing, as noted (at the right and center) by the dark stains and white mineral deposits.

and surface treatments) can split, curl, and absorb moisture more readily. Left untreated, these conditions accelerate roofing system failure.

Mechanical Abrasion

When tree limbs repeatedly strike the surfaces of a roof, when inappropriate snow removal methods are used, or when someone walks on or drags something across roofing material, the force created by the abrasion can remove or compromise protective barriers and coatings. As ultraviolet radiation from the sun combined with heat and freeze-thaw cycles work to degrade the material's structural integrity, the effect of this abrasion can become more pronounced. All roof systems are vulnerable to this problem. The best solution is to eliminate the source of the abrasion or provide supplemental protection such as pavers for walkways on low-sloped roofs.

Mechanical Failure

Beyond failure due to insects, fire, or rot, when a load exceeds the structural capacity of the roof, the structural system will fail and cause the roof to collapse. This failure can occur abruptly, as in a major snowstorm or wind event such as a hurricane. A roof collapse is an obvious failure, but less noticeable failures can occur where structural members are compromised but appear to be intact. Failure can also occur over an extended period as moisture penetration, insect damage, rot, ultraviolet radiation damage, or corrosion reduces the overall strength of the roof. These less catastrophic failures may be identified where cracking, bowing, or splitting of the structural members is revealed by sags and shifts in the structural framing and sheathing. This failure may occur when a roof or attic space is modified so that heat that previously escaped through the roof and melted the snow no longer can do so. In this instance, the increased weight of accumulated snow creates a load that had not existed previously. This weight may cause structural failure. This failure has been known to happen to low-sloped roofs on residential and light commercial buildings built in the mid-twentieth century where insulation systems were added in reaction to the rising energy costs of the late twentieth century.

Along with structural failure, material failure can occur within the sheathing and cladding systems. Prolonged cycles of freeze-thaw and thermal expansion and contraction can break down the composition of both water-absorbing and water-resistant materials. This is a particular problem in metals and other non-water-absorbent materials used for flashing and sealants that have a significantly different thermal expansion coefficient than the materials to which they are attached or that have different exposures to the sun that cause the sunlit portion to expand differently from the shaded portion of the same element.

Storm water collection systems are vulnerable to failure. Inadequate maintenance allows debris to clog the system. As debris, water, snow, or ice accumulates, the added weight can cause the gutter to break apart or detach from the roof. This problem, if left uncorrected, can cause damage from the backsplash that occurs along the eaves as water falls freely to the ground below.

Ice Dams

Ice dams are caused by freeze-thaw cycles combined with poorly insulated living spaces and inadequate roof drainage. As heat escapes from occupied parts of a building, the snowmelt collects in roof gutters or along roof edges that cannot drain adequately. When the air temperature drops below freezing, icicles form along the roof edge or solid ice forms in gutters and downspouts (see Figure 8-6). Over time, this process forms a layer of ice along the edge of the roof and, when left untreated, the ice penetrates beneath the roof sheathing. Continually exposed to heat escaping from the occupied space, the ice melts and causes a leak. Long-term exposure to this moisture can cause structural members and interior surface materials to rot, corrode, and grow mold. If this problem is left untreated, the roof, framing, and other building assemblies (e.g., soffit, fascia, cornice) adjoining the location of the ice dam eventually fail.

The area of damage may be confined to the roof edges and eaves. However, if left untreated for several years, damage can extend several feet upward from the roof edge. Once these paths open, moisture penetration can continue year-round.

Figure 8-6 Inadequate drainage combined with excessive heat loss through attic spaces can lead to the formation of icicles along the eaves of the roof. Left uncorrected, this can cause the formation of ice dams along the eaves of a roof.

The problem of ice dams has been recognized for several centuries, and various architectural solutions were developed. Two of the more common solutions were to incorporate snow guards or fences in roofs to keep snow from collecting over the eaves of a building or to install metal sheathing along the lower 3 feet of the roof along the eaves so that snow could slide off the roof more easily. The use of standing seam metal roofs in snowy areas is another solution to the problem.

Aside from the obvious moisture damage to exposed surfaces, as well as damage concealed within the wall and roof assemblies, several important contributors can be identified when checking for ice dams. The first is the lack of appropriately sized gutters and downspouts to handle water and ice buildup. The second is evidence of valleys, gutters, and downspouts that have not been regularly cleared. The third is lack of insulation or protection from heat escaping from penetrations (e.g., sealed gaps around light fixtures, piping, and venting) within the attic space. While none of these conditions individually indicates a potential ice dam problem, the presence of two or more of them is strong evidence of the cause of the problem when the damage along the edge of the roof has become apparent.

The contemporary solution to the ice dam problem is to eliminate the contributing factors. However, beyond these steps, one must recognize the importance of allowing the underside of the roof to remain cold in the winter months and dry year round if possible. For unoccupied attics, this is accomplished using gable vents. For occupied attics, a ventilation path from soffit vents at the eaves to a ridge vent along the top of the roof should be considered to allow drying and keeping temperatures below freezing.

RECOMMENDED TREATMENTS

The repair and replacement of roofing on historic buildings is a complex and often costly decision. Beyond the initial repair or replacement costs, both short- and long-term issues must be considered. When a roof is away from public view, a wider range of solutions are available compared to those for a roof seen by the public. In either case, the roof must be evaluated to determine if smaller individual repairs can correct the problem or if a comprehensive replacement of the roofing materials is needed. A series of smaller patch and repair projects may seem less expensive initially but may turn out to be more expensive in the long run, especially when the materials used in the repair are physically incompatible with the existing materials. Caution is needed when selecting replacement or repair strategies and materials. Likewise, selecting a roofing contractor who has experience with the historic roof system in question is imperative.

Compatible Materials

One contemporary trend that has caused vexation among historic preservationists nationwide is the use of incompatible roofing products. These products tend to be visually incompatible with the historic fabric and, in some cases, physically incompatible as well. For example, in the first instance, replacement asphalt shingle products are available with darker borders printed on them to imitate the shadow line from what should have been a thicker shingle material (see Figure 8-7). These products do not work well, particularly when viewed close up. Similarly, a number of recent composite plastic materials have been developed for new construction that may be inappropriate for older and historic buildings. Therefore, it is imperative to verify acceptable materials with the SHPO, the local historic landmarks commission, or other agencies with design review authority over the building.

A more significant problem arises, however, when the new roofing product is physically incompatible with the existing material. This may happen when attempting

Figure 8-7 The false shadow line printed on the edge of this asphalt shingle product does not adequately convey a sense of the thicker slate material that it replaced. This product would not be acceptable in many historic districts.

to cover or patch an existing roof with a newer roofing product (e.g., a single-ply membrane bonded to a BUR system). If the two separate systems do not expand and contract similarly, they can eventually tear themselves apart.

Repair Strategies

Care should be taken to select a contractor who is well qualified to repair the type of roof under consideration. This is especially true for historic roofing systems that vary significantly from contemporary construction practices. There are trade associations and preservation directories (see Appendix B) for traditional building crafts that can provide information in locating qualified traditional roofing specialists. If the roof has historic significance, these specialists should be brought in to do the investigation and confirm the extent of damage.

Whether due to inadequate maintenance or damage from storms, all roofs eventually will need some repairs. The first step in the repair process is to inspect the roof to determine the cause of the damage. This inspection involves looking at the exposed portions of the exterior as well as the interior structure and sheathing.

Since roofs are typically located well aboveground, access is the first limitation to overcome. Certain roofs, like those made of tile or slate, may not be safe when wet or may be too fragile to walk on. Preliminary inspections can be done from ground level using binoculars to verify such potential problems as broken or missing elements, or sagging or bulging structures, and to determine the overall condition of the roof. If the roof is safe to walk on, a hands-on visual inspection and a thermographic inspection can be conducted on the roof itself. Otherwise, it may be necessary to use mechanical lifts or construct staging over the roof to get a closer look. Beyond examining the overall cladding, specific attention should be given to the ridges, the eaves, and especially the flashing.

Leaks show up on the interior surfaces as darkened stains with, in many cases, a dried perimeter of chalky white mineral deposits. Metals, such as nails and flashing, may show signs of corrosion as well. Water flows by gravity downward from the point of a leak. However, gravity does not imply that water always drops straight down from the entry point. More likely, water will enter the leak and flow along structural members and the underside of the sheathing until it reaches a low point and drips onto the ceiling below. So, keep in mind that the source of a leaking roof may be several feet uphill and laterally offset from the point where the water enters the occupied space below.

When the final extent of needed repairs is determined, a decision must be made on whether to do a piecemeal repair or to replace the roof entirely. This decision should be based on the historic significance of the roof construction, the availability of matching compatible cladding materials for the repair, the cost to replace the roof versus patching it, the availability of qualified contractors to complete the work, and the long-term protection goals for the building and its contents. If the choice is to repair, then the materials used for the repair should be both visually and physically compatible with the historic roof. A careful interview with the potential contractor will determine what specific methods will be used to complete the repairs and their compliance with the applicable preservation standards or guidelines.

Roofing products, such as asbestos-cement shingles and boards, some asphalt shingles, roof felts, certain mastics, and adhesives (as well as other building products), are known to include asbestos to strengthen the material or to improve fire resistance. When the microscopic asbestos fibers are inhaled, they can become lodged in the lung and may eventually cause mesothelioma (a form of lung cancer). In recognizing this potential health risk, several consumer protection agencies enacted a number of restrictions and bans on the use of asbestos in the 1970s. In 1989, the United States Congress banned and scheduled a phase-out of asbestos-based products. However, much of this ban was later judicially vacated to allow many existing products to continue to be made and used, but the opportunity to introduce new asbestos-based products was eliminated. Legislation to ban the continued manufacture of existing asbestos-based products has not been enacted at the time of the writing of this book (Environmental Protection Agency [EPA] 1999; MesotheliomaHelp 2007). Listings of currently banned or restricted asbestos-based products and processes are available from the EPA.

If the presence of asbestos is suspected in an area to be disturbed by the repairs, then samples of the suspicious products must be sent to a testing laboratory for verification. If asbestos is known to exist, then asbestos abatement guidelines must be followed. While removal has been the normal course, encapsulating (physically isolating) the asbestos may be an option when the asbestos is nonfriable. If the asbestos is friable (loose or easily dislodged as dust) in the removal or repair of the product, then the asbestos must be abated. Verify the permitted course of action with local building and health officials.

Replacement Strategies

Inevitably, there comes a time when patching and piecemeal repairs may not be cost effective. When this happens, consider roof replacement. A complete replacement involves numerous possibilities and choices.

One choice is simply to cover the existing roof with an entirely new roof by attaching the new cladding directly to the existing roof. While this strategy is possible for wood or asphalt shingle roofs, most local building codes do not allow more than two layers of roofing on a roof. This strategy cannot be used for tile, slate, cement, or asbestos-cement roofs.

Another possibility is to replace the roof with similar materials. For fire safety reasons, many communities restrict the use of wood shingles. Asbestos-concrete shingles were manufactured well into the 1980s (National Park Service 1999). The asbestos fiber dust created during salvage presents a health risk in the unlikely event that salvaged materials are located. A number of slate and tile roof products are still available (either new or salvaged), in some cases from the original sources, but their availability and cost must be factored into the decision.

When suitable similar replacement materials are available, it is critical to hire a qualified contractor. As recommended in the previous section on repair strategies, it is important to interview potential candidates and ask specifically about their experience and their approach to installing the type of roof in question.

If the roof must be replaced, it may be possible to salvage existing roofing materials (especially tile and slate) from portions of the roof outside of public view and use these materials to replace damaged materials on portions of the roof that are visible to the public. Then the areas hidden from the public can receive the new replacement materials. This approach is not recommended for asphalt, wood, and metal roofs, where these materials may become damaged beyond reuse when they are removed.

Another issue with complete replacement is to verify that the structural framing can support the new roofing system adequately. This is especially important for replacement systems that incorporate increased insulation measures that would result in greater accumulation of snow loads on the new roof than on the previous roof. A qualified structural engineer should be consulted to assist in making this determination. Similar verification is needed to ensure that the underside of the roof is adequately ventilated to allow drying.

The removal of historic roofing systems must be done in accordance with local health and safety regulations. As noted earlier, certain roofing systems used materials, such as asbestos, lead, and PCBs, that are now known to present health hazards to people exposed to them. There are a number of licensed hazard abatement specialists who can assist in determining the appropriate means of mitigating risk in removing these hazardous materials.

References and Suggested Readings

Allen, Edward. 1999. *Fundamentals of building construction: Materials and methods.* 3rd ed. New York: John Wiley & Sons.

Arkansas Historic Preservation Program. 2004. Asbestos siding. http:// www.arkansaspreservation .org/historic-properties/national-register/ siding_materials.asp?page=asb

Association for Preservation Technology International. 1997. *Metals in historic buildings.* Chicago: Association for Preservation Technology International.

Belle, John, John Ray Hoke, Jr., and Stephen A. Kliment, eds. 1994. *Traditional details for building restoration, renovation and rehabilitation from the 1932–1951 editions of Architectural Graphic Standards.* New York: John Wiley & Sons.

Buchholdt, H. A. 1999. *An introduction to cable roof structures.* 2nd ed. London: Thomas Telford.

Cummings, Abbott Lowell. 1979. *The framed houses of Massachusetts Bay 1625–1725.* Cambridge, MA: Harvard University Press.

De Mare, Eric, ed. 1951. *New ways of building.* London: Architectural Press.

Gay, Charles Merrick and Harry Parker. 1943. *Materials and methods of architectural construction.* 2nd ed. New York: John Willey & Sons.

Griffin, C. W. and Richard Fricklas. 1995. *The manual of low-slope roof systems.* 3rd ed. New York: McGraw-Hill Book Company.

Grimmer, Anne E. and Paul K. Williams. 1992. *The preservation and repair of historic clay tile roofs.* Preservation Brief No. 30. Washington, DC: United States Department of the Interior.

Harris, Samuel Y. 2001. *Building pathology: Deterioration, diagnostics, and intervention.* New York: John Wiley & Sons.

Hobson, Vincent H. and Melvin Mann. 2001. *Historic and obsolete roofing tile: Preserving the history of roofing tiles.* Evergreen, CO: Remai Publishing.

Hornbostel, Caleb. 1991. *Construction materials: Types, uses, and applications.* 2nd ed. New York: John Wiley & Sons.

Jackson, Mike. 1995. Asphalt shingles. In Jester 1995, 248–253.

Jenkins, Joseph, 1997. *The slate roof bible.* White Rive Junction, VT: Chelsea Green Publishing.

Jester, Thomas C., ed. 1995. *Twentieth century building materials: History and conservation.* New York: McGraw-Hill Book Company.

Levine, Jeffrey S. 1992. *The repair, replacement, and maintenance of historic slate roofs.* Preservation Brief No. 29. Washington, DC: United States Department of the Interior.

Litchfield, Michael W. 2005. *Renovation.* 3rd ed. Newtown, CT: Taunton Press.

McCampbell, B. Harrison. 1991. *Problems in roofing design.* Stoneham, MA: Butterworth Architecture.

Merritt, Frederick S., ed. 1958. *Building construction handbook.* New York: McGraw-Hill Book Company.

MesotheliomaHelp. 2007. Asbestos bans in the U.S. and worldwide. www.mesothelioma-help.us/asbestos_bans.htm.

Michaels, Leonard. 1950. *Contemporary structure in architecture.* New York: Reinhold Publishing Corporation.

Nash. George. 2005. *Renovating old houses: Bringing new life to vintage homes.* Newtown, CT: Taunton Press.

National Park Service. 1999. From asbestos to zinc: Roofing for historic buildings. (Online version of the exhibit at The Roofing Conference and Exposition for Historic Buildings held in Philadelphia on March 17–19, 1999). http://www.cr.nps.gov/hps/tps/roofing exhibit/ complete.pdf.

Park, Sharon C. 1989. *The repair and replacement of historic wooden shingle roofs.* Preservation Brief No. 19. Washington, DC: United States Department of the Interior.

Peterson, Charles E., ed. 1976. *Building early America: Contributions towards the history of a great industry.* Radnor, PA: Chilton Book Company.

Pierpont, Robert N. 1987. Slate roofing. *APT Bulletin,* 19(2): 10–23.

Poore, Patricia, ed. 1992. *The Old House Journal guide to restoration.* New York: Dutton.

Rifkind, Carole. 1980. *A field guide to American architecture.* New York: Plume Books.

Rose, William B. 2005. *Water in buildings: An architect's guide to moisture and mold.* New York: John Wiley & Sons.

Scheffler Michael J. and James D. Connelly. 1995. Building sealants. In Jester 1995, 272–276.

Swanke, Hayden Connell Architects. 2000. *Historic preservation: Project planning and estimating.* Kingston, MA: R. S. Means.

Sweetser, Sarah M. 1978. *Roofing for historic buildings.* Preservation Brief No. 4. Washington, DC: United States Department of the Interior.

Tobin, Erin M. 2000. When the imitation becomes real: Attitudes toward asphalt and asbestos-cement roofing and siding. *APT Bulletin* 31(2/3): 34–37.

United States. Environmental Protection Agency (EPA). 1999. EPA asbestos materials bans: Clarification, May 18, 1999. http://www.epa.gov/asbestos/pubs/ asbans2.pdf.

Waite, Diana. 1976. Roofing for early America. In Peterson 1976, 135–149.

Watson, John A. 1979. *Roofing systems: Materials and application.* Reston, VA: Reston Publishing.

—— 1984. *Commercial roofing systems.* Reston, VA: Reston Publishing.

CHAPTER 9 Exterior Wall Cladding

Cladding for exterior walls is defined here as the outermost layer of material that encloses a building. Cladding is considered nonstructural in the sense that it is not intended to support the building structurally. Many of the decay mechanisms and principles guiding the remediation methods described for materials used as structural systems in Part 2 of this book are applicable to those same materials when they are used as cladding. The difference here is that the materials described in this chapter act as the skin of the building rather than as the building's main structural support.

The first shelters built by early colonists in North America were primitive and were intended to be temporary until permanent buildings could be constructed. Cladding was made from materials that were immediately available (e.g., bark, thatch, animal hides). Since then, various cladding systems have been developed to counteract problems caused by local climate, shortages of materials, and lack of expertise in construction methods, as well as to reduce construction time and cost. The development of exterior wall cladding was similar to that of roof cladding since these products involved the same materials and construction technologies. Many early cladding systems used wood or stucco (see Figure 9-1). Advances in construction technology led to the development of metal and asphalt products in the nineteenth century and composites, metal alloys, and plastic polymers in the twentieth century.

One common construction practice has been to use less expensive materials to imitate the appearance of more expensive ones. For instance, wood on houses and

175

Figure 9-1   Traditional wood cladding includes (a) shingles, (b) clapboards, (c) weatherboards, and (d) board and batten.

cast iron on commercial buildings were finished with paint mixed with sand to give the appearance of stone. Stucco was scribed to look like cut stone. Stamped metal panels were made to imitate three-dimensional textures, such as relief carvings, stone, and brick patterns.

Cladding on wood-framed buildings was typically fastened to sheathing that had been attached directly to the structural frame. Asphalt building felts and papers were introduced in the nineteenth century to control moisture and dampness. The use of boards for sheathing continued into the early twentieth century, when the first plywood and fiberboard sheathing products were introduced. Plywood was introduced in 1905. Fiberboard sheathing was first developed in England in 1772, patented in the United States in 1858, and became available for a range of interior and exterior uses under a variety of names (e.g., "masonite," "beaver board," "hardboard") in the 1910s and 1920s (Gould, Konrad, Milley, and Gallagher 1995). On load-bearing walls sheathing was not typically used, although stucco cladding could be applied directly to the exterior surface using a wood lath or wire mesh substrate to secure it in place.

With the transition to skeletal framing from the end of the nineteenth century to the early twentieth century, cladding "systems" became a popular way to enclose a building at a fraction of the cost of the materials they had replaced. This practice resulted in a number of innovative cladding systems (e.g., veneer masonry, curtain wall) in both the residential and nonresidential sectors. Although lack of demand, sufficient material resources, and the Depression of the 1930s hampered early acceptance in the construction market, these systems saw growing acceptance in the prosperous eras after World War I and World War II.

Veneer cladding and curtain wall systems became popular in this period, when the cladding formed the skin of the building but did not necessarily support other parts of the building. To eliminate moisture migration into the building, a drainage plane was incorporated into the construction. A drainage plane is a physical gap or separation between two layers of material. For example, an air gap between the cladding and the sheathing or masonry substrate was common in veneer-based cladding. The exposed exterior face of the wall was (or should have been) constructed so that water flowed down the surface to drip edges or to the ground below. Any moisture that seeped through the cladding material or joints would intercept the drainage plane and flow to the ground as well. To accommodate this drainage, weep holes or concealed gaps along the base of the wall were included. A well-constructed building included horizontal drip edges along joints and projecting surfaces, as well as flashing around the windows and doors, and along the tops and bottoms of the walls to direct water to the appropriate drainage path. Where an air gap was not possible, a moisture-resistant asphalt building paper or another impermeable membrane could be attached to the sheathing, over which the exterior cladding was then attached. Moisture would stop at these membranes and flow downward as well. These physical discontinuities and impermeable barriers had been incorporated into many of the better cladding systems construction practices in use by the mid-twentieth century.

Wood

The earliest permanent colonial constructions used timber framing, brick, and stone as structural systems. Traditional timber-framed buildings included wattle and daub to fill in the vertical spaces between the framing. However, the severe climate and temperature variations in the northern colonies almost immediately revealed that wattle and daub could not withstand the freeze-thaw cycles, winds, and moisture conditions present there. Wood, which was locally abundant, was a logical choice. Wood cladding that included clapboards, shingles, and boards was introduced in the early seventeenth century. While board siding is no longer common, the use of clapboards and shingles continues to this day.

Clapboard and Weatherboard

Clapboard and weatherboard are the quintessential American cladding systems. Originally, clapboards were wedge-shaped lengths of wood that were hand-split, or riven, along the grain of split logs and secured to the timber framing. This early riving process created an irregular edge rather than the more common straight edge of later periods. As cutting technology improved, the edges became more regular. In the southern colonies, a beaded edge was incorporated to act as a drip edge as well as to provide a certain degree of ornamentation. Mechanized clapboard production came with the use of the circular saw and the band saw in the early nineteenth century. The band saw was used to resaw a board along a bias to create two separate clapboards from a single board. The circular saw was used to riftsaw a log radially along its length to create the clapboards. Riftsawing produced better clapboards (Poore 1992).

Eventually, local traditions were established that distinguished clapboard installation regionally. In the Southeast, a variant of clapboards known as "weatherboards" were used that were wider and slightly thicker than the clapboards of the colonies in the North. Due to the colder temperatures that fostered snowy, rainy, and damp conditions, clapboards in the northern colonies were installed with a narrower reveal (typically 4–6 inches) to enhance their weather resistance. Meanwhile, in the mid-Atlantic and southern colonies, weatherboard boards had a reveal that could be several inches wider than those found in the North.

Clapboards and weatherboards were laid up in successive overlapping courses to enclose the wall. Because the board was cut along the grain of the wood, the exposed lower edge was more water resistant than the end grain located at the ends. To protect the boards from moisture exposure, the joints were commonly either butted (e.g., the ends of two boards were cut perpendicularly to the face and secured tightly next to one another) or scarfed (e.g., the ends of the boards were cut at complementary angles vertically into the face of the clapboard and overlapped one another). Clapboards and weatherboards were cut in assorted lengths to offset the joints so that one continuous vertical seam would not be created on the face of the wall. At the corners of the building, the ends could be mitered, staggered, or offset, or a corner board was attached to the respective face of each wall and the clapboards were butted up against it.

In one construction technique found on early clapboard systems, the clapboards were more closely overlapped at the lowest first few feet of the wall to provide better protection from backsplash of water dripping from the roof eaves where snow and damp conditions were more persistent. As the population expanded westward, clapboards became more uniform and simpler in design. The use of clapboards was most popular in the 1890s.

Shingle

Wood shingles were hand-split, trimmed to uniform dimensions, and attached in overlapping offset courses with the end grain exposed along the lower edge. Although

Figure 9-2 When the exposed portion of the shingle was cut into a variety of shapes, shingles became part of the exterior ornamentation of the building. They could be laid with a combination of shapes to provide additional visual interest.

changes have occurred in cutting technologies, shingle cladding has remained essentially the same for the past four centuries. Shingles were often cut in a variety of shapes (see Figure 9-2) that could be character-defining features of the exterior of the building. While used on various vernacular buildings, the use of wood shingles varied by architectural style, becoming most popular in the late nineteenth and early twentieth centuries.

As shingling practices evolved in the late nineteenth and early twentieth centuries, the use of plywood and building felts became more popular as sheathing. This combination reduced moisture penetration and air infiltration while providing protection for the shingles themselves.

Boards and Siding

Board siding typically could be mounted either vertically or horizontally. In its simplest form, it was oriented horizontally and the boards were nailed flush to one another to create a relatively flat surface. To reduce moisture and wind penetration, construction techniques evolved that resulted in profiles and textures now viewed as character-defining features.

Flat board siding was often painted to protect the wood. However, to give the appearance of a more substantial material, the practice of rustication was developed, wherein the otherwise flat boards were beveled along the horizontal edges and had a V-shaped vertical groove cut into them at regular intervals. The boards were then painted with a sand paint (paint to which sand had been added) to give the appearance of stone. In less expensive buildings this method was frequently used only on the facade that faced the street, while the other three facades were finished with less expensive siding and finish treatments. In other instances, the board siding was left flat but painted with sand paint and only the trim pieces, such as corner boards, water tables, and window trim, were rusticated. As an alternate to rustication, flat board siding was painted with regular paint or covered with stucco.

Early colonial builders hand-cut boards and planks for roofing and siding, which led to the early use of "board on board" and "board and batten" siding to both sheath and clad a building. Board on board siding consisted of boards that were mounted vertically on the side of the building in two offset alternating layers. Board and batten siding consisted of boards flushed mounted to each other with a joint between the boards running vertically where a batten (a narrow strip of wood) was secured over the joint to reduce air and moisture penetration. Either approach created a joint with multiple turns to block or significantly reduce moisture and air penetration. These methods reached the height of their popularity with the Gothic Revival style promoted by Andrew Jackson Downing and Alexander Jackson Davis in the mid-nineteenth century and its subsequent variants, the Carpenter Gothic and Stick styles, in the third quarter of the nineteenth century. However, board and batten siding was largely relegated to use on outbuildings as other cladding systems came into common use.

As the number of sawmills grew, sawn boards and clapboards of uniform size and appearance became more common. Improvements in cutting technology throughout the eighteenth century enabled the broader use of wood siding. Horizontal siding can be classified as beveled and nonbeveled. Beveled siding included siding that was attached in an overlapping manner to provide a drip edge that gave the overall surface of the wall a repeated beveled look.

Nonbeveled siding was typically secured flat against the framing without substantial overlap but did have a shiplap or tongue-and-groove joint along the bottom and top edges. A shiplap joint was typically a rabbet (groove) cut along the length of the lower edge that received a matching offset profile from the top edge of the board below it. Tongue-and-groove joints consisted of more complex complementary profiles that were cut along the top and bottom of each edge. Nonbeveled siding became popular in the late nineteenth century. Over time, the various forms of nonbeveled siding have been also known as "drop siding," "novelty siding," and "German siding." Novelty siding was often cut with a distinct profile consisting of bevels, rabbets, and concave shapes to create shadow lines, drip edges, and other details that added to both the functionality and decorative qualities of the siding.

Masonry Veneers

Before the introduction of masonry veneers at the turn of the twentieth century, load-bearing stone and brick were left exposed in lieu of exterior cladding or they could be clad in stucco. The transition to skeletal frames moved the use of load-bearing masonry to an increasingly smaller segment of the construction industry. On large commercial buildings, lighter-weight stone, ceramic, and terra-cotta veneers that simultaneously formed the exterior cladding, decorative finish, and ornamentation were introduced. In the residential market, brick and simulated brick veneers were developed. As a result, a variety of masonry-based materials were used throughout the twentieth century to simulate the appearance of masonry load-bearing systems at a fraction of their weight, cost, and construction time.

Stucco

Stucco is one of the oldest cladding materials in continuous use. Originally, it referred to the fine interior ornamental plaster, but over time it has come to denote, in the United States, the exterior cladding system. The term "rendering" has also been used to describe the stuccoing process. Early stucco was a combination of hydrated or slaked lime, sand aggregates, and water. In the 1820s, natural cements were also included in the mixture. A reinforcing material, such as plant fibers or hair, was sometimes added to make the stucco stronger. Other additives included animal blood, urine, eggs, keratin or glue size, varnish, wheat paste, sugar, salt, sodium silicate, alum, tallow, linseed oil, beeswax, wine, beer, and rye whiskey. The 1871 introduction of portland cement eventually replaced most of the lime content of the stucco mixture by the early twentieth century. Today, gypsum is used instead of lime (Grimmer 1990).

Stucco cladding consists of two or three layers of the stucco mixture applied to a keying system made from wood lath, wire mesh, or expanded metal lath attached to the substrate or sheathing. In some locations, cottonwood or willow saplings were used for this purpose. The typical three-coat method started with a base coat, also known as the "scratch coat." Next came the brown coat. The finish coat was last. The additives and binders were added to the first two coats, while the finish coat consisted of a very fine mesh grade sand, lime, and pigment mixture. In a two-coat stucco system, the first two coats were combined and the finish coat was then applied.

It should be noted here that stucco is similar to yet different from the traditional coatings that were applied to adobe buildings. Adobe buildings were coated in a mixture of fine sands, clays, and water that were parged onto the adobe brick. Both mixtures were applied wet and left to cure in place. Damaged surface areas were readily repaired on a continual basis. However, the inclusion of cements in the stucco mixtures that evolved through the twentieth century makes modern stucco incompatible and inappropriate as an adobe repair or replacement material because of the difference in strength and moisture permeability.

While the stucco was still wet or uncured, a finish treatment could be worked into the surface. A variety of finishes provided the desired texture. In some cases, stucco was finished with a smooth texture and lines that imitated masonry joints were scribed into the surface in a process known as "sgrafitto" (see Figure 9-3). A number of rough-cast surface textures were also available, and varied by region and architectural style. These finishes commonly included pebble dash (created by

Figure 9-3 Sgrafitto is the process of scratching or etching stucco to imitate stone or brick. A layer of smooth, wet stucco was scribed to imitate joints. For added realism, the "joints" were filled with putty or colored to replicate the mortar that would have normally been found between the bricks or stones.

throwing or troweling pebbles against the wet stucco) and spatter dash (created by throwing diluted plaster or whitewash against the stucco). Stucco could also be colored and painted to imitate various natural stone colors. Finishes could be varied on differing elements of the facade to differentiate them. For example, the overall facade may have had a smooth finish, while the water table and foundation portions were textured to appear as though made from a different type of stone.

While constructing a stucco cladding system has long been a manual process, a pneumatic concrete application system (e.g., Shotcrete) has been in use for surface construction and repair applications since shortly after it was introduced in 1910. In recent years, synthetic stucco products, also known as "exterior insulation finishing systems" (EIFS), have been created for use in place of stucco. Originally developed as a retrofit system for existing masonry buildings in Europe in the 1950s, EIFS was introduced into the United States in 1969. This system is a combination of a stucco/acrylic polymers/pigment mixture and rigid insulation panels and has been adapted for use on skeletal framed buildings in new construction (Pencille 2000). It is not yet considered an historic material. EIFS has been plagued with a number of moisture-related problems when not properly installed or protected with flashings.

One common issue with stucco is the need for continual maintenance. Historically, stucco required periodic application of whitewash or paraffin wax. These applications provided continued protection of the stucco and filled in fine hairline cracks before they became a substantial problem. Stucco is vulnerable to water damage not only from exterior exposure but also from water leaking into the wall from the roof or from poorly flashed and sealed penetrations, such as windows and doors, as well as condensation from water vapor escaping from interior living spaces. Stucco is also vulnerable to the many problems associated with stone, masonry, and concrete described earlier in this book.

Masonry Veneer Panels

The late nineteenth century was a transition period between load-bearing and skeletal framing systems. During this period, the concept of a curtain wall still included a heavy masonry infill (e.g., brick, concrete block, clay tile), but this soon began shifting as masonry products, such as stone veneer and its subsequent imitators (e.g., terra-cotta, ceramic veneers, hollow cast stone and terra-cotta sculptural forms), were introduced in efforts to make walls as lightweight and self-supporting as possible. These veneering products were cut or cast to permit easy handling and installation on the exterior of the building or attachment to a supporting frame.

As these masonry and concrete-based materials were replaced by cladding materials, exterior masonry cladding systems became significantly thinner and lighter. At the turn of the twentieth century, stone veneer was nominally 2 to 4 inches thick; however, manufacturers continually reduced this thickness, and stone veneers used in the 1950s were nominally 1 to 2 inches thick. Experience in the late twentieth century began to reveal that certain stone veneers (e.g., marble) cannot tolerate the effects of weathering with such a reduced thickness.

Masonry veneering products were either mounted in mortar applied to a substrate material or secured to a supporting frame or shelf angle typically made from ferrous metals. As problems with corrosion in the fastening systems became apparent, efforts to minimize failures brought about the use of nonferrous fastening products, such as bronze or stainless steel. Alternatively, corrosion-resistant coatings were used to keep the fastening system intact. Holes were formed or cut into these veneering products to receive tie rods or mounting clips. Gaps between the veneer material and the fasteners were filled with grout or mortar. Similarly, since the joints between the panels needed to accommodate thermal expansion of the panels while keeping out moisture, joint sealants, such as caulk, plaster, grout, and, by the 1950s, silicone-based products, were used for the purpose. Many of these practices and sealants have since been found to be problematic, as weathering of the veneer material, corrosion on the connecting materials, and sealant failure acted together to break apart the assembly.

Brick and Simulated Brick Veneers

The introduction of balloon framing and the continuing evolution of stick-framing methods through the early twentieth century brought the eventual recognition that full load-bearing brick wall construction could be done with a combination of wood framing and brick veneering systems. In brick veneer construction, the building was framed and sheathed in wood; then a single wythe of brick was constructed as a cladding material to give the appearance of a brick building. The result was a building seemingly constructed of brick but at a lower cost. By the turn of the twentieth century, brick veneer walls had come into use. These walls can be identified by their lack of header courses in the bond pattern on the exterior wall.

Simulated brick veneer masonry systems were introduced in 1929 (McKee 1995). In these systems, concrete was applied to the face of a wall in a manner similar to that of stucco. However, the mortar was colored and stamped to resemble various masonry surfaces (see Figure 9-4). These surfaces included an assortment of brick patterns but also cut stone and other masonry-based systems.

Figure 9-4 Simulated masonry veneers (shown at the left) were introduced for new construction in 1929 and were also used to cover facades of existing buildings, much like aluminum siding, to provide an updated appearance and reduce maintenance work.

Asbestos-Cement Shingles and Boards

Since siding versions of these products did not occur until the mid-twentieth century, early asbestos siding was simply the roofing product installed on the vertical walls. By the 1940s, a variety of siding products had been developed that imitated clapboards, shingles, shakes, and thatch. Asbestos-cement boards were also used for exterior siding, as well as for insulation or fireproofing in utility areas and interior walls (Arkansas Historic Preservation Program 2004). While these products reached the height of their popularity in the first half of the twentieth century, their production continued well into the 1980s until safety concerns and economic competition from other siding products (e.g., vinyl siding) caused them to be discontinued (Tobin 2000).

Metal Cladding

Metal cladding has taken on several forms in the past 200 years. While early systems were typically created by castings that were meant to provide a less costly alternative to ornamental stone, many metal cladding systems were also devised as enhanced systems to fireproof the exterior of otherwise vulnerable buildings. As the stamping and rolling methods of the nineteenth century evolved, thinner and lighter sheet metals were introduced that similarly could be used to imitate ornate stone and wood carvings. The twentieth century saw the use of extrusion production systems that led to the creation and refinement of the curtain wall systems that persist to this day.

Cast Metals

In the early nineteenth century, cast metals were used in a variety of ways as structural, ornamental, or both uses in buildings. Among the earliest facades that employed cast iron cladding were those constructed by Daniel Badger in the mid-1820s in New York and the Miners Bank in Pottsville, Pennsylvania, constructed in 1830 by John Haviland. By the mid-nineteenth century, various cast and rolled metals (e.g., iron, bronze) were in use as cladding on exterior walls as increased production capabilities made these materials more readily available. While the early contributions of Badger and Haviland are noteworthy in the introduction of cast iron architecture, James Bogardus became the major figure in this field after he erected in 1849 the Edgar Laing Stores, a warehouse in New York that has been recognized as the "first structure with self-supporting, multi-storied exterior walls of iron" (Waite and Gayle 1991, 2). The facades consisted of an assembly of smaller pieces that were bolted together and attached to the structure of an otherwise plain building.

One compelling feature of cast iron facades was that they could be crafted in a foundry and shipped via the burgeoning network of railroads to remote locations, where they were assembled on the building site. When completed, the facade could be finished with sand paint to give the building the look of stone. Cast iron facades remained popular until just after the turn of the twentieth century.

Curtain Walls

The early twentieth century saw the introduction of numerous metals and metal alloys, including steel, aluminum, and stainless steel, to systems that were crafted and assembled in a manner similar to that of cast iron. These systems were the early forerunners of the modern curtain wall assemblies that became common in the mid-twentieth century. The first true modern curtain wall building in the United States is credited to the architect Willis Polk, whose Hallidie Building was completed in 1918 in San Francisco (see Figure 9-5).

Figure 9-5 The Hallidie Building in San Francisco has been recognized as the first true modern curtain wall building in the United States.

Historically, curtain walls consisted of an assembly of many repetitions of framing, cladding panels, glazing, connections, and structural supports. Over time, curtain wall systems have been classified into four categories: stick, unit, unit and mullion, and panel. "Stick system" installation begins with the mullions, and then the horizontal rails are installed. The insert panels go next, followed by the glazing. This system is the earliest form of the modern curtain wall construction and is still in use. "Unit systems" are composed entirely of a repeated sequence of modular units preassembled at the factory. The units include spandrel glass panels (nontransparent glazing introduced in 1935) (McKinley 1995) and sometimes regular vision glazing to serve as a window. The vertical and horizontal edges form the mullions and rails, respectively, between adjoining units. One unit can be up to three stories tall. "Unit and mullion systems" are a combination of the stick and unit systems. The mullions are installed first, and the preassembled units are then installed between them. The unit can be a single story in height or can consist of smaller separate spandrel and vision glazing subunits. "Panel systems" are similar to unit systems except that the panels are not preassembled units but instead homogeneous sheet metal or castings. The panel system comes in two types—architectural and industrial. Architectural panels are designed for a specific building, while industrial panels are uniformly made and distributed for simpler repetitive facades (Architectural Aluminum Manufacturers Association 1979).

In addition to vision and spandrel glazing, infill panels could consist of metal alloys and, by the 1950s, composite materials based on fiberglass and polymers. Other panel types were also composed of combinations of perforated panels, mesh materials, or more sculptural elements, such as louvers. As the effects of weathering were recognized, aluminum became increasingly popular due to its light weight and its self-sealing patina of oxides that helped to protect it from the effects of exterior weathering.

In 1924, porcelain-enameled metal panels were introduced. Based on earlier technologies from Germany, these panels consisted of ceramic frit or enamel paint fused to a metal substrate. Originally, these panels were available only in black or white and were used primarily on gas stations and food service buildings before World War II. Postwar manufacturers significantly expanded the color palette and marketed these panels to a broader architectural audience. Unfortunately, while the enamel protected the exposed surface, the base metal substrate was vulnerable to corrosion if exposed to moisture. Although marketed as durable, low-maintenance products, the panels could be buckled by the force of winds or dented by objects striking them. When this occurred, the porcelain enamel could crack or chip and expose the substrate beneath it. These forces could also act to open joints at the edges of the panels.

By their very nature, curtain walls include many joints between the various components. Early systems used a variety of materials to seal these joints, but by the mid-twentieth century, the forerunners of modern sealants were being used. In the 1950s, the increased use of metal curtain wall construction, with its greater need for expansion and contraction tolerances, brought about the introduction of elastomeric sealants that could accommodate those tolerances. By the early 1960s, a range of sealants were being used that were derived from butyl (synthetic rubber), polysulfide, acrylic, and silicon-based formulations. Weathering and aging in some of these products became increasingly evident by the late twentieth and early twenty-first centuries and led to the refitting of many early curtain wall systems. One of the more notable projects of this type was the Lever House, designed by Skidmore Owings and Merrill and constructed in 1952 in New York City. After years of piecemeal repair and replacement, the decision to reclad the entire building was made. The curtain wall system was removed and replaced with a system derived from the original plans using contemporary curtain wall technologies. Although the new appearance is identical to the original, this practice begs the question of whether the building is still the original building or just a replication of it (Curtis 2002).

Sheet Metal

Metals fabrication technologies of the late nineteenth century allowed a variety of decorative patterns to be stamped into metal panels and strips that were later fastened either directly to exterior sheathing or mounted on a support framework. These patterns ranged from simulated brick to ornamental carvings. As discussed previously, metals were used for ornamental cornices to simulate carved stone. The cornice was built up from a frame covered with stamped, rolled, or brake-formed metal cladding. Due to the hollow nature of the final assembly, these features were vulnerable to corrosion if moisture penetrated to the interior.

Edward Charlebois patented a sheet metal clapboard in 1903, and John. J. Muryn patented sheet metal weatherboards in 1928. However, the 1939 patent of Frank

Hoess led to the widespread use of aluminum siding, which was introduced in 1946 (Lauber 2000). The burgeoning housing market after World War II led to the demand for a cladding system that was readily available, faster to install, and less expensive than traditional clapboard or shingle systems. While aluminum siding was used on both new and existing houses, one of its key selling points was reduced maintenance, particularly painting. Unfortunately, as home owners have since discovered, the paint that comes on the original aluminum siding can decay and wear off. When this happens, the repainting cycle for the siding becomes similar to that of regular wood siding.

Asphalt

The introduction of asphalt roofing products created the possibility of adapting asphalt-based building papers to siding applications. In some instances, roofing products were directly adapted to siding by nailing them to the sheathing. More commonly, rolls of asphalt paper were fastened to the sheathing. Like roofing products, asphalt siding included stone granules on the exposed surface to protect it from ultraviolet radiation decay. These granules could be configured to imitate a wood grain, stone, or brick bond pattern. In addition to the granules, asbestos may have been used to make the product more durable and heat resistant. Asphalt siding was extremely popular during the Depression and as a siding material for low-cost housing.

Asphalt siding that is exposed to long periods of sunlight or is in close contact with humans and animals is prone to more rapid decline in performance. Like its roofing counterpart, asphalt siding is vulnerable to granule erosion and tearing from wind and mechanical contact, as well as failure due to sunlight and moisture exposure.

TYPICAL PROBLEMS

Although a structural component, such as a load-bearing wall, can support *and* enclose a building, as in the case of logs, bricks, and masonry, the function of cladding is to provide protection from exterior environmental forces. Cladding materials are vulnerable to many of the same decay mechanisms (e.g., rising damp, differential settling, insects, rot, and corrosion) that plague structural components made of those same materials. Therefore, inspection and testing methods similar to those used to assess the condition of structural materials can be applied, as appropriate, to cladding systems as well. The primary problems in cladding systems result from the direct and indirect effects of moisture, sunlight, and wind.

Moisture Penetration

Rain, melting snow, and groundwater all present problems for cladding assemblies. The two most vulnerable areas are the joints between discontinuous materials and the protective coatings. Lack of maintenance for either of these two entry points begins the decline of the cladding system. The next area of concern is the deterioration of the roof and the eaves or parapet above the wall to which the cladding is attached. While the joints and surfaces may be intact, moisture penetration can occur within the wall as the roof and eaves begin to fail. Lastly, the construction practices and the quality of the materials used need to be assessed to determine if a problem was built into the wall assembly. The construction problems (e.g., poor or no flashing, poor vapor barrier installation, inferior materials) may arise within the original construction or as part of more recent modifications.

Visible signs of early moisture problems include open joints, corrosion or rot, and failed exterior finishes (see Figure 9-6). As deterioration proceeds, the substructure of the cladding within the wall begins to fail, causing sagging, bulging, and cracking to become increasingly prominent. Moisture penetrating the wall from

Figure 9-6 Untreated moisture penetration has caused noticeable decay in this fascia and soffit.

within the cladding assembly is indicated by peeling paint or finishes and soiling stains emanating from joints. The prolonged continuation of the decay mechanisms will cause them to interact and escalate the rate of deterioration until the situation is remedied or the cladding fails.

In some instances, the source of the moisture penetration can be quite obvious but present only seasonally, such as when snow accumulates and remains for a long period of time along the base of the wall during the winter months or when there is a continual backsplash of water dripping from the roof, striking the ground below, and rebounding against the lower portion of the wall. In other cases, the construction or changes to the construction may be to blame, as when drainage systems are compromised or removed and water is allowed to run freely down the face of the wall across joints. If inadequate or no drip edges are incorporated into and along the horizontal joints, water can flow by surface tension or be drawn by capillary action into the joint and, in turn, into the interior of the cladding assembly itself.

Less obvious may be the water that enters the envelope from an otherwise unidentified leak in the roof, cornice, eave, or parapet construction located above the failing cladding. With the advent of the concept of drainage planes, cladding systems were designed to permit moisture migration within the wall to flow downward to the base of the wall and be relieved through weep holes and other openings. These weep holes may have been inadvertently plugged by being overpainted, allowing soiling products to clog them, or improperly designed or constructed in the original construction or later modifications.

In low-rise buildings, one moisture source comes from landscaping practices that can cause problems. Many buildings may not have originally had plantings along the foundation. As landscaping practices changed, plantings were introduced in this location. Whether by intent or due to inadequate care, climbing plants (e.g., ivy) seek out sunlight and moisture and can result in vines growing into and through joints and gaps in the siding. When this happens, moisture and other decay mechanisms (e.g., insects) further compromise the cladding. Both climbing and non-climbing plants located along the wall can trap moisture and prevent drying sunlight and wind from entering the space between the building and the plants.

Inadequate maintenance (e.g., raking, pruning, and trimming) of these plants, their beds, and adjoining sections of grass can also allow the accumulation of decayed vegetation to eventually build up the soil, which further traps or fosters moisture problems. Lastly, the modern use of sprinklers may deposit water directly on the cladding if sprinkler heads are located too close to the building. While this staining mimics the appearance of rising damp, the water deposits can usually be differentiated by the hard water stains left after repeated cycles of wetting and drying have occurred.

Sunlight and Heat

Exposure to sunlight and heat affects all types of cladding in various ways. The variations in sunlight on each facade cause the south and west facades to experience more decay due to ultraviolet radiation. Facades oriented to the east receive a moderate amount of ultraviolet energy, and those oriented to the north receive the least amount. The ultraviolet energy in sunlight can cause decay on exposed unfinished wood surfaces or on wood surfaces whose finishes have been allowed to fail. Ultraviolet energy can also cause protective finishes to shift in color or deteriorate (e.g., turn chalky or flake).

Sunlight and the daily thermal cycles (especially freeze-thaw cycles) cause materials to expand and contract as sunlight shifts and temperatures vary throughout the day and night. For cladding assemblies composed of thermally compatible materials (e.g., with similar thermal expansion rates and weather-resistant, flexible joints), these variations should be readily withstood. However, if one material does not react similarly to the material adjoining it or if thermal expansion is restricted, internal stresses can build up that can eventually break the materials apart. This problem can become evident when nail heads are pushed out from wood siding, brick and mortar spall or disintegrate, and infill panels buckle or flex. However, while thermal expansion will occur to some degree in all materials, this problem becomes more pronounced in construction assemblies that use metals in combination with sealants and gaskets. When metal cladding expands and contracts (or flexes due to wind pressures) repeatedly over time, metal fatigue can develop. This is noticeable when splits, creases, corrosion under fractured finishes, and unusual bulging or concave surfaces have occurred.

When joint sealants age and deteriorate, they lose their flexibility and tend to break apart. Sealant and gasket failure then admits moisture into the joints, which can accelerate the decay of the wall cladding. This moisture can then attack the assembly from the inside, where it is less noticeable until exposed pieces fail or fall off the building. Stamped metal cornices, terra-cotta, stone, and brick veneers, and metal curtain wall assemblies secured in place with ferrous metal anchors are particularly vulnerable to this problem.

Wind and Mechanical Abrasion

Wind pressure varies continuously, which causes the cladding to flex repeatedly and can cause fatigue within the cladding as well as additional stress on the cladding where it is secured to the building. Low-tensile-strength materials, such as asphalt shingles and metal siding, can fail around these connections. With sufficient continuous wind pressures (e.g., unusually strong winds, hurricanes, tornados), the uplift pressure can tear the cladding from the building.

More common, however, are the effects of windblown objects, such as tree limbs or wires, repeatedly striking the cladding surface (see Figure 9-7), which can cause dents, cracks, abrasion of the protective finishes, or outright failure of the cladding.

a   b

Figure 9-7 Wind can initiate a number of decay problems on exterior cladding: (a) wind-scoured aluminum siding and (b) mechanical abrasion from a light cord.

Similarly, the continuous scouring effect of airborne dust, sand, or ice pellets can abrade the protective finish of the cladding, leaving it exposed to potential decay from atmospheric sources. Smooth or otherwise glossy surfaces can become pitted. If the underlying material absorbs moisture, the material can become vulnerable to moisture-related decay mechanisms.

Material Losses

A building will move in response to the forces of gravity, wind, and seismic activity. This reaction can cause the building to settle, sway, or move substantially based on its construction technology and local soil conditions. The long-term effects of settling include cracked surfaces, spalling masonry, and out-of-plumb door and window frames, which all contribute to moisture penetration.

The stress and failure caused by wind and seismic events can be similar to those due to settling, but they are more readily noticed after a significant wind or seismic event. Connections can fail, joints can open, and cladding can sag or fall from the building. As corrosion or rot begins and the cladding begins to decay without consistent maintenance, these components fail and allow decay mechanisms to further destroy the cladding and the materials beneath it. Similarly, wind events can strip away small portions of the cladding, which also accelerates the decay beneath the cladding. Once the cladding is compromised in this manner, moisture enters the assembly and eventually affects the interior spaces of the building.

As decay sets in and cladding begins to fail, precautions must be taken to protect public safety. Protective netting, screening, and hard enclosures over adjoining sidewalks and public assembly areas must be installed to ensure that materials falling from the building cannot potentially cause harm to people passing below.

Soiling and Finish Failure

The combination of soiling, moisture, and inadequate maintenance is a recipe for accelerated failure of cladding. Air pollution, dust, and bird droppings can be deposited directly on the cladding or carried onto the cladding as runoff from poorly maintained gutters and roofs. Beyond the aesthetic issues of the soiling itself, the accumulation of these materials can invite biological growths and chemical reactions that accelerate decay. Biological growths, such as moss, lichens, and small plants, can take root. If left in place, their roots can grow sufficiently to begin breaking apart joints between the cladding elements. When this occurs, moisture and its related problems can penetrate beneath the cladding. For materials like wood and ferrous metals, this moisture can pose a substantial problem. Chemical reactions from air pollution and acid rain can react with surface finishes and cause them to fail more rapidly than normal. Painted surfaces can fail, and finishes can be dulled or become etched from continued corrosive contact. Rather than be allowed to accumulate unchecked, the soiling materials must be removed as part of an ongoing maintenance program and, wherever possible, the source of the contamination must be removed or mitigated.

Substitute Siding

While many cladding systems are now considered historic if they were installed as the original cladding when a building was first built fifty or more years ago, substitute siding is any exterior cladding material that is placed directly over or is used to replace preexisting exterior cladding material installed in an earlier historic period. The most common recent forms of substitute siding on residential buildings are aluminum (and, subsequently, vinyl) siding and imitation masonry veneers. Deteriorated and soiled cladding was often considered too difficult to maintain, and these substitute siding systems were simply installed over the existing siding or cladding materials. In larger buildings this practice has been a growing concern to preservationists, as many cladding systems of the recent past are rapidly deteriorating. In the past decade, a growing number of larger buildings have had new replacement cladding installed in an effort to alleviate this problem. Surprisingly, however, the use of substitute siding systems is not just a recent phenomenon.

By the mid-seventeenth century, clapboards were used over the exposed existing wattle and daub of the earliest colonial buildings to reduce maintenance and improve their appearance. Furthermore, evidence of this continued philosophical approach can be traced back 300 years as new materials and construction systems evolved. For example, cast iron and stamped metal panels were used to modernize the appearance of existing buildings in the early nineteenth century (Brand 1994). This practice has continued down to the present whenever the "modernization" of a building's exterior is contemplated using the then "modern" material system on an older building.

Two major problems can occur when cladding is installed over the original cladding of a building built in an earlier period. First, the new siding may not correct the decay mechanisms acting on the original cladding. Second, the installation method used may irreversibly damage or remove historic fabric.

Figure 9-8 To properly install aluminum siding on an existing building, ornamental elements were cut back or removed to create a flat plane so that the siding could be properly attached. In many instances, this process removed or concealed character-defining features of the original exterior cladding and surfaces.

An example of the first case is when water from a leaking roof or an inadequately maintained gutter system has entered the wall assembly internally. The installation will not necessarily eliminate the source problem. Furthermore, the new cladding will conceal this continuing decay and decay from other subsequent decay mechanisms (e.g., mold, rot, corrosion, or insect damage) from direct view. Ultimately, this decay can result in the complete failure of both the original and new cladding and the continued penetration of moisture into the interior cavities and materials beneath it.

The second problem that substitute siding creates is in its installation approach. Many substitute siding systems require a flat plane to which the new cladding is attached. The installation often requires the removal of existing ornamental and character-defining features of the building to establish the necessary flat plane. Often, these items are cut back or removed altogether and discarded (see Figure 9-8). Similarly, the attachment method (e.g., masonry mortars, adhesives, or mechanical fasteners) may permanently damage the original cladding. The irreversible nature of these approaches has created numerous problems when the goal of a project has been to restore the original exterior appearance.

RECOMMENDED TREATMENTS

Due to the evolving nature of cladding systems, acceptable treatments are extremely varied. However, they generally fall into three categories: conserve, repair, and replace. In certain instances, the cladding may no longer be manufactured for new construction and is available only through salvage. More commonly, the existing material can simply be conserved, repaired, or, if need be, replaced by using techniques in common use today, adapting current techniques to earlier practices, or reusing traditional building methods. Lastly, it may be necessary to replace a large portion, if not all, of the existing cladding if it is too badly deteriorated or poses a life safety threat.

Cladding made of the traditional materials described in Part 2 can use the same remediation treatments described there for each type of material; therefore, for the sake of brevity, those treatments will not be repeated here. More problematic and idiosyncratic are the numerous curtain wall systems built in the mid-twentieth century. While they were made from many of these same otherwise historic materials, their fabrication and construction methods were sometimes more experimental in nature. As a result, these systems require closer evaluation to determine how well they are performing.

Conservation Strategies

Many cladding systems will not last more than a few decades without some maintenance to keep them viable. Annual and seasonal maintenance provides the needed attention to catch and eliminate decay mechanisms before they become substantial. In this regard, a maintenance log of all activities related to the cladding should be developed to enable a long-term view of recurring issues that may point to a more serious problem. Each cladding system is vulnerable to specific decay mechanisms, as described above. Overall, however, the careful assessment and conservation of joints, surface finishes, and overall physical integrity is important in maintaining the existing historic fabric.

As part of the ongoing maintenance of the building, care must be taken to monitor conditions. While sudden natural and man-made events can cause readily noticed damage, long-term effects of weathering, aging, and decay may not be readily or immediately apparent unless specifically checked. Two aspects of the inspection of the overall condition of the cladding are important. First, it is important to assess the integrity of the surface materials (e.g., finishes, structural integrity, continuity, and visual appearance), the joints between materials (e.g., missing caulk, sealants, flashing, or gaskets), and connections (e.g., missing, damaged, or deteriorated). Second, the overall cladding should be evaluated to see if it remains in its proper orientation and is not bulging, bowing, sagging, or showing other signs of internal failure of its connection systems to the substructure. When problems are discovered, one should note them and develop an overall assessment of the existing conditions. Nominal problems can usually be readily fixed. On the other hand, a review of the overall system may indicate a larger, more serious problem (e.g., corrosion has compromised the connections and joints or moisture penetration is occurring on a broader scale).

As buildings age or as the goal of conserving the historic integrity of the original materials becomes more important, conservation strategies allow the retention of as much historic fabric as possible more economically. Preservation has been referred to as long-term applied maintenance, and that concept is valid here. What is important is to look at the detail and to gain a broad overview and understand the implications of both. If ongoing maintenance needs are continually deferred, or if the decision to forgo an expensive conservation procedure would result in the loss of the complete building, then appropriate repair and replacement treatment strategies must be selected. These strategies must be chosen with care.

Repair Strategies

A step up from conservation occurs when deferred maintenance has caused cladding materials to fail. Here still, though, the need to take both a detailed view and a broad view remains critical, as a series of numerous small repairs may mask a larger overall problem. At this point, initiating repairs becomes necessary. When done annually, as part of the maintenance program, these repairs may be relatively nominal. These problems should not be neglected for long periods of time since they have typically compromised the overall integrity of the cladding systems, which in turn can accelerate decay problems.

Many techniques for roofing repairs can be directly applied to similar vertical cladding products. Shingles, clapboards, and other cladding veneers can be replaced with exact replacements, if available, or with comparable products of similar size, material, and construction. Appropriately formulated epoxies and repair adhesives for wood, cementitious materials, and stone can be used to reinforce and consolidate failing materials. Dutchman and other spliced inserts can typically be used so long as they have compatible physical characteristics, can be differentiated from the original materials, and are reversible.

For sophisticated cladding systems, such as curtain walls and thin veneer masonry systems, consultation with the original manufacturers, if they are still in business, is an almost certain requirement. Original records, such as shop drawings, can illustrate how the cladding system was fabricated and give clues to how it was installed. If necessary, strategic removal of individual units, panel components, or interior portions may be needed to ascertain the condition of the construction materials and the construction practices used in the original installation. When this information is known, an appropriate repair strategy can be implemented. By reviewing the long-term maintenance logs, one can decide whether to continue repairs on a piecemeal basis or pursue more significant replacement strategies.

Replacement Strategies

The *Standards* permit in-kind replacement of materials when the original materials are missing or too deteriorated to retain. However, while replacement of common materials (e.g., wood shingles or siding) and product assemblies (e.g., brick veneers or stucco) is allowed, this frequently creates a dilemma when synthetic materials (e.g., terra-cotta, cast stone) and complex assemblies (e.g., curtain walls, thin veneer cladding) are in question. Cutting, casting, and other replication techniques may provide the opportunity to fabricate suitable replacements (e.g., GFRC, cast aluminum, fiber-reinforced polymers, cast stone, precast concrete, epoxy concretes, and polymer concrete) directly from readily available matching natural or synthetic materials (Park 1988). However, as construction technologies advance from the period of the original installation, many of the fabricated products and the companies that made them cease to exist. Stone quarries close, a certain grade and species of wood becomes expensive, and companies drop product lines or go out of business. Modern construction practices include many contemporary products and methods that may appear to replicate the original installation, but caution must be used when selecting them. As noted in earlier chapters, substitute materials available today may appear appropriate when first installed but subsequently may decay (e.g., due to wind and sun exposure, physical incompatibility, or poor installation practices). Replacement materials should be chosen from among the materials approved by local preservation oversight agencies that are physically compatible and best serve the long-term preservation needs of the building. For example, medium density fiberboard (MDF) is frequently an acceptable choice, while vinyl siding is not.

As the curtain wall systems constructed in the mid-twentieth century age and in some cases deteriorate or no longer meet contemporary performance demands, a question has been raised: what actually constitutes preservation and keeping a building authentic? With many curtain wall systems, there have been opportunities to deconstruct the cladding and replace it. While the visual appearance and detailing of the original construction were achieved, the question remains as to whether this replacement was a preservation project or a reconstruction project. The image remains the same, but the authenticity is no longer there. This question pervades all levels of preservation technology as original materials are discarded due to poor maintenance practices or the misperception that they cannot be restored economically. Unless there is a serious underlying decay problem like extensive rot or corrosion, much of the damage may actually consist of only superficial finish problems.

Simulated masonry veneers and aluminum siding (and, subsequently, vinyl siding) were frequently installed on older residential buildings to "enhance" their appearance and reduce maintenance costs. In nonresidential buildings, curtain wall systems introduced in the early twentieth century as an exterior cladding system on new buildings were, by the 1950s, being installed on many commercial buildings built in the late nineteenth and early twentieth centuries as owners began "updating" their buildings. This trend continued into the late twentieth century but began to be reversed in some locations as the unique architectural character of the buildings was recognized for its historic value and appearance. In a number of projects, previous replacement cladding systems were removed in the hope of recapturing the original appearance of the building, only to reveal the substantial damage done to the original historic building fabric when the replacement or substitute siding was installed. When any of these replacement cladding systems were used, many historic buildings also suffered additional damage as decay mechanisms continued or subsequently started in the areas concealed by the substitute siding.

If removal of the substitute siding is an option, evaluate the condition of the cladding beneath it. Using an approach similar to that used in investigating potential repair needs, selected portions of the siding may be removed to gain access to the original cladding beneath it to determine its condition. As the siding is removed, note the extent of damage caused by the installation of the siding and attempt to identify locations where previous protruding features (e.g., trim, brackets, and building ornaments) may have been located. The best approach to determine what might be missing is to look for the physical evidence of an element's having been cut back, the "witness marks" of something that has been removed (e.g., outlines of trim profiles where paint was applied to the trim piece or subsequent coats of paint have built up along the edge of a removed ornament), or to interpret historic photographs if they are available. Replacement features must be based on physical evidence.

References and Suggested Readings

Allen, Edward. 1999. *Fundamentals of building construction: Materials and methods.* 3rd ed. New York: John Wiley & Sons.

Architectural Aluminum Manufacturers Association (AAMA). 1979. *Aluminum curtain wall design manual.* Chicago: AAMA.

Arkansas Historic Preservation Program. 2004. Asbestos siding. http://www.arkansaspreservation.org/historic-properties/national-register/siding_materials.asp?page=asb

Ashurst, John and Nicola Ashurst. 1988. *Practical building conservation, Vol. 4: Metals.* New York: Halsted Press.

Association for Preservation Technology International. 1997. *Metals in historic buildings.* Chicago: Association for Preservation Technology International.

Belle, John, John Ray Hoke, Jr., and Stephen A. Kliment, eds. 1994. *Traditional details for building restoration, renovation and rehabilitation from the 1932–1951 editions of Architectural Graphic Standards.* New York: John Wiley & Sons.

Brand, Stewart. 1994. *How buildings learn: What happens after they're built.* New York: Viking Press.

Cummings, Abbott Lowell. 1979. *The framed houses of Massachusetts Bay 1625–1725.* Cambridge, MA: Harvard University Press.

Curtis, Wayne. 2002. No clear solution. *Preservation* 54(5): 46–51, 118.

De Mare, Eric, ed.1951. *New ways of building.* London: Architectural Press.

Erdley, Jeffrey L. and Thomas A. Schwartz, eds. 2004. *Building facade maintenance, repair, and inspection* (ASTM STP1444). West Conshohocken, PA: American Society for Testing and Materials.

Fisher, Charles E. and Hugh C. Miller, eds. 1998. *Caring for your historic house: Preserving and maintaining structural systems, roofs, masonry, plaster, wallpapers, paint, mechanical and electrical systems, windows, woodwork, flooring, landscape.* New York: Harry N. Abrams.

Foulks, William G. ed. 1997. *Historic building façades: The manual for maintenance and rehabilitation.* New York: Preservation Press.

Friedman, Donald. 1995. *Historical building construction: Design, materials, and technology.* New York: W. W. Norton & Company.

Gayle, Margot and Carol Gayle. 1998. *Cast-iron architecture in America: The significance of James Bogardus.* New York: W. W. Norton & Company.

———, David W. Look, and John G. Waite. 1992. *Metals in America's historic buildings: Uses and preservation treatments.* Washington, DC: United States Department of the Interior.

Gould, Carol S. 1997. Masonite: Versatile modern material for baths, basements, bus stations, and beyond. *APT Bulletin,* 28(2/3): 64–70.

Gould, Carol S., Kimberly A. Konrad, Kathleen Catalano Milley, and Rebecca Gallagher. 1995. Fiberboard. In Jester 1995, 120—125.

Grimmer, Anne. 1990. *The preservation and repair of historic stucco.* Preservation Brief No. 22. Washington, DC: United States Department of the Interior.

Harris, Samuel Y. 2001. *Building pathology: Deterioration, diagnostics, and intervention.* New York: John Wiley & Sons.

Hornbostel, Caleb. 1991. *Construction materials: Types, uses, and applications.* 2nd ed. New York: John Wiley & Sons.

Howell, J. Scott. 1987. Architectural cast iron: Design and restoration. *APT Bulletin,* 19(3): 51–55.

Hunt, Dardley. 1958. *The contemporary curtain wall.* New York: F. W. Dodge.

Jester, Thomas C., ed. 1995. *Twentieth century building materials: History and conservation.* New York: McGraw Hill Book Company.

Johnson, Paul G., ed. 2003, *Building performance of exterior building walls* (ASTM STP1422). West Conshohocken, PA: American Society for Testing and Materials.

Kelley, Stephen J. 2001. Conflicts and challenges in preserving curtain walls. *APT Bulletin* 32(1): 9–11.

Lauber, John. 2000. And it never needs painting: The development of residential aluminum siding. *APT Bulletin* 31(2/3):17–24.

Litchfield, Michael W. 2005. *Renovation.* 3rd ed. Newtown, CT: Taunton Press.

McKee, Ann Milkovich. 1995. Simulated masonry. In Jester 1995, 174-179.

McKinley, Robert W. 1995. Spandrel glass. In Jester 1995, 206–11.

Melander, John M. and Albert W. Isberner, Jr. 1996. *Portland cement plaster (stucco) manual.* Skokie, IL: Portland Cement Association.

Merritt, Frederick S., ed. 1958. *Building construction handbook.* New York: McGraw-Hill Book Company.

Milley, Kathleen Catalano. 1997. Homasote: The "greatest advance in 300 years of building construction." *APT Bulletin*, 28(2/3): 58–63.

Mitchell, Robert A. 1991. Whatever happened to Lustron homes? *APT Bulletin* 23(2): 44–53.

Myers, John H. (revised by Gary L. Hume). 1984. *Aluminum and vinyl siding on historic buildings: The appropriateness of substitute materials for resurfacing historic wood frame buildings*. Preservation Brief No. 8. Washington, DC: United States Department of the Interior.

Nash. George. 2005. *Renovating old houses: Bringing new life to vintage homes*. Newtown, CT: Taunton Press.

Park, Sharon. 1988. *The use of substitute materials on historic building exteriors*. Preservation Brief No. 16. Washington, DC: United States Department of the Interior.

Pencille, Doug. 2000. EIFS Facts. http://www.dspinspections.com/eifs_facts.htm

Peterson, Charles E., ed. 1976. *Building early America: Contributions towards the history of a great industry*. Radnor, PA: Chilton Books.

Poore, Patricia, ed. 1992. *The Old House Journal guide to restoration*. New York: Dutton.

Rose, William B. 2005. *Water in buildings: An architect's guide to moisture and mold*. New York: John Wiley & Sons.

Simpson, Pamela H. 1999. *Quick, cheap and easy: Imitative architectural materials, 1870–1930*. Knoxville: University of Tennessee Press.

Scheffler, Michael J. 2001. Thin-stone veneer building facades: Evolution and preservation. *APT Bulletin*. 32(1): 27–34

——— and James D. Connelly. 1995. Building sealants. In Jester 1995, 272–276.

Schwartz, Thomas A. 2001. Glass and metal curtain-wall fundamentals. *APT Bulletin*, 32(1): 37–45.

Slaton, Deborah and William G. Foulks, eds. 2000. *Preserving the recent past 2*. Washington, DC: Historic Preservation Education Foundation.

Sullivan, Anne. 1995. Shotcrete. In Jester 1995, 102–107.

Swanke, Hayden Connell Architects. 2000. *Historic preservation: Project planning and estimating*. Kingston, MA: R. S. Means.

Tobin, Erin M. 2000. When the imitation becomes real: Attitudes toward asphalt and asbestos-cement roofing and siding. *APT Bulletin* 31(2/3): 34–37.

Van den Branden, F. and Thomas Hartsell. 1984. *Plastering skills*. Homewood, IL: American Technical Publishers.

Waite, John and Margot Gayle. 1991. *The maintenance and repair of architectural cast iron*. Preservation Brief No. 27. Washington, DC: United States Department of the Interior.

Weaver, Martin E. 1997. *Conserving buildings: A guide to techniques and materials*. New York: John Wiley & Sons.

Yeomans, David. 2001. Origins of the modern curtain wall. *APT Bulletin* 23(1): 13–18.

CHAPTER 10 Windows

Windows are one of the most distinctive character-defining features of any building. Beyond use for enclosure, light transmission, and thermal protection, windows may provide a view both out from and into a building. In this chapter, generic window systems will be discussed as part of the exterior building fabric. Art and stained glass windows, which present a more visual statement than the simple windows described here, will be discussed as part of the interior finishes and ornament later in the book.

From an architectural aesthetic point of view, windows are integral to the overall composition of a building. Unfortunately, window replacement can be among the most contentious undertakings when rehabilitating, restoring, or adaptively using a building. Many replacement decisions are arbitrary and often little related to the actual condition of the window. Many deteriorated but still functioning windows have been replaced due to lack of recognition that surface conditions did not truly represent the potential remaining utility of the window and the economic benefit that could be achieved by nominal repairs rather than wholesale replacement.

Historic Overview

Windows have served various functions throughout recorded history. Recommendations for the use and placement of windows have been recorded as early as the first century BC by Vitruvius, a Roman architect, in his *Ten Books on Architecture* (Morgan 1960). The original purposes of windows were to provide a view, to admit light and fresh air, and to provide protection from outdoor elements. The Romans used various translucent materials to fill in the window opening to make their rooms more temperate. These materials included mica, alabaster, linen, and various types of glass set in frames to reduce air flow while still admitting light. Although the size, shape, configuration, and number of windows varied through time as architectural

Figure 10-1 Windows are significant character-defining features of residential and nonresidential buildings alike. Shown here is nearly a century of window evolution in Chicago.

styles and building technology changed, the window is considered one of the most significant character-defining features of any building (see Figure 10-1). The National Park Service, which administers the reviews for compliance with the *Standards*, states in *The Repair of Historic Wooden Windows* that these important characteristics include the pattern and size of the openings, proportions of the frame and sash, configuration of windowpanes, muntin profiles, wood type, paint color, glass characteristics, and the associated ornamental details (Myers 1981).

As architectural styles evolved, changes in window technology often simultaneously introduced the use of building construction systems and materials that varied dramatically from those of previous styles. Over time, aspects of these newer windows were incorporated into older buildings in an attempt to "update" them. Nowhere is this trend more evident than in the treatment and use of windows whose replacement has been driven not only by aesthetic concerns but also by concerns about maintenance and energy costs. Hence, we see the ongoing updating to what is considered to be the most contemporary window system. Casement windows were replaced by sliding windows. Sliding windows evolved from single-hung to double-hung as suitable window hardware technology emerged. Refinements in cutting-edge technology allowed the creation of finer woodwork details, while advances in the industrialization of wood products fabrication and glassmaking processes led to the availability and widespread distribution of windows across the nation. Progress in materials fabrication methods led to the creation of sash made from steel, aluminum, and eventually vinyl that competes with wood sash today. And lastly, the demand for efficient energy performance led to the development of multilayered glass systems.

In the earliest stages of a proposed project, it is important to determine the condition of the existing windows and verify the opportunity for their continued use. The importance of doing so cannot be understated, as again noted in *The Repair of Historic Wooden Windows*:

> Evaluating the architectural or historical significance of windows is the first step in planning for window treatments, and a general understanding of the function and history of windows is vital to making a proper evaluation. (Myers 1981, 1)

Often, although unsightly due to lack of proper maintenance, existing windows can be repaired, upgraded, and reused for less money than a complete replacement. Both wood and metal sash can be repaired and reused to preserve the historic appearance of a building without resorting to their complete replacement. Energy performance needs seemingly tip the balance toward replacement; however, a number of strategies for maintenance, performance upgrades, and cost reductions exist that do not make complete replacement the only course of possible action. Retention and reuse of the historic fabric is an important aspect of complying with the *Standards*. The cost of appropriate repairs for existing windows can qualify for historic preservation tax credits.

Window Components

Historically, most windows have been composed of the same standard elements—glazing, sash, and hardware. The basic components have remained constant, although the methods used to fabricate them were updated through mechanization practices introduced beginning in the early nineteenth century. Depending on when they were installed, the major variations in windows stem from the available glazing materials and the available stock of construction materials. The appearance of windows was influenced by the currently fashionable window accessories and ornamental details associated with a particular architectural style. Window accessories included shutters and storm windows. Ornamental details included trim, casework, framing, and sills.

In the earliest colonial settlements in North America, many buildings either did not include windows or did not use glazing in the window openings. Due to technology limitations and the expense of glass itself, early glazing consisted of an assembly of small (e.g., 2–4 inches per side) glass diamond-shaped panes, known as "quarrels" or "quarries," held in place by a lead came or wooden muntins. Later, windowpanes or "lights," as they came to be called, were larger in size (e.g., 6 by 8 inches or 8 by 10 inches) and rectangular in shape due to improved glassmaking practices.

The largest variations in windows were not in the window sashes themselves but in the way the sashes were installed. The basic sash was composed of horizontal top and bottom rails, two vertical side stiles, and glazing. When the first colonists settled along the eastern seaboard, the most common window type was either the fixed sash window or the casement window with glazing comprised of quarrels (see Figure 10-2).

Although fixed sash windows could not be opened, casement windows were hinged along one vertical side. By 1670, horizontal and vertical sliding windows had been developed in Europe and had been introduced to the colonies in North

Figure 10-2 A grouping of casement windows with quarrel glazing.

America by the early 1700s. While both sash arrangements were initially present, the vast majority of later buildings incorporated the vertical orientation. The vertical sliding windows were originally single-hung and later evolved into the common double-hung window system still used today. A single-hung window consisted of two separate window sashes mounted in a vertical arrangement in a single framed opening. In its original form, the top sash was fixed in place and only the bottom sash was operable. The bottom sash was held open by pegs or notched pieces of wood. Refinement of these windows led to the use of various counterbalancing systems, such as cord-mounted sash weights and retractable spring mechanisms that held both sashes in place.

As demand grew for windows composed of larger glass panes than the earlier quarrels, the muntin was developed to divide the opening in the sash into smaller divisions that each held a single glass pane or light. This configuration formed what is known today as the "true divided-light" window. In describing windows, the practice has been to count the number of panes or lights in the sash. Hence a twelve-light sash held twelve individual panes of glass, most likely in three rows of four panes. An eight over twelve or "8/12" window consisted of an upper sash containing eight lights over a lower sash containing twelve lights. Like other decorative components of a window, the number of lights in the window was a character-defining feature of a particular architectural style.

The most common material used for window sash from the early seventeenth to the late nineteenth centuries was wood. Early window sash was crafted locally by a joiner or carpenter who typically fashioned the profile of the sash by hand using a hand plane and then joined the stile and rails with a mortise-and-tenon joint fastened with wooden pegs or wedges. This wood window construction practice remained constant for several centuries. Metal sash was first developed in the mid-1800s but did not achieve market success until the 1890s, when the steel rolling technologies were optimized. Unlike wood sash, metal sash was fabricated in a factory from stock sections of rolled steel that were welded or mechanically joined to construct the sash and framing systems. Steel was commonly used originally, but later materials included aluminum and other alloys. The popularity of steel sash was led by its use in many industrial, commercial, and institutional buildings of the early twentieth century as well as the numerous residential architectural styles of the 1920s and 1930s.

Window hardware and accessories varied by architectural style but included a number of elements still common today. The earliest windows had only minimal hardware—latches and hinges or hinge pins. Hinges and other metal accessories, such as latches and catches, were originally made from wrought iron or cast iron by a blacksmith. In the early 1800s, mechanized production was introduced. Window hardware products used from the nineteenth century through the late twentieth century commonly included a variety of window latches, sash locks and stopping mechanisms, and counterweight systems to hold the window sashes open.

Shutters commonly provided security and privacy, and blocked unwanted winter wind and summer sun (see Figure 10-3). For security purposes, early shutters were solid panels; however, as climate varied by region, the construction of a shutter was adapted to suit it. Shutters in colder climates tended to be solid panels or "blinds," whereas in warmer, humid climates they were more likely to be louvered. Likewise, architectural fashion often dictated whether the shutter was located on the exterior of the building or formed part of the interior enclosure surrounding the window. Storm shutters were used in cold climates on the exterior of the building but later were supplanted by demountable storm windows that came into increased use with the introduction of the heating systems of the nineteenth century (Garvin, 2001).

Figure 10-3   Shutters were used for a variety of purposes that promoted comfort, privacy, and security.

Originally, storm windows were fixed to the exterior of the window during the winter and removed or propped open in the warmer months for ventilation. The eventual use of wire mesh screen windows, introduced in 1890, led to the semiannual ritual of removing the storm windows and replacing them with screens and vice versa. As time progressed, shutters became less functional or served only as reminders of their uses in earlier styles. Likewise, demountable storm/screen window systems were supplanted by aluminum combination windows to reduce maintenance labor and add convenience for the homeowner.

The largest change in window making evolved as the Industrial Revolution of the early 1800s introduced mechanized methods of cutting and joining separate pieces. As the process of creating stock millwork, window sash, and hardware became more mechanized and as the expanding railroad transportation network reduced shipping times, window manufacturers were able to create manufacturing supply systems that forced many of the smaller independent joinery carpenters out of the window-making business. The introduction of free mail delivery to rural areas led to the use of catalog sales for both wood and metal windows and set in motion an overall mass production ethos that still exists. These basic elements of glazing, sash, and hardware led to the diversity of window types found in the United States today. All windows fundamentally work on the same operating principles and are vulnerable to similar failure mechanisms that will be described below.

Glassmaking

From Roman times through the nineteenth century, glazed windows represented wealth or power. History reveals a number of instances in which monarchies have imported skilled workers or have encouraged the growth of glassmaking in their own countries. However, by the eleventh century, the finest glass came from Venice and manufacturing methods were closely held secrets in the glassmaking guilds that flourished there. To protect the economic wealth gained from the glass trade, local governments enacted steep penalties against any craftsperson wishing to leave the region. These penalties were eventually stiffened to include the death penalty (Elliott 1993). Despite these penalties, the glassmaking industry eventually developed in Northern Europe. Overall, however, the expense of obtaining glass limited its use to buildings owned by the wealthiest people and religious institutions.

The seventeenth century brought recognition of the abundant natural resources available in North America and encouraged the founding of the early colonies. Demand for wood fuel sources, hastened by the demand for glass in England, was causing English forests to be cut down at an alarming rate. In 1615, based on this deforestation and its threat to shipbuilding, King James I issued a *Proclamation Touching Glass* banning the use of wood for making glass (Davison 2003). This proclamation forced a shift to coal as a fuel source, and glassmakers began looking to the American colonies for solutions to their production problems.

Several early colonial settlements were established with the partial intent of making glass to serve the demand in Europe. These settlements included Jamestown, where glasshouses were established in 1608 and 1621 but failed, and New Amsterdam in 1638, which was only marginally successful. Archaeological research along the eastern seaboard shows evidence of a number of other failed ventures during this period. Attempts to establish glassmaking operations in the early colonies met numerous obstacles. While raw materials and fuel were abundant, the demands of establishing a working colony, poor management practices, disease, and Indian uprisings were common causes of failure. Therefore, in seventeenth-century North

America, window glass was viewed as a rare commodity and colonists used oiled paper, animal skins, or wooden shutters to cover window openings. As a community prospered, those who could afford glass usually imported it from Europe. Later emigrating colonists were advised to bring windows or glazing with them. In Europe, tenants of a building often owned their window sashes since many building owners of the time did not provide it.

Despite many attempts to establish glassmaking ventures, the eighteenth century saw little increase in the availability of window glass, as only a handful of these attempts were even marginally successful. The first successful glass factory has been attributed to Caspar Wistar in Salem County, New Jersey (near Philadelphia), in 1738 (Davison 2003). Despite slow, steady growth in glassmaking efforts in the first half of the nineteenth century, window glass was still largely imported from Europe. This trend continued well into the twentieth century. The 1850–1920 period saw growth in the industry particularly in the Pennsylvania-West Virginia-Ohio region, where there were abundant coal resources, convenient transportation, and an assortment of small glasshouses that had united into a handful of larger conglomerates. Despite this successful period, the majority of window glass was and still is imported from Europe.

The process for making window glass dates back before Roman times. The main ingredients were silica, lime, and soda, with variants in the mixture that included potash and other minerals. These ingredients were mixed in a furnace until they fused together into a homogeneous mass. A portion of the mixture was removed using a blowpipe. Window glass consisted of two types: crown glass and cylinder glass. For crown glass, a mass of molten glass was gathered at the end of the blowpipe and the glassblower blew a sphere of glass. The sphere was then attached to a pontil (a metal rod used to handle the molten glass), and the blowpipe was cut away to create an open-ended hollow bowl of glass. This ball was spun until it formed a crown-shaped element that was reheated, spun, and reheated until it formed a flat circular disk. The disk was then removed from the pontil and allowed to cool. Depending on the skills of the glassblower, the disk reached approximately 4 feet in diameter. The disk was then cut into a series of standard sizes, and the scraps (including the "bullet" from which the pontil had been attached to the disk) were returned to the furnace and remelted. Window glass made by this method can be recognized by the concentric circular optical distortion lines (see Figure 10-4) created as the disk was spun and annealed while being flattened.

The process for making cylinder glass started in a similar fashion, except that the sphere was continuously worked into an elongated cylinder. The cylinder could reach 4 or 5 feet in length. It was removed from the blowpipe, and both curved ends were removed. The cylinder was then cut lengthwise and through an annealing process was flattened into a sheet of glass. This sheet was allowed to cool and was cut into a series of standard sizes. Window glass made by this method can be recognized by the parallel optical distortion lines created by the flattening process as the cylinder was annealed and cooled.

Of the two, crown glass was more desirable and therefore more expensive. Depending on the mixture and cleanliness of the soda, lime, ash, and other materials used and the heat of the furnace, both of these glass types display air bubbles and other imperfections that are not present in machine-made modern glass and distort the optical performance of the glass. This difference should be noted when considering replacing the glazing or repairing a broken windowpane since the

Figure 10-4 The annealing process of making crown glass creates concentric optical distortion lines in the window glass. These lines are visible in the upper portion of the large pane of glass in this window when one looks at the roof of the neighboring garage.

replacement will noticeably alter the optical performance characteristics of a window, particularly if replacement occurs adjacent to windows still containing handblown glass. Replacement panes crafted from handblown glass are still available from manufacturers.

In the late seventeenth century, the French developed a new process for making sheet glass. This product is known as plate glass. At first intended for use in making mirrors, plate glass eventually found its way into the window glass market, especially for retail uses. Plate glass was formed by pouring molten glass on a flat surface and smoothing it to a uniform thickness. After the glass cooled, it was finely polished on both sides. The clarity and uniform thickness of plate glass were considered vastly superior to those of handblown window glass made using the crown or cylinder methods (Elliott 1993; Konrad et al. 1995).

Improvements in glassmaking over the next 250 years focused largely on mechanization to increase production as demand for window glass grew when people migrated westward across North America. By the early twentieth century, glassmaking had become much more of a science as the nascent automobile industry fostered the advancement of even greater mechanization. This period also saw the introduction of tempered glass and laminated glass. Originally designed for use in the automobile, these glass products saw limited use in buildings constructed before the mid-twentieth century. The use of tempered glass became required by many building codes in the late twentieth century. In recent years, laminated glass has been increasingly and successfully used to upgrade single-pane windows, usually without extensive alterations to the window sash and muntins. An early double-paned window assembly in a single sash called the "Thermopane" window was developed in 1934. These windows later came into widespread use as an energy conservation product in the late twentieth century.

In 1952, a major shift in glassmaking came from early experimentation with the process of making float glass. In this method, the molten glass was supported on a vat of molten tin and, simultaneously, the exposed upper surface of the glass was "fire-finished" with an open flame to create a polished surface akin to the finished surface of crown glass. The glass sheet was then placed in an annealing oven to cool and was later cut or reheated and molded into the desired curvilinear window shapes. This process significantly sped up production capacity by reducing fabrication time and breakage losses incurred in the then standard polishing processes. Float glass was introduced into the window glass market in the late 1950s (Elliott 1993; Jackson and Day 1992).

Any number of window and sash combinations can be found in a variety of applications within existing historic and older buildings. While wood was the material of choice through the mid-nineteenth century, a variety of metal window frames and sashes have been introduced in the past 150 years. All have gone through varying levels of maintenance over the years, leaving some in relatively good operational condition and others in extremely poor physical and visual condition. A number of repair and upgrade methods have been developed to allow the retention of historic windows. Many window products were produced in the late twentieth century for new construction, and their use as replacement units for existing buildings formed a secondary market. When replacement windows were recognized as a viable product for historic buildings, a number of window manufacturers developed products specifically for that market segment, while others sold their existing products with little regard to compatibility with historic buildings. Some manufacturers adopted custom-made windows as part of their sales operations. Therefore, it is important to verify that the windows sold by a specific vendor meet the requirements for the historic building being rehabilitated after determining that the existing windows cannot be made usable. Described below are several key strategies and processes available for remediating problems with existing windows.

WINDOWS AND SASH

Applications

In residential construction, windows installed prior to the turn of the twentieth century were traditionally fabricated from wood. After 1890, metal sash and frames were introduced and competed directly with wood windows. Until the mid-twentieth century, most windows included single-paned glazing systems with a variety of storm window options to accompany them. Many windows were operable and served as the primary nonmechanical means to provide cooling and ventilation prior to the advent of air conditioning. Double-hung windows were viewed as particularly useful because, by lowering the top sash, warm air trapped at the ceiling could be removed. Conversely, the lower sash could be raised to admit relatively cooler air.

Windows were also intended to admit daylight to the interior spaces of the building. Some windows were fixed in place (nonoperable) and used only to admit daylight. These windows exist in the form of fanlights and sidelights near entrance doors and as glass panels in interior doors. To aid in light and air flow into interior spaces while maintaining the privacy and security of a closed door, operable transoms were located above interior and exterior doors.

Early windows were small and few in number, in part due to technology limitations and in part due to taxes on glass. As technology improved and glass became more readily available, windows grew in size and number. The finest examples of eighteenth-century architectural styles feature an exuberant display of windows. The nineteenth century saw a multiplication of windows placed in groupings of twos and threes, as well as the introduction of what was later referred to as the "picture

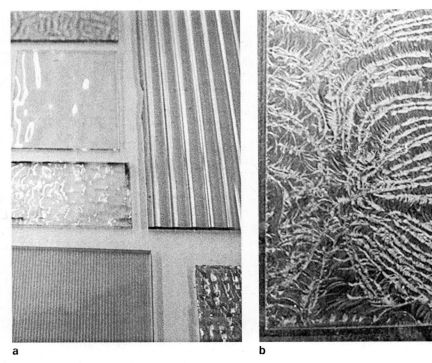

a   b

Figure 10-5 A variety of processes were used to create "obscure" glass to enhance privacy in buildings, including (a) several types of pressed or rolled glass and (b) glue chip glass.

window." As the century progressed, the number of individual lights in the window sash was reduced as the size of the windowpanes increased.

Ironically, after the American Centennial in 1876, when manufacturing could produce larger expanses of glazing, the revival of housing styles of bygone days reintroduced the smaller-paned casement and double-hung window. By the turn of the twentieth century, steel windows were also available and were worked into the period revival styles as well as appearing in the International, Art Deco, and Art Moderne styles popularized in the 1920s and 1930s. This period also introduced the use of clear and prismatic glass blocks as glazing. Post–World War II America found itself on the now well-documented path to suburbia. The metal fabrication processes perfected during the war were now directed at the American homeowner. New construction of the 1950s featured windows made of aluminum. This era also introduced the widespread use of aluminum storm window/window screen combination windows that were used on new and older buildings alike, much to the aesthetic detriment of many historic buildings.

Historically, commercial construction for offices used many of the same ventilation and daylighting practices applied to residential construction. Tall double-hung windows enabled daylight and a modicum of ventilation comfort. Interior glazing panels with "obscure" glass composed of embossed or etched panes allowed light to enter interior spaces while providing limited privacy (see Figure 10-5). Commercial offices and industrial work spaces were the first major users of steel-framed windows at the turn of the twentieth century. A more detailed discussion of glazing in retail storefronts is presented in Chapter 12.

Institutional buildings, such as hospitals and schools, also promoted the use of operable windows and daylighting and followed similar usage patterns in materials. Like commercial and industrial applications, institutional construction practices adopted the use of steel frames and sashes early in the twentieth century.

Along with the problems associated with wood and architectural metals identified in earlier chapters of this book, windows experience problems related to structural failure, missing features, and insensitive replacement.

Structural failure most frequently occurs as a result of moisture and subsequent corrosion, rot, or insect damage. Window sash is highly vulnerable to moisture problems due to the condensation and freezing that occur at and around single-paned glass and steel sash that has no thermal breaks incorporated into it. Due to the higher thermal conductivity of single-paned glass and steel window sash, the inside surface temperature of the window materials is significantly lower than that of other, better-insulated building materials. Warm, moist air inside the building cools when it touches the colder glass surface. The moisture condenses onto the window, sash, and frame and subsequently collects at the horizontal surfaces of the window assembly. Combined with deteriorated surface coatings, this moisture penetrates the wood and accelerates deterioration (see Figure 10-6).

This problem can occur even when storm windows are present. In this instance, two factors are at work. First, as caulking, weather stripping, and window putty deteriorate, moist air can move through the resultant gaps and cause condensation between the window and the storm window. Concurrently, poor maintenance of the storm window can cause the weep holes provided to allow moisture relief between the windows to become blocked and trap the moist air. The condensation and deterioration processes also affect metal frames and sash, particularly those made of ferrous metals. As the corrosion or rot progresses, the deteriorated components eventually fail and the window framing system falls apart. One common misguided attempt at maintenance that contributes to this problem has been to repeatedly paint

Typical Problems

Figure 10-6 Condensation on single-paned windows can cause long-term moisture-related problems, as shown by the paint failure here.

the surfaces of the sash and muntin bars without properly preparing and repairing the surfaces of the damaged areas or repairing the paths of moisture penetration. Long-term moisture condensation also can cause early ferrous metal lintel systems above a window to corrode, deteriorate, and fail in place. This subsequently causes the structure above the window to fail and the shifting materials of the building envelope to impinge on the window itself.

A second problem with windows is the loss of character-defining features due to vandalism or poor maintenance that resulted in their decay and removal. Commonly, when a matching window element was not readily available, a substitution was made that did not match the appearance of the original window. In the 1950s and 1960s, when aluminum siding or simulated masonry products such as Perma-Stone was applied to older buildings to reduce maintenance, and in the 1970s, when new siding was installed to reduce energy costs, the decorative trim elements and accessory items were removed from windows to provide a flat surface to which the siding was attached. This practice continues today with the installation of vinyl siding.

The third problem, insensitive replacement, is both a stand-alone condition and one indirectly related to the previous problem. Often throughout the eighteenth to twentieth centuries, the original window no longer seemed appropriate to the owner's aesthetic needs or maintenance preferences. Hence, for example in a home, the wood double-hung windows were removed and replaced with a picture window or an aluminum slider of different dimensions. The window opening was altered to accommodate the new window and the remaining open portions of the original opening were filled, often with visually incompatible materials (see Figure 10-7). While this practice was seen as a fashionable updating of a building window system, it has resulted in a variety of later-period window types being added to earlier-period buildings. The *Standards* recognize that these later replacements may have become part of the historic fabric and do not mandate a return to the earlier window systems. However, if such a return is contemplated, the *Standards* do require

Figure 10-7 Insensitive replacement of the original windows has significantly altered the visual and historic integrity of this building in Ann Arbor, Michigan.

that the replacements be based on documented evidence (e.g., photographs, written descriptions, or physical evidence).

The movement toward the reduction of maintenance needs in the 1950s and 1960s and the energy crises of the 1970s produced the trend of replacing larger wood-framed and steel-framed single-paned windows designed to admit daylight with other fenestration systems. For wood windows, this replacement initially involved the use of aluminum-framed and subsequently vinyl windows. For steel-framed windows, it involved a variety of translucent fiberglass-insulated sandwich panel systems, used in modern fenestration systems. As it turned out, none of these approaches were found to comply with the *Standards* and consequently caused a furor in the preservation community.

Since windows play such an important role in the visual historic integrity of a building, the *Standards* are quite specific on the subject of how to approach their repair or replacement. Many deteriorated windows can be repaired or have replacement sashes, frames, and casements installed. When replacements occur, care should be taken to match the sash profiles, the opening sizes, and the number of lights per sash. These strategies will be discussed in detail below.

Recommended Treatments

Repair Strategies

A long-held philosophy in preservation practice first stated by A. N. Didron in 1839 stated that "it is better to preserve than to restore and better to restore than to reconstruct" (Murtagh 2006, 4). This statement implies that the best strategy is to preserve an object rather than to repair, restore, or replace it. While applicable to any number of historic materials, this statement is especially true for windows. For buildings that have been well maintained, rehabilitation may not need to be extensive. However, in these buildings, there may still be problems underlying many coats of paint, for example, that are not immediately detectable by visual inspection alone. Paint and other coatings can often act as a binder that holds the surface together while concealing significant decay inside the window itself. Thus, any inspection of even a well-maintained building should include a systematic probing of the components of the window sash to determine their true structural integrity.

More common are the rehabilitation projects that are initiated well after routine maintenance has fallen by the wayside. A cursory visual inspection alone may give the impression that the windows need to be replaced simply because of their poor aesthetic appearance. Total replacement need not be necessary. Many projects have proceeded with the repair and reuse of existing windows after a systematic physical inspection revealed that beneath the multiple coats of paint, the structure of the windows was relatively intact. This is not to say that some windows do not need repair or replacement but rather that a careful evaluation of the windows can lead to their retention and repair at a lower cost than complete replacement.

Routine maintenance on windows should be done annually. The first step is to remove loose and flaky paint to expose the surface beneath. Mechanical stripping systems or hand-scraping tools that do not damage the underlying materials are generally acceptable. Removing all coats of paint is not necessary, but multiple coats of flaking paint may need to be sanded to remove the rough edges before priming and repainting. Next, one should examine the integrity of the sash and window itself. This process includes checking the integrity of the sash, muntins (if present), and frame as well as the glass and window putty. Look for signs of rot in wood or corrosion in metals. Minor areas of decay can be treated in place. If the materials are intact, then the glass and window putty can be replaced as needed. For metal windows, exposed metal should be cleaned thoroughly and treated with an anticorrosive primer. Similarly, exposed areas on wood windows should be primed appropriately and repainted. Care should be taken not to overpaint the gap between the sash and frame, as too much paint will result in the window being "painted shut." Examine the window hardware and adjust or repair it as needed. Examine and adjust or replace the weather stripping and any caulking as needed.

For regular maintenance, only nominal scraping, repairs, and repainting are generally required. However, for less frequent maintenance cycles, the extent of needed repairs may increase substantially. This prolonged cycle may lead to more extensive preparation for the repairs. Buildings constructed before 1978 are likely to have been painted with lead paint, so plan to remove the paint accordingly. If preliminary inspection reveals that large areas of paint are to be stripped, verify the local health department removal and disposal requirements before starting the work. Attempts to accelerate the stripping process should be made cautiously. Sandblasting wood windows is not acceptable. Although low-pressure wash systems seem benign, they introduce a sizable quantity of moisture, which can be detrimental. Stripping using heat guns is a possibility, but caution is needed since the heat may vaporize lead paint and damage the underlying and adjoining materials. Stripping using an open flame is not advised. Chemical strippers may be acceptable, but pay careful attention to the manufacturer's directions for their use. Careful consideration should also be given to the impact of using chemicals and other stripping processes on deteriorated areas, as they may exacerbate the original problem and complicate the repairs. If lead paint is present, sanding and heat or chemical stripping methods will require the use of appropriately filtered air masks and protective clothing.

The next step is to repair damage found during routine maintenance. Cracked, broken, or missing panes can be placed without removing the sash. Attention should be paid to the type of glass and whether it is appropriate to install modern glass or use crown or cylinder glass.

For repairs to wooden sash, remove the sash, temporarily cover and secure the window opening, and repair the sash at a workbench where the window can be securely clamped in place. In removing the sash, care should be taken to disconnect the sash weights, when present, and secure them for reuse. The weight pockets are located along the vertical sides and can usually be accessed by removing the cover located near the bottom of the pocket. Frayed or broken sash cords should be replaced. Removing the window putty and the glass panes, while not necessary, is advisable, particularly if heat stripping is planned, since this removal will forestall possible glass breakage as the stripping and repairs are completed. There are two approaches to consider when attempting the repair. The first approach is to stabilize the weakened material and then use epoxy fillers to re-create the surface profile of the affected area(s). This process includes several steps once the surface has been cleared. The wood must be dried and a fungicide applied to the affected areas. A penetrating consolidant is applied next. Once the consolidant has cured, epoxy fillers are applied to the remaining void. The epoxy is then sanded or carved to re-create the original surface profile. The second method is to remove the decayed material and to splice in a replacement piece of compatible material (see Figure 10-8). The affected area is cleared of decayed material until only healthy material remains. The damaged area is further cut into a regular three-dimensional shape, and

Figure 10-8 A dutchman repair in this window has been shaped to match the contour along the edge of the window sash.

replacement material is cut to fit that shape. The replacement piece is then fastened in place and trimmed to match the surrounding surface. The repair is primed and painted to match the rest of the window.

For windows with steel sash and frames, the inspection processes are similar. However, the repair process can vary dramatically. Steel sash windows can have a different fabrication and installation method than the simple wood window, as the frames may have been welded to the adjoining steel framing of the building or embedded in adjoining masonry. This fabrication method may eliminate the opportunity to remove the sash and framing due to the potentially high costs of removing and replacing the materials holding the window in place. Along with the glazing and the sash itself, inspect the condition of the paint, all hardware, the integrity and wear of all connections and hinges, and the level of corrosion. Based on the level of prior maintenance and as described in *The Repair and Thermal Upgrading of Historic Steel Windows* (Park 1984, 8–9), the maintenance and repair process may be composed of the following steps:

- Remove dirt and grease from the material.
- Remove rust and corrosion.
- Remove flaking paint.
- Align bent and bowed metal sections.
- Patch depressions.
- Splice in new metal sections.
- Prime exposed bare metal sections.
- Replace missing screws and bolts.
- Clean, lubricate, or replace hinges and other hardware.
- Replace broken or missing glass and glazing compound.
- Caulk masonry surrounds.
- Repaint windows.

As with wood windows, peeling paint and unsightly surface corrosion may influence the initial reaction to pursue complete replacement. The condition of the paint may not be a true indicator of the window's structural integrity, and the corrosion needs to be assessed for its effects. Corrosion can be classified in three ways. The first is "light" corrosion located on the surface of the metal. This corrosion can easily be scraped away and sealed. The second is "medium" corrosion that has penetrated beneath the surface. In this case, the surface appears blistered but the metal still retains its structural integrity. The third is "heavy" corrosion that has penetrated the surface and has caused structural failure. The two latter conditions may require new material to be spliced into the deteriorated section or, if extensive, require its complete replacement (Park 1984). Individual windows may exhibit all three conditions simultaneously. The extent of the damage will dictate whether to repair or replace the window.

Methods for stripping paint and corrosion from metal sash vary from those of wood windows in that certain metals, particularly steel, can withstand more abrasive techniques than are appropriate for stripping paint from wood. Depending on its severity, corrosion may be removable by simply using solvents or rust removers and a wire brush. Heavy corrosion may need to be sandblasted; however, note that sandblasting can damage the surrounding materials and the glass. Therefore, care should be taken to protect these adjoining elements (e.g., glazing and other exposed surfaces) during the sandblasting process. If large portions are to be stripped, removing and safely storing the glazing is advisable. After the corrosion is removed, the bare metal should immediately be primed using a corrosion-inhibiting primer and then repainted. Do not leave bare metal exposed to further corrosion. Verify with the local landmarks commission or the SHPO as to what guidelines and restrictions exist for sandblasting steel sash. Also, check with local authorities to confirm any specific requirements for dust control, lead abatement, and disposal practices. Other cleaning practices include flame cleaning or dipping in iron phosphates, phosphoric acid, or sulfuric acid (Waite 1980). These practices are best used if the damaged sash can be removed, since heat from the flame may ignite surrounding materials and the dipping processes typically require that the piece be fully immersed in the solution.

Once the metal has been stripped, repairs can be made. Any bowed sections must be straightened and realigned. Bowing can occur as a result of mechanical impact or from the expansion that occurs as metal corrodes. Determine the cause of the bowing and correct it. If not already done, the glazing may need to be removed and safely stored during this straightening process. Depressions and pitted areas can be filled with metal putties, and damaged sections can be removed and replaced with a splice or dutchman repair. Here again, care should be taken when welding in the splices to prevent ignition of adjoining materials.

For both wood and metal windows, it is also important to verify the conditions of the frame, lintel, sill, and other associated features. Sills should slope to allow water to flow away from the window and be watertight to keep water out of the wall below the window. As with any maintenance or repair process, care should be taken to confirm the initial source of the decay problem and correct it appropriately. Repairing and returning the window to its original location without correcting the source of the problem(s) will simply become an ongoing and recurring cycle.

Replacement Strategies

After a thorough course of inspections and evaluations has been completed, the decision to retain and repair the existing windows or replace them can be made. Part of the decision will be based on retaining the historic appearance of the building, while other factors, such as energy operating expenses and maintenance, may come into play as well. The vinyl windows and translucent fiberglass panel replacement systems mentioned earlier that were installed as maintenance reduction and energy conservation measures bore no physical resemblance to the original window systems that they replaced. As a result, significant debate has arisen from the unsympathetic replacement of windows with contemporary products that were not visually compatible with the original windows. For windows, two key factors in the appearance are the relationship of the sash width to the glazing opening and the shadow profiles created by the muntins and mullions that define the individual window openings (see Figure 10-9).

Since vinyl sash is much wider than the typical wood or metal sash, the glazing area is smaller. Many common vinyl windows may have fake muntins applied to the outside of the window, no muntins at all, or do not use true divided lights. These aspects make vinyl windows inappropriate choices for replacement windows, according to the *Standards*. Also problematic is the ability to repair and replace vinyl windows in the future, since many of the original manufacturers have left the market. If window replacement is deemed the only viable alternative, many window manufacturers do carry a line of stock windows or can custom build appropriate replacement windows that use wood, a combination of vinyl- or aluminum-clad wood windows, and aluminum windows that match or replicate the look of the historic wood or steel windows that they are intended to replace. The appropriateness of these windows should be confirmed by any oversight agencies with review authority over the

a   b

Figure 10-9   Notice how the visual character of the window system changes between the original windows (a) and the replacement windows (b).

project. While these windows are more expensive than stock windows, the expense may be offset, if applicable, by the use of historic preservation tax credits.

Modern translucent fiberglass sandwich panels have not been acceptable replacements for traditional historic windows, since they do not retain the historic appearance of the original windows, although they have been used where retaining the integrity of the historic appearance was not required. However, like exterior wall cladding products developed in the 1950s, the translucent fiberglass-insulated panels of that era are becoming viewed as historic materials in their use as original construction. Some of these original fenestration systems and skylights from the 1950s and early 1960s have already begun to fail. Unfortunately, advances in product design in the late twentieth century may make exact replacement difficult, if not altogether impossible, since many of the original manufacturers, if they are still in business, may no longer be able to provide in-kind replacement.

Improving Thermal Efficiency

Historically, the thermal efficiency of windows, especially residential windows, had been enhanced by installing storm windows on the exterior of existing window openings. Using modern exterior storm windows raises aesthetic issues if the appearance of the storm windows does not match that of the storm windows used in the period when the building was constructed. To offset this problem, storm panels can be installed on the interior side of the window. These panels provide the benefit of retaining the exterior appearance while reducing heat loss.

The energy crises of the 1970s created a clamor for replacement windows with higher thermal performance. While intended for new construction, these windows were also sold to owners of existing buildings who were hoping to reduce their energy costs. Ironically, subsequent studies showed that, especially in wood windows, rather than installing new double-paned windows, greater energy efficiency could be gained by adding storm windows or the proper maintenance of caulking and weather stripping. Other studies also found that, since heat rises vertically, greater energy savings could be achieved less expensively by increasing attic insulation rather than by upgrading to double-paned windows. Despite these findings, demand remains strong for multipaned, high-performance windows.

Thermopane windows that included two layers of glass enclosing an air space were developed in 1934, but significant demand for them did not occur until the energy crises of the 1970s, when the resultant marketing efforts of window manufacturers led to the widespread removal of single-paned windows nationwide. Although removal of existing windows and the associated repairs increased the price of an already expensive product, windows were often replaced with little understanding of how long it would take for the annual energy cost savings to pay for the cost of the retrofit (Sedovic and Gotthelf 2005). However, after the aesthetic failures of these replacement windows, the need for double-paned windows that displayed an aesthetic equivalence to historic true divided light windows soon became evident.

Early attempts to use some type of muntin system in conjunction with double-paned windows were aesthetic failures. The muntins were composed of three-dimensional strips of plastic applied to the exterior and interior of the sash to approximate the appearance of a true divided light window. Unfortunately, the gap between the exterior and interior muntins created by the double-paned glass panel resulted in a visually odd appearance, and the plastic strips frequently failed or were removed. The numerous historic preservation tax credit projects of the 1980s fostered a growing niche market for new and replacement windows that provided a more visually sensitive appearance. Many window manufacturers developed what has come to be known as "simulated divided light" windows. In this type of window, the muntins are built from the same materials as the window sash and are integrated into the fabrication of the window assembly. To eliminate the gap between the interior and exterior, metal strips are inserted between the two layers of glass in locations corresponding to the muntins. The muntin profiles can be custom made to match the original windows. When fully assembled, the muntins appear to be continuous through the glass and provide a more substantial appearance creating the traditional shadow profiles common to the historic older window they replaced.

Early double-paned glazing units were dimensionally narrower in thickness and in some instances could be retrofitted directly into the existing sash by cutting an appropriately sized rabbet into the existing sash and muntins to accept it. However, this approach meant that in windows with numerous lights, there would need to be a single small double-paned unit in each space created by the muntins. While possible, this method is quite expensive and involves careful evaluation to determine if the windows being retrofit can withstand the loss of structural material needed to insert the window units. Experiments with creating window units that held inert gas and mylar films between the glazing panels resulted in even more increased thicknesses of the window panel. As these variations were developed, the thicknesses began to exceed what could easily be retrofitted into an existing window sash.

A variety of other measures were used to improve the thermal performance of windows. They included the development of low-emittance or "low-E" coatings that could be applied directly to the glass as it was made. These coatings block and reflect radiant heat so that, depending on the orientation, the coatings could keep warmth inside in the winter and radiant heat from the sun outside in the summer. As a retrofit application, various after-market adhesive films were produced as well. Both of these methods presented problems for historic preservationists; the former required the removal of historic window glass, and the latter was pretty much irreversible since the adhesively attached film was meant to be permanently installed. Some early films also suffered from poor installation and subsequent thermal expansion-contraction problems that caused the film to bubble, wrinkle, or tear.

A recent alternative to replacing the window itself has been the use of laminated glass consisting of multiple layers of glass with one or more layers of plastic film sandwiched between them. This film reduces ultraviolet light and heat gains from the sun. Laminated glass can be used directly in place of original single-pane glazing that is not a character-defining feature of the building (e.g., cylinder or crown glass has optical distortions that create wavy lines and ripples). This approach combines the retention of original window sash and muntins with the growing need to enhance sustainability in existing buildings (see Figure 10-10).

Although improved thermal resistance was provided by the glass panel of the double-paned windows, studies found that heat loss also occurred through the sash. While not as severe in wood windows due to the insulation properties of the wood, this heat loss was particularly evident in metal sash windows, which were fabricated in such a way as to provide a continuous path for the heat to flow through. This loss can be seen when observing the condensation or freezing that occurs on the inside face of metal sash. Window manufacturers solved this heat loss problem by changing the way the window is fabricated. Instead of forming a continuous path, the interior and exterior facing sections of the metal sash are extruded or rolled separately. The

Figure 10-10 Laminated glass was used in this window upgrade to retain the historic appearance of the window while improving energy efficiency and reducing sound transmission from the street.

sections are then assembled incorporating a gasket or other material discontinuity that forms a thermal break that significantly reduces heat flow. Many replacement windows fabricated to replace historic metal sash windows now include this feature.

Other improvements have centered on the reduction of untreated cold air, also known as "infiltration," that enters through gaps between the sash and the frame and between the frame and the building structure. For gaps between window components themselves, a variety of measures can be used. Upgrading or adding weather stripping can significantly reduce infiltration. Compression metal weather stripping mechanisms can now be installed in existing windows on the frame next to the sash that will maintain a constant seal at that location. Also successful is the installation of weather stripping in the adjacent meeting rails in double-hung windows. Neither of these methods significantly harms the overall appearance of the window, and they greatly improve its thermal performance. For use between the frame and the building itself, a variety of resilient caulking materials have been formulated, as well as a group of expandable foams that can be sprayed into voids and gaps in the window construction. These foams then cure in place and form an impenetrable barrier to moisture and heat. Caution is advised in using these expanding products since they can create pressures inside the wall cavity or leak out of the walls in unexpected places.

References and Suggested Readings

Belle, John, John Ray Hoke, Jr., and Stephen A. Kliment, eds. 1991. *Traditional details for building restoration, renovation, and rehabilitation: From the 1932–51 editions of Architectural Graphics Standards.* New York: John Wiley & Sons.

Byrne, Richard O. 1981. Conservation of historic window glass. *Bulletin of the Association for Preservation Technology,* 13(3): 3–9.

Calloway, Stephen, gen. ed. 1996. *The elements of style: A practical encyclopedia of interior architectural details from 1485 to the present.* Rev. ed. New York: Simon & Schuster.

Carmody, John, Stephen Selkowitz, and Lisa Heschong. 2000. *Residential windows: A guide to new technologies and energy performance.* New York: W. W. Norton & Company.

Davison, Sandra. 2003. *Conservation and restoration of glass.* 2nd ed. Oxford: Butterworth-Heineman.

Elliott, Cecil. 1993. *Technics and architecture.* Cambridge, MA: MIT Press.

Fisher, Charles. E., III, Deborah Slaton, and Rebecca A. Shiffer. 1997. *Window rehabilitation guide for historic buildings.* Washington, DC: Historic Preservation Education Foundation.

Garvin, James. L. 2001. *A building history of northern New England.* Hanover, NH: University Press of New England.

Hull, Brent, 2003. *Historic millwork: A guide to restoring and re-creating doors, windows and moldings of the late nineteenth through mid-twentieth centuries.* New York: John Wiley & Sons.

Jackson, Albert and David Day. 1992. *The complete home restoration manual: An authoritative, do-it-yourself guide to restoring and maintaining the older house.* New York: Simon & Schuster.

Jester, Thomas C. 1995. *Twentieth century building materials: History and conservation.* New York: McGraw-Hill Book Company.

Leeke, John. 2004. *Save your wood windows.* Portland, ME: Historic Homeworks.

Litchfield, Michael W. 2005. *Renovation.* 3rd ed. Newtown, CT: Taunton Press.

Louw, H(enti). J. 1983. The origin of the sash window. *Architectural History,* 26:49–72, 144–150.

—— and Robert Crayford. 1998. A constructional history of the sash-window c. 1670–c.1725 (part 1). *Architectural History,* 41:82–130.

—— and Robert Crayford. 1999. A constructional history of the sash-window, c. 1670–c. 1725 (part 2). *Architectural History,* 42:173–239.

Konrad, Kimberly A., Kenneth M. Wilson, William J Nugent, and Flora A. Calabrese. 1995. Plate glass. In Jester 1995, 182–187.

McKearin, George and Helen McKearin. 1941. *American glass.* New York: Crown Publishers.

McKearin, Helen and George McKearin. 1950. *Two hundred years of American blown glass.* Garden City, NY: Doubleday & Company.

Meany, Terry. 2002. *Working windows: A guide to the repair and restoration of wood windows.* Guilford, CT: Lyons Press.

Minor, Joseph E. 2001. Focus on glass. *APT Bulletin,* 32(1): 47–50.

Morgan, Morris Hickey, trans. 1960. *Vitruvius: The ten books on architecture.* 1914 reprint. New York: Dover Publications.

Murtagh, William J. 2006. *Keeping time: The history and theory of preservation in America.* New York: John Wiley & Sons.

Myers, John. H. 1981. *The repair of wooden windows.* Preservation Brief No. 9. Washington, DC: United States Department of the Interior.

Nash, George. 2003. *Renovating old houses: Bringing new life to old houses.* Newtown, CT: Taunton Press.

Neumann, Dietrich. 1995. Prismatic glass. In Jester 1995, 188–193.

——, Jerry G. Stockbridge, and Bruce S. Kaskel. 1995. Glass block. In Jester 1995, 194–199.

Park, Sharon C. 1984. *The repair and thermal upgrading of historic steel windows.* Preservation Brief No. 13. Washington, DC: United States Department of the Interior.

Peterson, Charles W., ed. 1976. *Building early America: Contributions toward the history of a great industry.* Radnor, PA: Chilton Books.

Poore, Patricia, ed. *The Old House Journal guide to restoration.* New York: Dutton.

Sedovic, Walter and Jill H. Gotthelf. 2005. What replacement windows can't replace: The real cost of removing historic windows. *APT Bulletin,* 36(4): 25–29.

Smith, Baird. 1978. *Conserving energy in historic buildings.* Preservation Brief No. 3. Washington, DC: United State Department of the Interior.

Swiatosz, Susan. 1985. A technical history of late nineteenth century windows in the United States. *Bulletin of the Association for Preservation Technology,* 17(1): 31–37.

Waite, John G. 1980. *Metals in America's historic buildings: Part II: deterioration and methods of preserving metals.* Washington, DC: United States Department of the Interior.

CHAPTER 11 Entrances and Porches

OVERVIEW

Porches have a long and varied history in their use and placement in American residential buildings. When the first permanent homes were constructed, porches did not exist. Many of the earliest colonial buildings did not have porches since the medieval precedents from which they were drawn did not typically have them. However, porches became increasingly popular social spaces in the eighteenth century as economic prosperity created more leisure time for socializing. The attraction of protected spaces for utilitarian activities further enhanced the popularity of the porch as part of the daily routine of food preparation, cleaning, and other domestic pursuits. By the early nineteenth century, the porch had become a fixture in American culture. With the exception of buildings based on medieval precedents, their subsequent revival styles, and the early-twentieth-century modern styles, porches have been a major character-defining fixture of many architectural styles.

The porch survived well into the twentieth century but after World War II retained only a vestige of its former importance. Lack of maintenance or increased demand for living space led to the removal or conversion of porches to interior living space, much to the detriment of the historic appearance of the buildings they served. The late-twentieth-century neo-traditionalism and new urbanism trends led to the rediscovery of the porch as a social and utilitarian space and have promoted a wave of porch remodeling on existing historic houses.

Figure 11-1 Porches have varied in size, purpose, and location: (a) stoop, c. 1744, (b) stoop, c. 1826, (c) social porch, c. 1890, (d) sleeping porch, c. 1900.

As social customs changed, porch sizes fluctuated accordingly, reaching their peak at the turn of the twentieth century. The number of porch types is as varied as their specialized uses (see Figure 11-1).

TYPES

Each entrance evolved to serve the needs of its users. The front door was the formal entrance where invited guests and visitors entered the home. Servants and tradespeople entered the home via side and rear entrances. As utility needs evolved, sleeping porches and other small porches and balconies also emerged as part of the American home.

The original stoop (believed to have been brought to America as the Dutch "stoep") was the simplest form of a porch. In its most primitive form, a stoop was a small raised wood or stone platform adjacent to an entrance. Other than perhaps an overhanging second story, there was no roof (although some buildings had small projecting or recessed entryways and vestibules that today could be perceived as porches). Steps were added as needed to reach the sidewalk.

Stoops

By the turn of the eighteenth century, the stoop had grown larger in direct proportion to the wealth of the homeowner and the size of the house. Refinements included a projecting hood or canopy that evolved into a roof to protect visitors from rain. The porch roof was supported by wall-mounted brackets, posts, or columns. Ornamentation followed the architectural tastes of the time and ranged from plain wooden posts to highly decorated stone columns that followed the classical orders of architecture.

As prosperity increased, trade with the Caribbean colonies demonstrated how a shaded, well-ventilated porch could provide relief from the sun and heat and serve as a social space. This design concept took hold in the southernmost colonies and eventually began to spread. In New England, however, the desire to take advantage of the warming sunlight forestalled the use of the larger porch common in the South, and only stoops and smaller porches came into use.

The "Social" Porch

In the nineteenth century, the Gothic Revival style and the work of A. J. Downing introduced the more exuberant use of the porch. Originally providing shelter from the rain and sun, porches had until then a utilitarian nature. However, continuing their evolution from the simple stoop, porches grew much larger and more ornate as they became recognized as spaces for social interaction. Porches served as both formal and informal spaces that acted as transitional zones between the house and the street and the house and the garden.

As the popularity of porches grew, porches were added to buildings that previously had only a stoop or no porch at all. Existing porches were expanded. New buildings were constructed with numerous types of porches, including wraparound porches, multistory porches, side galleries, verandahs, and piazzas (see Figure 11-2). Porches became wider, deeper, and more elaborately decorated. Shutters, windows, and wire-mesh screen (when it became available in the late nineteenth century) were installed to enhance comfort, privacy, and year-round enjoyment. All of these changes provided opportunities for social interactions with passersby on the street and as part of the homeowner's social gatherings.

The heyday of the porch in America has been defined as the period from the late nineteenth century to the eve of World War II (Jackson 1985), when it became a symbol of the center of the community. This perception began to change with the

Figure 11-2   The porch became an important social gathering space across the nation: (a) Charleston, South Carolina, (b) Salt Lake City, Utah, (c) Bastrup, Texas, and (d) Cape May, New Jersey.

introduction of air-conditioning, the automobile, and television. Originally invented in 1906, air-conditioning was introduced to residential buildings by the late 1920s and reached the middle classes by the 1950s. This began the decline in the popularity of the porch as a social center of activity. In the air-conditioned comfort of the living room, people turned to watching television. The automobile allowed them to seek out activities rather than waiting for passersby on the front porch and contributed further to the demise of the porch by the noise and pollution it generated on the street. As the automobile gained popularity, so did the garage, which moved from its original location behind the house to the side and then to the front of the house. Meanwhile, the front porch dwindled in size as its former social uses moved to the rear patio or deck behind the house.

Other Special-Use Porches

Prior to the advent of air-conditioning, porches were used to escape the trapped heat and humidity often found inside the house. Service porches emerged to facilitate household cleaning activities. Drawing, in part, from the English conservatory precedent, sun porches became a common feature. The porte-cochere emerged as a means of entering the house without being exposed to the rain or sun.

Sleeping Porch

The sleeping porch was based on several cultural phenomena. The two primary ones were the need to escape the stifling heat of enclosed living spaces and the perceived medical benefits of sleeping in fresh air. These porches were located adjacent to bedrooms and were especially popular in the late nineteenth and early twentieth centuries.

Service Porch

Service porches were located at the rear and side entrances to permit domestic activities, such as food preparation, cleaning, and deliveries (e.g., ice for the ice box), to occur away from the more formal entrances to the house. These porches were often stoops or smaller, narrower porches than those used for formal social gatherings.

Sun Porch

Attached greenhouses or conservatories were common in many of the largest houses of the late nineteenth century. By the early twentieth century, they had become scaled down to form the middle-class equivalent—the sun porch. During this era, earlier porchless building precedents (e.g., the Colonial Revival styles) had been adapted to include the sun porch as an amenity. Typically located along the south or west side of the home, these porches provided a means to enjoy available sunlight in an enclosed space during the winter months. In many smaller homes, these spaces were later winterized to provide additional year-round interior living space.

Porte-Cochere

One specific type of porch combined the way that people arrived at and departed from the house. The porte-cochere was an oversized porch roof that extended over the driveway where people could embark and disembark from a carriage and enter the house without being directly exposed to rain, snow, or direct sunlight.

Site Features

In addition to porches, other structures that used the same materials and construction practices used on porches could be found on the site. These features included gazebos, arbors, and other shelters constructed for leisurely enjoyment or utilitarian purposes. While more common on larger, more formal estates, many of them can be found in a scaled-down version on smaller properties.

MATERIALS AND COMPONENTS

The entrances of the earliest colonial houses commonly were composed of a wooden or stone sill and a door. As communities prospered or as the need to construct a basement or ventilated crawl space became clear, entranceways were often elevated above the adjoining sidewalk or path leading to the entrance. Construction materials included wood, stone, brick, masonry, and metal; however, the basic components of the porch remained consistent, although the level of ornament varied with current architectural tastes.

The basic components of the porch include the foundation and substructure, flooring and steps, roof support, balustrades, roof, and ornamentation. These components are vulnerable to decay when not well maintained. However, when a component is absent or has been replaced, it may never have been there or was removed after it failed. A replacement component should be based on physical evidence and, if available, early photographic images.

Foundation and Substructure

Early wooden porch foundations consisted of individual wood posts or horizontal wood beams placed either directly on the ground or supported by a continuous stone footing or stone piers. For entrances located close to ground level, the wooden porch could enclose a modest crawl space that allowed the vulnerable wooden components to be separated from the ground and its related decay mechanisms. As entrance levels increased, this crawl space became more visible and, in some instances, was later modified for storage and other accessible spaces. The crawl space was typically, but not always, enclosed by an apron or skirt composed of latticework, boards, or other screening materials. All crawl spaces required some form of ventilation openings to relieve the moisture that resulted from drainage from the porch, condensation, or groundwater conditions. These openings consisted of small slots or groups of louvers, gaps created by ornamental cutwork, or decorative grilles and screening materials.

In masonry-based construction, the visible exterior perimeter could serve both load-bearing and ornamental purposes. The enclosure formed by the perimeter masonry was filled with rubble, gravel, sand, or concrete placed directly on the ground or supported by vaults and flat arches to enclose a crawl space (or other accessible space) while supporting the porch above.

Flooring and Steps

When a crawl space was present, wood floorboards or decking planks were supported directly by the wood joists or masonry structural frame. Masonry flooring (e.g., stone, brick, or tile) was supported by structural masonry construction methods. By the early twentieth century, metal pan construction with concrete decking had also been introduced. Where terraces adjoined or were used instead of a porch, sand, stone dust, crushed gravel, or a solid concrete or stone slab could be used as a substrate upon which stone, brick, or tile could be directly placed.

In general, the steps leading to the porch were most frequently constructed of the same materials as the decking. Depending on the architectural style and materials used, a step could be made of individual treads and risers; formed from a single piece of material; or fabricated from an assembly of smaller units. Wood and cast metal (e.g., iron or bronze) steps are examples of the first type, while stone, brick, and concrete are examples of the latter two types. Wood or cast metal steps were supported by stringers at each end, with additional stringers included on wider stairways that needed intermediate support. The riser between each tread could be a continuous solid piece of material or could be left out altogether. Metal risers could also include decorative cutwork and openings to enhance crawl space ventilation. When cast iron porch and stairway assemblies were introduced in the early nineteenth century, the decking and stairs came as a disassembled kit of parts from the foundry that produced them. These kits were assembled on-site and attached to the building.

When monolithic stone was used, it could serve as both tread and riser simultaneously. Otherwise, the tread and riser could be built up from individual masonry units or a combination of stone or cast masonry slabs. Mid-twentieth-century post–World War II housing saw the significant use of concrete stoops constructed in place and sometimes covered with brick veneer.

The columns, posts, or piers supporting the roof were made from wood, stone, brick, or metal (see Figure 11-3). Early load-bearing construction methods for wood columns or posts typically resulted in the use of solid wood timbers or posts. Later refinements led to the use of a construction system where columns were built up from wood boards or staves. The result was a column with a hollow center that still could support the load imposed on it but was lighter to transport and erect. Over time, it became apparent that moisture (from condensation or leaks) trapped inside the hollow core could cause the column to rot from the inside out. As a result, weep holes and small gaps at the bottom of a column were an essential feature that allowed the relief of this moisture before damage could occur. This principle was later followed on built-up cast iron and other prefabricated assemblies for similarly hollow yet moisture-sensitive components of the porch. Unfortunately, excess paint buildup and other debris can block these openings and trap moisture inside the column.

Similarly, large masonry columns could be constructed by building them up from smaller pieces. What appears to be a substantial stone column may in fact be a brick column rendered and detailed with stucco or concrete. True solid stone columns were typically used only in the most expensive and largest buildings as a display of power and wealth, while less expensive imitation products could have been used in less public entrances.

The use of load-bearing masonry, brick, or stone piers instead of traditional columns and posts also came into fashion with the Arts and Crafts and Bungalow styles at the turn of the twentieth century. In some cases, the piers are true load-bearing construction; in others, they may simply be a veneer concealing concrete or steel posts.

Columns, Posts, and Piers

a   b   c

Figure 11-3   Three forms of column construction: (a) wood columns are typically hollow and constructed using segmented vertical sections fastened together; (b) stone columns are solid monolithic components secured together using metal rods or by simple gravity; (c) brick columns could be left exposed or were often covered with plaster and finished to give the appearance of stone.

Balustrades

Balustrades are the railing system found along the perimeter of a porch and along the stairs. A balustrade is composed of balusters, a handrail, and sometime a bottom rail. The balusters are the vertical pieces that provide screening for privacy while also supporting the top rail. In larger formal houses, these balusters were constructed of stone and were connected using cast iron and wrought iron fasteners secured in place by grout or mortar. Wood balustrades used nails, and cast iron balustrades were either bolted or welded together.

Balustrades were supported by attaching them to a newel post (an ornamental vertical post typically found at the top or bottom step of the stairway) or a porch column. Due to its vulnerability to damage and decay, an original balustrade may have been removed rather than replaced. The replacement, if any, may be of a much simpler design, such as metal pipe, or made from more modern materials than were available when the house or porch was built. If this condition is suspected, evidence often exists in the form of assorted nail and bolt holes, obvious patched infill, remaining remnants, and paint buildup. In some cases, however, the porch may simply have been constructed without balustrades of any sort.

Roof and Ceiling

The support structure for the roof could be either a separate assembly that was secured to the vertical wall of the house and rested on columns, posts, or piers located at the perimeter of the porch or the roof could be a cantilevered extension of the horizontal framing for an upper floor. Porch roof construction materials and practices were similar to those used on the roof of the main house. The major difference was that the porch was exposed to the annual cycle of sun, wind, and moisture from above and below. Unlike roofing on the main house, however, leaks may have gone unnoticed or may have been tolerated for longer periods of time. This tolerance allowed moisture penetration and decay to become much further advanced than might otherwise have been tolerated inside the home. The same precautions noted for roof drainage systems on the main house also apply to the porch roof.

In certain informal and rustic architectural styles (e.g., Carpenter Gothic, Arts and Crafts, or Bungalow), the underside of the roof structure was left exposed to view. In more formal styles (e.g., Queen Anne), a ceiling was constructed to provide a more finished appearance. Ceilings for porches were attached to the underside of the roof framework on porches that projected from the face of the vertical wall of the house. For recessed porches or porches under cantilevered portions of an upper floor, ceilings were attached to the overhanging framing construction. Wood ceilings were usually constructed with flat flush-mounted boards. The use of tongue-and-groove bead board (a board with a small bead cut along one edge and sometimes along the center of the board face) for porch ceilings became popular in the mid-nineteenth century as a means of obscuring individual joints. Early beads were spaced as much as 4 inches apart. As aesthetic tastes changed, the spacing between the beads was reduced such that by the early twentieth century a 2-inch or narrower spacing was the norm.

Ornamentation

The degree of ornamental decoration has varied over the past 400 years. The presence or absence of ornament was largely dictated by architectural tastes of the time, the economic prosperity of the building owner, and the availability of materials and skilled craftspeople. Although stone has always been the ultimate sign of power and wealth, porches often were built from lesser materials and then finished to look like stone. Early ornamentation was achieved using wood shaped by cutting, carving, and turning and then painted with sand paint. Later advances in technology introduced casting and stamping, particularly with metals.

By the early nineteenth century, details resembled ornamentation found on ancient Greek and Roman buildings as columns, brackets, and balusters were "sculpted" to emulate classical architecture. The introduction of the scroll saw fostered many variations on ornamental elements. Instead of fully turned three-dimensional carvings, a silhouette of the turning was substituted on less expensively constructed buildings. This cutwork became increasingly mechanized and was supplemented by a wide variety of preassembled panels and elements composed of turned spindles, ball and dowel, and other fretwork combinations that could be ordered from a catalog and shipped to the site. This phenomenon reached its peak with the Queen Anne and related styles of the late nineteenth century. Similarly, cast iron and stamped metals were introduced in the middle and late nineteenth century for this purpose.

Variations in architectural styles added, extended, or eliminated entirely the various flourishes provided by the ornamental trim. The many revival styles of the early twentieth century reintroduced simpler forms of ornamentation, as these styles recalled a precedent that did not originally include large porches but perhaps an elaborate stoop. The Art Deco, Art Moderne, and International styles of the mid-twentieth century eliminated much of the architectural ornament entirely and instead promoted clean, streamlined, and industrial rectilinear forms. The use of ornamentation continued to decline and was virtually eliminated in the period after World War II as low-cost, mass-produced house construction took hold and converted the front porch back to a simpler, less expensive-to-build stoop.

TYPICAL PROBLEMS

Due to the exposed nature of much porch construction, moisture penetration, finish failure, and the decay of the connections and of the construction materials themselves are all potential problems. As moisture penetration and the assorted decay mechanisms attack the various materials from which the porch is constructed (see previous chapters for descriptions), the components of the porch will begin to fail. Paint will peel, roofing and decking will begin to disintegrate, and exposed wood or metal will begin to rot or corrode. Unless an annual inspection and maintenance program is in place, the seasonal variations in temperature, sunlight exposure, and moisture will act in a cyclical manner to eventually cause the structural failure of portions of the porch, if not the entire porch itself. This failure often results in removal of the porch or selected portions of it.

Structural Failure

In most instances, structural failure is the result of external forces acting on the material. Wood on porches is particularly vulnerable to absorption of moisture through its end grain and exposed horizontal surfaces, which can initiate rot. Likewise, ferrous metal connections (e.g., nails, bolts, threaded iron or steel rods) that have been exposed to moisture from open joints can be weakened by corrosion and ultimately fail. Stone and metals can be damaged by deicing salts.

In addition to decay mechanisms introduced by moisture acting on the materials, groundwater fluctuations and soil subsidence (settling) can compound the effects of these problems. Expansive clay soils swell and shrink in reaction to seasonal groundwater fluctuations. Activities such as tree removal or excavations, which change the drainage patterns, can affect the groundwater. Decayed footings and porch columns, inadequately sized footings (especially after an open porch is enclosed to create interior space), extreme snow loads, and differential settling between the house and the porch can all cause vertical displacement that results in additional strain on the structural connections between the porch and the house. In some instances, where the house and porch have significantly different weights (e.g., a masonry house with a wood porch), the house may settle more or less than the porch.

In areas of the country where seismic activity or extreme wind conditions exist, the porch can be damaged by a single event or by the cumulative effects of a series of smaller separate events. After each such event, the porch should be inspected for damage.

Altogether, decay, differential settling, seismic activity, and extreme winds can ultimately cause sufficient structural failure to the porch so that it requires drastic measures to protect life safety. The porch can be repaired, removed, or replaced. These actions can threaten the historic integrity of the house. Insensitive repairs and removal can have a major impact on the visual appearance of the house. Replacement with an inappropriate porch of a different size, configuration, or materials can also have a negative effect on the historic integrity of the house.

Missing Features

Often, lack of adequate maintenance, fire, or vandalism have caused portions (if not all) of the porch to lose key structural or ornamental features. In some cases, the desire to eliminate the need for maintenance or the perceived expense of replicating a

a   b

Figure 11-4 The porch has been removed from this building (a) and the facade has been resided, but the missing ornamental detail of the porch is evident by (b) the witness marks left by soiling and overpainting around the profile at its prior location.

Figure 11-5 As owner-occupied properties were transformed into rentals or when additional space was required, porches were often enclosed, a process that frequently seriously diminished the visual appearance of the house.

porch has resulted in its removal and replacement with a porch made from other materials. In the 1950s and later, replacing wood posts or columns with cast aluminum or other metal products or with simple steel piping became common. When a porch has been replaced or removed completely, visual clues to its original configuration and location may be found on the building itself. These clues often consist of outlines of soiling or overpainting along the edges of the previous ornament, discontinuities in ornamental details on the main building where the porch roof had been located, or indications of where the porch was connected to the house (see Figure 11-4).

Unsympathetic Enclosures

As air-conditioning became available and the family's social spaces moved indoors, the porch began to lose its prominence as both a social space and a place to escape overheated interior spaces. Increasingly, it was seen as inconsequential space. As a result, the perceived need for additional enclosed interior living space often meant the enclosure of a previously open porch.

In some instances, the desire to create a year-round usable space prompted homeowners to enclose the porch with a glazing/screening system composed of aluminum combination windows or similar products. In others, when more living space was needed, porches were enclosed and used for a variety of interior uses (see Figure 11-5). This practice also commonly occurred as former single-family owner-occupied homes were converted to multiple-unit rental apartments or as growing families expanded their living space.

These enclosures generally significantly altered the overall appearance of the building. Some enclosures were created by simply enclosing the columns and ornamental details within the wall. Restoration-minded building owners should not be surprised to find that some portions, if not all, of these features remain intact within the exterior wall. Unfortunately, and far more likely, these details were removed. Sometimes they were stored in a basement, attic, shed, or barn; in other cases, they were discarded.

RECOMMENDED TREATMENTS

Porches are constructed from the collective materials and assemblies described earlier in this book. While treatments for each of those materials and assemblies have already been described, the treatments presented below provide an overview of the repair and replacement strategies that can be employed when working on porches.

Repair Strategies

For porches that have been reasonably well maintained, repairs often imply simple further maintenance (e.g., painting, caulking joints, and repairing minor surface defects). As the level of annual maintenance declines, the level and extent of repairs increase. Many repair and maintenance activities associated with the roof, exterior cladding, windows, and foundations apply to porch materials as well. Repairs imply nominal intervention focusing on specific components and locations on the porch. When a component needs to be repaired, there is usually sufficient evidence remaining on the porch to give a general indication of what it should look like when intact. As with all repairs, the source(s) of deterioration should be identified and corrected. Repairs generally can be classified into the following types: structural, roofing, drainage, flooring, cosmetic, and finishes.

The most important yet least accessible repairs are those involving the structural systems, particularly those portions concealed within the roof and substructure of the porch. Inspection and repairs of structural components are the most invasive and ultimately affect many other types of repairs. When maintenance has been poor, structural elements, such as footings and roof framing, need to be visually and physically inspected to determine their condition. These inspections may require removal of ceiling materials, porch apron, or nominal excavations to access and assess the suspected problem areas. When a porch is well maintained, a structurally stable porch will have structural components that are level (or uniformly sloping for drainage) and plumb. Nothing has sagged or twisted from its original position. Ideally, this pristine condition is a result of no water penetration, no insect damage, and appropriate structural capacity. Rot and insect infestations are problems for wood components, especially where destructive insects are known to be common (see Figure 11-6).

What otherwise might appear to be a visually intact component may in fact have been seriously compromised internally. Closer inspection may reveal splitting and cracking. Often, the exterior paint acts as a consolidant holding the component

Figure 11-6 Installing termite shields is a common method to protect wood from attack. In any repair involving termite damage, it is advisable to install these shields as a deterrent to future attacks.

together. Simple probing with a pointed object can test its integrity. If the surface remains firm under moderate pressure the component may be structurally sound, but further pressure may indicate more serious underlying damage. Similarly, if sounding the component (rapping with knuckles or a rubber mallet) causes a solid resonant response, then the component is likely to be structurally intact. In materials other than wood, evidence of bending, breaking, cracking, spalling, erosion, and loss of physical integrity indicates structural problems.

Structural repairs also include inspecting and possibly repairing the method by which the porch is attached to the main building. Differential settling can cause stress on the connections and cause the joints between the porch and the main building to open, admitting water. Inspect the flashing and sealants used along this joint for watertight continuity. As structural components such as columns and footings fail, the porch is likewise pulled down from the face of the main building, putting further strain on the connections to the building. Once the damaged components are identified, they can be repaired or replaced as needed. If necessary, jacking the porch up to relieve sagging may be possible; however, this should only be done under the supervision and consultation of a licensed professional familiar with structural repairs. As noted previously in Part 2, there are many repair strategies that do not necessarily result in the complete removal of these structural components if the damage is minor. In all cases, the porch should be structurally stabilized both before proceeding with the inspection and during the repair.

Roofing on porches is similar to that of the main building except that a porch is typically exposed to the outdoor elements year-round. The inspection and repair of porch roofing involve the same processes as other roof types.

Drainage repairs have three aspects. First, there are the obvious problems involving missing, inadequate, or improperly maintained roof drainage. Clogged or missing gutters cause the water to flow in an uncontrolled manner: it can randomly drip off the roof, causing backsplash problems; water can migrate to the interior of the porch roof; and water can freeze into ice dams. All drainage components should be checked for clogs and physical integrity. The second aspect is drainage from the flooring. Wooden decking should have nominal gaps between each floor board or should slope toward a controlled drainage path. Other flooring should readily allow water to flow toward drains or scuppers along the porch perimeter. (Note: weep holes or drainage paths from hollow columns should be open to allow any moisture inside the column to drain.) Lastly, drainage on the ground around the perimeter of the porch should accommodate moisture relief from runoff from the roof and porch. If no roof gutters and downspouts are present, then coarse gravel or a French drain may be needed. Plantings should be pruned, and built-up soil and decayed plant materials should be removed. This is particularly important if porch footings are made from wood or contain ferrous metals.

Flooring is the most common component, particularly on wooden porches needing repair. As flooring and stair components experience rot, corrosion, or settling, the resulting compression of materials can result in loss of physical strength and in sloping and uneven surfaces, which in turn present safety hazards. The unpainted end grain of wood draws water into the board, which can then begin to rot (see Figure 11-7). Similarly, porous masonry can absorb moisture and succumb to the ravages of freeze-thaw cycles. Exposed or unpainted ferrous metals can corrode and fail, particularly at connections and along joints. When salts for snow removal are introduced, surface decay can occur that can evolve into serious material losses.

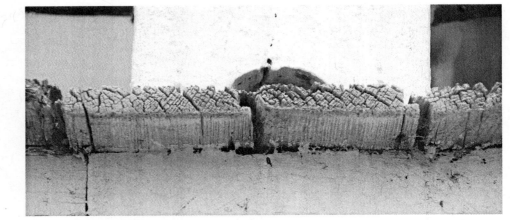

Figure 11-7 Exposed end grain on wood porch floor boards makes them easily susceptible to rot.

Therefore, protective finishes should be well maintained to keep water and harmful salts away from the raw flooring materials. As described earlier in discussing their respective materials, there are consolidants and sealants that allow retention of the historic fabric at a nominal cost compared to complete replacement of the damaged component. Wood is probably the easiest material to replace in a piecemeal repair process. When this is done, all surfaces, including those to be concealed from view, should be coated with a suitable primer. Cracked and missing masonry flooring may be replaced using similar new or salvaged masonry material from other sources or salvaging it from less prominent portions of the floor (and putting the new stone in those less important locations). The bedding substrate should allow the repaired flooring to lie smoothly and permit water to drain adequately. Repairing metal flooring and stair components may be completed by removing decayed pieces and fitting in replacement components. An identical replacement for the entire component may only be available through salvage or as a custom casting.

Cosmetic repairs include the repair of torn screens, broken glazing, shutter louvers, and other ornamental components (e.g., awnings). These items are typically part of a larger component assembly, which is otherwise intact. These infill repairs can involve using contemporary materials (e.g., modern float glass) that match the existing materials in the larger component or, depending on the historic accuracy desired, custom-made materials (e.g., crown glass) similar to those in the existing component.

Lastly, but probably best known, is the repair of finishes. As noted throughout this book, protective surface finishes are the major defense against such problems as moisture penetration and ultraviolet radiation decay. When these finishes fail, decay of the exposed material begins. The distortions brought on by this decay open joints, permitting additional moisture decay to occur from within the assembly as well. Two essential components of repairing these finishes are surface preparation and sealing open joints. All surfaces to be refinished should be clean and free of anything that will hinder a sure bond of the finish to the existing surfaces. Biological growths, soiling, oils, salts, and other contaminants must be removed. Failing finishes should be sanded or scraped to remove loose debris. Any areas where the finish is firmly in place can remain, and any areas with exposed raw materials should be primed. If lead paint is present, it will need to be removed and disposed of following local health department guidelines. Once the surface is prepared, open joints should be sealed with care not to seal weep holes or other intentionally constructed drainage paths. Application of the finish should follow the manufacturer's guidelines.

Replacement usually occurs when the entire porch or major components of it are missing or have deteriorated beyond the point of simple repair. When there is physical evidence of the original components, these components can be used to directly model the replacements. For example, brackets and other portions of ornamental trim may have been removed previously rather than being repaired (or perhaps just repainted). In some instances, there may be other brackets or ornament remaining in place or these removed items may have been stored on-site. In either case, these components can serve as a template to (re)construct replacements. In other cases, there may be several severely damaged components that can be disassembled and reassembled into one (or more) reconstituted versions of the original component. Many modern materials (e.g., structural foams, plastics, and recycled rubber) used in new construction are intended to simulate the appearance of a more traditional material. Like other modern synthetic materials, these potential replacement elements must be evaluated for physical compatibility, long-term durability, and expected maintenance requirements.

Replacement Strategies

Unfortunately, there are often cases where the original porch has been completely removed or remodeled so that the direct physical evidence of the original materials, ornamental features, and construction methods has been lost. In this case, use other research methods to determine what the porch previously looked like. Photographs may be obtainable from the local tax office, historic archives, neighbors, and previous homeowners (see Figure 11-8). Other methods include interviewing previous residents and neighbors, reviewing local historical society sources, and, if the house is prominent, reviewing local newspaper articles and photographs. If no such documentation is available, then the design can be based, in part, on comparable porches in the immediate neighborhood.

When the subject building is in a historic district or otherwise subject to the *Standards*, documentation of how the proposed replacement porch will conform to

a   b

Figure 11-8 Porches are character-defining features of a building. Replacement of missing original components must be based on physical, archival, and photographic evidence. This porch was reconstructed based on a combination of all three types of information: (a) before reconstruction and (b) after reconstruction.

the historic appearance of the original construction must be presented for review. In many instances, an existing porch that has been enclosed or extensively remodeled can remain in place without violating design guidelines. However, if the existing porch is removed, the replacement porch must meet the design and material compatibility requirements of the reviewing agency. The documentation prepared should support the design decisions made in arriving at the design of the replacement porch. One common issue in the design of replacement porches is the height of the railings. Modern building codes require minimum railing heights based on modern safety parameters. For houses located in historic districts, this restriction may possibly be waived to ensure the visual historic integrity of the house.

References and Suggested Readings

Belle, John, John Ray Hoke, Jr., and Stephen A. Kliment, eds. 1994. *Traditional details for building restoration, renovation and rehabilitation from the 1932–1951 editions of Architectural Graphic Standards.* New York: John Wiley & Sons.

Clark, Clifford, Jr.1986. *The American family home: 1860–1960.* Chapel Hill: University of North Carolina Press.

Cole, Roy F., Jr. 1981. Restoring a porch: A fast inexpensive method that doesn't require house jacks. *Fine Homebuilding* (April/May):: 14–17.

Cummings, Abbott Lowell. 1979. *The framed houses of Massachusetts Bay 1625–1725.* Cambridge, MA: Harvard University Press.

Dolan, Michael. 2002. *The American porch: An informal history of an informal place.* Guilford, CT: Lyons Press.

Downing, A. J. 1969 (reprint). *The architecture of country houses.* New York: Dover Publications.

Ewen, Ted. 1981/1982. Classical style in a porch addition: Tips for a restoration expert on deck construction, column building, and weatherproof design. *Fine Homebuilding* (December/January): 22–25.

Gowans, Alan. 1989. *The comfortable house: North American suburban architecture 1890–1930.* Cambridge, MA: MIT Press.

Harris, Samuel Y. 2001. *Building pathology: Deterioration, diagnostics, and intervention.* New York: John Wiley & Sons.

Historic Preservation Office Department of Planning and Urban Development. 2001. *Guidelines for porch repairs and replacement.* Portland, ME: Department of Planning and Urban Development.

Hunter, Christine. 1999. *Ranches, rowhouses and railroad flats: American homes: How they shape our landscapes and neighborhoods.* New York: W. W. Norton & Company.

Jackson, Kenneth T. 1985. *Crabgrass frontier: The suburbanization of the United States.* New York: Oxford University Press.

Jester, Thomas C., ed. 1995. *Twentieth century building materials: History and conservation.* New York: McGraw-Hill Book Company.

Kahn, Renee and Ellen Meager. 1990. *Preserving porches.* New York: Henry Holt and Company.

Leeke, John. 1991. *Practical restoration report: Wooden columns.* Portland, ME: Historic Homeworks.

———. 2004. *Practical restoration report: Wood-epoxy repairs for exterior woodwork.* Portland, ME: Historic Homeworks.

Litchfield, Michael W. 2005. *Renovation.* 3rd ed. Newtown, CT: Taunton Press.

Mahoney, Kevin M. 1993. Porches that won't rot. *Fine Homebuilding* (April/May): 44–47.

McAlester, Virginia and Lee McAlester. 1986. *A field guide to American houses.* New York: Alfred A. Knopf.

Nash. George. 1992. Renovating old porches: Common problems can be solved with simple repairs. *Fine Homebuilding* (June/July): 73–74.

———. 2005. *Renovating old houses: Bringing new life to vintage homes.* Newtown, CT: Taunton Press.

O'Brien, Tom. 1997. Restoring wood with epoxy. *Fine Homebuilding* (February/March): 60–65.

Phillips, Morgan. 1982. A stone porch replicated in wooden blocks: The Morse-Libby House. *Bulletin of the Association for Preservation Technology,* 14(3): 12–20.

Poore, Patricia, ed. 1992. *The Old House Journal guide to restoration.* New York: Dutton.

Rifkind, Carole. 1980. *A field guide to American architecture.* New York: Plume Books.

Sullivan, Aleca and John Leeke. 2006. *Preserving historic wood porches.* Preservation Brief No. 45. Washington, DC: United States Department of the Interior.

Vogt, Lloyd. 1985. *New Orleans houses: A house-watchers guide.* Gretna, LA: Pelican Publishing Company.

Wilkes, Kevin. 2000. Building a grand veranda. *Fine Homebuilding* (June/July): 106–109.

CHAPTER *12* **Storefronts**

Historically, the central business district (CBD) of a community was formed where the initial settlement of the town occurred, although subsequent prosperity or disaster often prompted the migration of the CBD to other sections of the community. Originally, shops were often the first floor or a front room of a building, with the shop owner's family occupying the remainder. Economic cycles have continuously reverberated throughout the country over the past four centuries. In times of prosperity in well-established communities throughout the country, CBDs intensified their land use or expanded into adjoining residential neighborhoods. This growth, for the most part, resulted in removal of the earliest buildings in the CBD to build new ones or in the extensive remodeling of earlier storefronts.

By the eighteenth century, residential or commercial space above the first floor had become customary. This upward growth was repeated periodically as communities prospered and continues today. There have also been numerous cases in which the CBD was abruptly obliterated by fires, wars, floods, earthquakes, and other natural disasters. Building codes were enacted to protect life safety, and a number of buildings surviving from earlier periods were then seen as economically undesirable and were replaced or abandoned. Lastly, severe economic decline forced businesses to close or relocate, causing the abandonment of a building. Today, as a result of these factors, many storefronts from the earliest settlement periods of large cities exist mainly as reconstructions or restorations based on physical or documented evidence. Meanwhile, those storefronts in smaller, less prosperous towns have survived due to earlier lack of economic opportunity to replace or modernize them.

The economic success of a community has always been closely tied to the transportation system of the time. The earliest towns were established along bodies of water to enhance access to trade with Europe and other colonies. As the transportation

239

networks evolved from pedestrian and animal-powered systems to trolley car, interurban, and transcontinental train systems and finally to the automobile and the interstate highway system, so too did the dispersal of residential and commercial buildings. Smaller shopping districts were created by adapting existing buildings or constructing new ones to serve the outlying "streetcar" neighborhoods of the late nineteenth and early twentieth centuries and then the mid-twentieth-century automobile-oriented suburbs. The massive migration to the suburbs after World War II and the rise of the shopping center in the 1950s, with the subsequent transition to enclosed shopping malls in the 1960s, left many urban and small-town retail building owners struggling to attract customers. The interstate highway system of the 1950s, coupled with the urban renewal programs of the 1960s, destroyed many inner-city residential neighborhoods and commercial districts.

The commercial buildings of the postindustrial landscapes of the nineteenth-century westward expansion and the twentieth-century suburban sprawl demonstrate how storefronts have or have not survived. Throughout the past four centuries, store owners who struggled to remain in business during prolonged economic downturns often attempted to attract or retain customers by updating the physical appearance of their buildings. Additionally, as a result of either man-made or natural disasters, those building owners who could afford to do so rebuilt, relocated, expanded, or updated their business establishments using the building styles and materials in current fashion. Recognizing these trends as a severe threat to the continued existence of many historically significant buildings and the cultural landscape, the National Trust for Historic Preservation (NTHP) created the "National Main Street Center" in 1980 to assist communities and store owners in attracting people back to the CBDs (NTHP 2006).

The success of the National Main Street Center program and other revitalization efforts has generated renewed interest in both urban and small-town commercial districts nationwide. With the realization of the importance of the historic architectural character that many small towns and urban commercial districts possess, the historic buildings within them are increasingly being recognized as potentially important physical assets that can attract new customers. While not specifically required by the *Standards*, many storefronts are being restored to their earlier configurations. In other cases, the subsequent modernizations themselves may have acquired historic significance and can remain in place as is or can be rehabilitated.

TYPICAL COMMERCIAL BUILDING FORMS

Since commercial buildings were constructed to meet local economic conditions, many different building forms were developed (see Figure 12-1). In *The Buildings of Main Street*, Richard Longstreth (1991) lists the following types of commercial buildings:

- Two-part commercial block: the most common form of commercial building, this building features a first-floor storefront with one to three upper floors above the main floor used for residential or commercial purposes.
- One-part commercial block: this form is a one-story building with only the commercial storefront as the facade.
- Enframed window wall: this form imitates the previous two but is typically much wider and uses the glazing and skeletal framing systems available around the turn of the twentieth century to allow more daylighting and display window space.

- Stacked vertical block: a taller version of the two-part commercial block, this form includes a first-floor storefront and then repeated multiple layers of a similarly detailed architectural facade to define each of the upper floors.

- Two-part vertical block: this form includes the first-floor storefront, but the upper floors are more unified to enhance the vertical lines of the building.

- Three-part vertical block: this is a two-part vertical block capped with a third zone consisting of one or more stories. This form was used in early skyscrapers that followed the classical ordering of a Greek column (e.g., base, shaft, and capital).

- Temple front: this form imitates a Greek temple, with a series of columns capped by a large pediment, and can be as many as three stories tall.

- Vault: this form includes a large two- to three-story central vault flanked by small subordinate vaults. It was commonly used for banks and theaters.

- Enframed block: this form includes a classically detailed central section of columns, pilasters, or an arcade framed by narrow end bays.

Figure 12-1 Several common storefront types: (a) two-part commercial block (the most common), (b) one-part commercial block, (c) enframed block, and (d) arcade.

- Central block with wings: this form includes classical element of the enframed block that was projected from less ornamentally detailed wings.
- Arcaded block: this form includes evenly spaced, round-arched openings in a two- to three-story facade.

There are also storefronts that defy classification and may actually be a combination of two or more of the above types. This combination could have occurred as part of the original construction of a new building or as earlier buildings were remodeled (Longstreth 2000).

STOREFRONTS

Traditionally, the storefront was located at street level. However, in more prosperous cities and towns, variations in storefronts emerged that sometimes included a mezzanine level and a second floor or several floors of a low-rise building as the storefront. The street-level storefront typically covered the first floor and/or complemented the facade (if any) above the first floor. This arrangement provided a sense of identity in smaller commercial districts and maintained a human scale in districts composed of taller buildings. As storefronts were updated, this visual connection was often ignored or the entire facade was concealed beneath contemporary cladding. In many other instances, only the first-floor storefront was updated since it was the one most visible to the street-level passerby. Lastly, fires or other catastrophes may have caused the loss of one or more upper floors while the lower floors remained reusable. In these instances, the damaged upper floors were either rebuilt or demolished and the remaining lower floors continued to be used. These updates often created visually disjointed results, not only on the building itself but also in relation to the buildings alongside it.

Components

Like houses, early stores tended to be small by today's standards. Over time, the typical floor space and volume became increasingly large. Storefronts evolved to include

Figure 12-2 The typical components of the storefront of a one-story commercial building.

several standard components: display windows, recessed doorways, transoms, bulk-heads, signage, and cladding (see Figure 12-2).

Display Windows

Perhaps the most notable characteristic of any storefront is the display window system. In the first commercial storefronts, unlike those of today, there were no large display windows to differentiate the shop from the nearby residential buildings. However, as glazing technologies improved throughout the next four centuries, shops and stores used increasingly large windows. Early windows consisted of multiple panes of glass (e.g., 4 by 6 inches or 8 by 10 inches) constructed to form a large display window that eventually grew larger as technology improved. The introduction of plate glass in the early nineteenth century enabled store windows to be made from larger sheets of glass. Naturally, the industrialization occurring at this time had begun to change and expand the commercial districts of larger American cities, and these were the first to adapt to this new technology.

By the early twentieth century, large display windows were considered the norm in urban areas and display windows essentially spanned a large portion, if not all, of the first-floor storefront (except, of course for the doorway and supporting load-bearing walls). The skeletal frame and the curtain wall allowed an even greater expanse of glass, with only the vertical mullions that held the glass in place preventing a continuous span of glass across the front of the building. For stores in smaller or less prosperous communities, the use of large multipaned windows persisted well into the early twentieth century.

Doors

The earliest storefront doors were constructed for security from invasive attacks and protection from the weather. They did not typically include windows. As a community became more established and security concerns were reduced, doors on newer buildings featured window glazing within the door, as well as windows (sidelights) flanking the sides of the door and/or a window above the top of the door. This window could be a series of small windowpanes or a more ornamental fanlight.

Early doors were constructed from wood but eventually were also made from or augmented by wrought, cast, stamped, or rolled metals. Doors for storefronts of the early twentieth century were made from bronze, nickel-silver, aluminum, stainless steel, or other metals. These doors were often ornamentally finished to match other exterior and interior elements that complemented or were part of the storefront.

As commercial activities grew, doorways became recessed to provide increased merchandise display space. Several variations on this arrangement also permitted the use of "islands," which were locked, full-height display cases containing wares that were used to entice passersby into the store. Coincidentally, these recessed entrances also provided some shelter from the sun, wind, rain, or snow.

Transoms

Prior to the introduction of modern lighting systems, many buildings were constructed to take advantage of natural daylighting. To do so, ceilings were much higher than those in many modern buildings. The taller ceiling allowed deeper penetration of usable daylight. To supplement the light coming in through the display windows, transoms (panels of glass) were installed over windows and doors.

Figure 12-3 Daylighting played an important role in lighting a store. The combination of high ceilings and a transom filled with prismatic glass was believed to extend daylighting to the rear of the store.

The use of prismatic glass to refract light and extend it further into an enclosed space was extremely popular in the nineteenth century. Introduced to storefronts in 1896 and manufactured until 1940 (Neumann 1995, 189), these prismatic panes were often assembled into panels and installed in transoms and other windows. Windows using prismatic panes are a character-defining feature of buildings constructed in this period (see Figure 12-3). Although they are no longer in production, replacements can sometimes be obtained from salvage companies. Unfortunately, due to the variety of prism styles and colors, finding exact matches may be difficult.

Along with admitting daylight, transoms could enhance ventilation. Transoms above doorways and display windows could be opened to allow fresh air to enter the building or hot, humid air trapped at the ceiling to be expelled. Later, after electric lighting and air-conditioning became preferable to natural daylight and ventilation in the twentieth century, transoms were often covered by signage or curtain walls and concealed above lowered ceilings. In some instances, they were removed; in others, they simply were covered.

Bulkheads

As display windows evolved, the portion of the wall below the window that supported the window installation became known as a "bulkhead." Bulkheads varied in size as display windows continued to get larger. By the mid-twentieth century, they had all but disappeared (and by the late twentieth century they had done so). Older storefronts can be identified by the larger size of their bulkheads and consequently smaller display window size. In efforts to update them, bulkheads were often simply reclad with contemporary materials, such as thin stone veneer or a curtain wall assembly. In cases of extremely overzealous updating, bulkheads and their display windows were demolished and completely replaced with contemporary materials.

Ornamental Elements

Many storefronts featured ornamental elements (see Figure 12-4) that served decorative or functional purposes. A flat signboard located above the transom of a display window advertised the name of the store. In later modifications, this area was sometimes expanded over the transoms when interior ceilings were lowered and interior electric lighting was added. Other signage could include a "blade" sign projecting perpendicularly from the building, identifying the purpose of the store. In less literate times, signs included figural forms to identify the services and goods without using words.

Along the top of the facade typically were a cornice and some form of pediment that included the name of the original owner or the purpose of the building. Over time, the cornice and pediment were fabricated from the entire range of materials described earlier in this book, beginning with wood and stone and eventually succeeded by terra-cotta and stamped metal as they became available. Lack of maintenance and concerns for public safety often resulted in their removal without being replaced.

One ornamental feature that was developed for thermal and lighting control was the awning. By the late nineteenth century, awnings had become a primary means to control sunlight and provide shade. They could be adjusted daily and were extended or retracted as needed. Awnings, like transoms, fell out of favor as air-conditioning

Figure 12-4 Store owners originally used figural signs to enable illiterate people to visually identify products or services. The early twentieth century saw the emergence of the building itself as an advertisement, as shown in this building in Boston.

came into use in the mid-twentieth century. Careful inspection may reveal the signs of earlier connections that held an awning assembly in place long after the awning had been removed. Similarly, more permanent protection was sometimes provided by a canopy over the entrance that also served as a sign and a means of getting to the curbside while being protected from the weather.

On buildings with upper floors above the storefront, the ornamental detailing followed the architectural style of the building. A common practice was to leave the facade of the upper stories untouched even though the storefront had been "updated." For this reason, much of the original ornamentation often survived to modern times.

Cladding Systems

Beyond the window and door configurations, cladding systems are an important character-defining feature of many storefronts. As noted previously, the material of which the load-bearing wall was constructed also formed the exterior cladding. For earlier storefronts with smaller windows, cladding followed systems similar to those found on residential buildings, with the most important and prosperous buildings using stone or constructions that emulated stone.

Early cast iron storefronts were a conceptual forerunner of the skeletal frame/ curtain wall combinations that emerged in the late nineteenth century and increased the expanse of display window glazing. This expansion reduced the cladding locations to bulkheads, signboards, and other framing elements around the windows. With the continuing desire to modernize older storefronts came the inevitable replacement or covering of existing storefronts with curtain walls, thin stone panels, or ceramic veneers of various types. As previously discussed, the early twentieth century saw the introduction of numerous cladding and curtain wall systems. Among the most unusual cladding systems was structural glass (also known by the trade names Vitrolit™ and Carrara Glass™), which was introduced in 1900 and produced

Figure 12-5 Vitrolite and Carrara Glass were products that imitated more expensive marble.

until 1962 (Dyson 1995, 201). This product was typically nearly opaque and simulated the appearance of polished marble (see Figure 12-5).

The aluminum rolling and extruding processes developed during World War II established a relatively low-cost means of manufacturing curtain walls. In the postwar era and beyond, numerous storefronts were remodeled to include these systems. Former bulkheads, transoms, and signboard areas were removed or covered with panelized systems of spandrel glass, structural glass, plastic infill, or decorative metals. This process often resulted in loss of or serious damage to the historic buildings materials.

By the mid-twentieth century, improvements in glass fabrication technology had resulted in the expansion of display windows to nearly the full height of the storefront itself. A nominal amount of space was required for framing to support the glass and seal joints between the glass and adjoining surfaces. Cladding had been reduced to a minimum.

Typical Problems

The most common problem with many storefronts is that the one currently on the building may not be the original one or may be the accumulated result of several earlier updates that may or may not be compatible visually and physically with the original construction and style. Beyond that, the range of decay processes affecting the specific materials of the storefront comes into play. Because of their public nature, storefronts are vulnerable to additional problems due to security, vandalism, and graffiti (see Figure 12-6).

Updates and Incompatible Repairs

The exterior cladding and storefront systems are the first materials to investigate to determine if they are original or later additions. This examination is particularly important where placing a curtain wall system over the original facade may have irreversibly damaged the original or earlier storefront or where poor maintenance practices have degraded the overall condition of the building.

Figure 12-6 Insensitive updating was common during the mid-twentieth century, when contemporary curtain wall construction was used to cover the upper facades of three of these stores. The store at the left-center shows the original facade.

Previous repairs may not have been completed in a professional manner and subsequently failed or are accelerating decay of the storefront. This decay should be noted and temporarily stabilized until the final treatment is undertaken. These temporary measures include securing tarpaulins or installing more rigid plywood coverings over the damaged portions of the storefront.

Material Decay Mechanisms

All of the decay mechanisms described in earlier chapters can act on the materials used to construct the current storefront and any underlying remnants of earlier storefronts. Identifying the materials used on the storefront construction and determining the status of the decay is critical. The assessment of the storefront's condition can include visual inspections, nondestructive testing, and possibly the selective removal of surface materials to access the interior assemblies.

Security

As certain business districts became less frequently used after business hours, concern for security of the buildings' contents increased. Many store owners installed security gates, shutters, and roll-down overhead doors over the display windows and entrances. While they provided security, these devices substantially altered the appearance of the buildings and perhaps damaged historic building fabric in the process of being installed. Even in revitalized areas, removing the anchoring systems for these devices may have further damaged the building fabric or left holes that allow moisture penetration into the wall.

Many communities consider unoccupied buildings a security and public safety threat. Therefore, unoccupied buildings create an attractive nuisance that may fall under local ordinances regarding public safety and liability. Without proper security, a variety of undesirable visitors (e.g., vandals, arsonists, illegal salvagers, transients, or animals) may find their way into these buildings. Building entrances and windows must be secured in accordance with local ordinances to prevent unwanted entrance.

Vandalism and Graffiti

The mechanical defects caused by deterioration and soiling of storefronts can invite further damage through vandalism and graffiti, especially in many less prosperous business districts. As these problems occur and nothing is done to correct their effects, a cycle of repeated vandalism and graffiti can accelerate the downward spiral of decay. Earlier repairs made with incompatible or substandard materials cause the building and its environment to appear further blighted and considerably less attractive to customers or potential new businesses. Unstable elements of the storefront should be secured in place or removed and stored until a decision on how to treat the storefront is made.

Cleaning processes used to remove or cover graffiti must be compatible with the material on which the graffiti is located. The medium (e.g., pencil, ink, or paint) used to create the graffiti should be identified and an appropriate removal system used to remove it. Care must be taken with porous surfaces, such as masonry, that can absorb the medium used to create the graffiti and the cleaning solvents used to remove it. On masonry surfaces, graffiti can be removed using a variety of methods, including erasers, solvents, poultices, water and detergents, alkaline compounds, bleaches, abrasives, and lasers (Weaver 1995).

For nonporous surfaces (e.g., metals, glass, or plastics), the graffiti may be removed with detergents or nonabrasive cleansers. Many solvents and chemicals are formulated for specific applications. Directions for applying these cleaning products must be followed to prevent additional damage to the surface or endangering the personnel using the product. When concerns about safety arise, consult the manufacturer's MSDS prior to using the product. To determine the suitability of the cleaning system, test it first on an inconspicuous portion of the affected area and assess the results before proceeding.

On painted wood or metal surfaces, the graffiti can be painted over using the same color of paint as the existing surface. However, this is successful only when the paint exactly matches the color and texture of the undisturbed surfaces adjoining the damaged site.

The proposed treatment decision will depend on the results of the investigation of the existing building as well as the discovery of documentation on the earlier, if not the original, appearance of the storefront. Successive economic cycles will most likely have had a substantial effect on the storefront as owners periodically updated their buildings or let them decay.

All treatment choices should include a review of physical and documented evidence to confirm the earlier appearance. This process will identify which components that were added later might be removed or altered. At this point, the decision to remove the later modifications can be made and the storefront may be restored to its original appearance. However, if the subsequent alterations have achieved historic significance in their own right, then they may be left in place and simply rehabilitated as needed.

Unfortunately, insensitive updating (e.g., a twentieth-century installation of a metal curtain wall over a nineteenth-century facade) may have substantially removed or destroyed the underlying character-defining features. This condition could prompt a reconstruction or replacement treatment that would fall under local design guidelines. In all cases, the approval of oversight agencies (e.g., planning department, landmarks commission, and state and federal review boards) must be obtained to achieve the required levels of appropriateness and conformance to applicable standards.

Restoration Treatments

Restoration is perhaps the optimal treatment when so many updates have occurred that they have significantly altered the appearance of the storefront. When the original facade has been damaged through insensitive updating and other modifications, it may be necessary to restore the storefront (or the overall facade) to a specific period of its existence based on well-established physical and documented evidence. When this is done, later alterations that were added after the restoration period and have not achieved significance of their own should be documented prior to their alteration or removal. Remaining materials from the restoration period should be preserved and reused. Deteriorated features from the restoration period should be repaired rather than replaced. Severely deteriorated or missing features can be reproduced based on documented evidence. New additions outside of the restoration period must be differentiated from the existing building and must be able to be removed without harming the portions of the building from the historic period. In addition, any materials used in new construction must be physically compatible with the historic materials.

Recommended Treatments

Rehabilitation Treatments

When restoration to a specific historic period is not desired, rehabilitation can be pursued. "Rehabilitation" refers to the comprehensive repair of the facade as it currently exists. The original historic character-defining features, materials, finishes, and construction techniques of the facade will be retained and preserved, along with any changes that have acquired historic significance in their own right. Deteriorated historic features will be repaired rather than replaced but, when necessary, severely deteriorated features can be replaced with new features that match the original in design, color, texture, and, if possible, materials. This issue becomes important when rehabilitating a storefront constructed of materials that are no longer manufactured. Treatments used for cleaning, repair, and replacement should be the gentlest ones possible and should be reversible. New construction should not destroy the historic materials, features, and spatial relationships that are character-defining features of the building. As with restoration, new work should be reversible, be differentiated from the old, and be compatible with the historic materials, features, size scale, proportion, and massing to protect the historic integrity of the property and its setting.

Reconstruction Treatments

When the original storefront has been extensively damaged, removed, or allowed to severely deteriorate, reconstruction may be more economically feasible. Exactly duplicating the materials and methods used to construct the original facade is possible. However, the *Standards* specifically state that the new construction must be clearly identifiable as a contemporary construction so as not to give a false impression of being the original.

In many cities with historic commercial districts, guidelines may dictate the design parameters of the (re)construction (Pregliasco 1988). While the parameters vary by municipality, they are meant to maintain the overall context that has been

Figure 12-7 Local design guidelines dictate that new storefronts be compatible with regulating lines, solid void rhythms, street setbacks, and height restrictions. In this store located in Camden, Maine, a national corporation accepted the local design guidelines rather than use its typical corporate design.

identified as historically significant for the particular commercial district (see Figure 12-7). These parameters typically include:

- Height: the average height of buildings in the streetscape, generally intended to maintain a sense of scale.

- Width: the average width of a building lot. Buildings covering two or more lots typically visually divided the facade into smaller units that simulated that width.

- Setback: storefronts conformed to established setbacks to accommodate sidewalks and create a block face. While awnings and canopies projected over the sidewalk, the facades were typically uniform along the block.

- Proportion of window openings: over time, window openings achieved a consistent height and width, particularly on the floors above the street-level storefront, which contained significantly more glazing.

- Horizontal rhythms: the pattern created by the width of the window openings relative to the amount of solid material between them.

- Massing and scale: the overall composition of the facade and how the detailing of the surface based on the previous parameters came together to avoid large, monolithic surfaces that detracted from the overall visual scale of the streetscape.

Additional parameters vary by community but sometimes require materials that are visually compatible with adjoining buildings, awnings, signage, and lighting. One contentious parameter is color. Many review agencies comment on it only if it is part of a permanent material like stucco or brick, while others attempt to control paint colors. Color selection should be done carefully to complement colors found on adjoining buildings but should not be taken to an extreme of creating a bland visual setting. Many architectural styles had period colors associated with them.

References and Suggested Readings

Auer, Michael, J. 1991. *The preservation of historic signs.* Preservation Brief No. 25. Washington, DC: United States Department of the Interior.

Belle, John, John Ray Hoke, Jr., and Stephen A. Kliment, eds. 1994. *Traditional details for building restoration, renovation and rehabilitation from the 1932–1951 editions of Architectural Graphic Standards.* New York: John Wiley & Sons.

Dane, Suzanne, G. 1997. *Main street success stories: How community leaders have used the Main Street approach to turn their towns around.* Washington, DC: National Main Street Center.

Dyson, Carol J. 1995. Structural glass. In Jester 1995, 200–205.

Fleming, Ronald Lee. 1982. *Facade stories: Changing faces of Main Street and how to care for them.* New York: Hastings House Publishing.

Jackson, Mike. 1991. Preserving what's new. *APT Bulletin,* 23(2): 7–11.

Jandl, H. Ward. 1982. *Rehabilitating historic storefronts.* Preservation Brief No. 11. Washington, DC: United States Department of the Interior.

Jester, Thomas C., ed. 1995. *Twentieth century building materials: History and conservation.* New York: McGraw Hill Book Company.

Konrad, Kimberly Kenneth M. Wilson, William J. Nugent, and Flora Calabrese. 1995. Plate glass. In Jester 1995, 182–187.

Liebs, Chester. 1985 *Main Street to miracle mile.* New York: Little, Brown and Company.

Longstreth, Richard. 1991. The significance of the recent past. *APT Bulletin*, 23(2): 12–24.

———. 2000. *The buildings of Main Street*. Walnut Creek, CA: Alta Mira Press.

National Trust for Historic Preservation. 2006. History of the National Trust Main Street Center. http://www.mainstreet.org/content.aspx?page=1807&section=1.

Nelson, Lee H. 1977. The 1905 catalog of iron store fronts, designed and manufactured by Geo. L. Mesker & Co., architectural iron works, Evansville, Indiana. *Bulletin of the Association for Preservation Technology*, 9(4): 3–40.

Neumann, Dietrich. 1995. Prismatic glass. In Jester 1995, 188–193.

———, Jerry Stockbridge, and Bruce Kaskel. 1995. Glass block. In Jester 1995, 194–199.

Patterson Teller, de Teel, ed. 1984. *The preservation of historic pigmented structural glass (Vitrolite and Carrara glass)*. Preservation Brief No. 12. Washington, DC: United States Department of the Interior.

Pregliasco, Janice. 1988. *Developing downtown design guidelines*. Sacramento: California Main Street Program.

Randl. Chad. 1991. *Repair and reproduction of prismatic glass transoms*. Preservation Tech Notes: Historic Glass No. 1. Washington, DC: United States Department of the Interior.

———. 2005. *The use of awnings on historic buildings: Repair, replacement and new design*. Preservation Brief No. 44. Washington, DC: United States Department of the Interior.

———. 2006. *The preservation and reuse of historic gas stations*. Preservation Brief No. 46. Washington, DC: United States Department of the Interior.

Stovel, Herb. 1985. Scrape and anti-scrape: False idols on Main Street. *Bulletin of the Association for Preservation Technology*, 17(3/4): 51–55.

Thomas, Bernice L. 1997. *America's 5 & 10 cent stores: The Kress legacy*. New York: John Wiley & Sons.

Utah Pioneer Communities Program. 1999. *Restoring a historic commercial building: A workbook for downtown business and property owners*. Salt Lake City: Utah Department of Community and Economic Development. http://history.utah.gov/historic_preservation/documents/restoringahistoriccommercialbldgfinal2.pdf.

Weaver, Martin E. 1995. *Removing graffiti from historic masonry*. Preservation Brief No. 38. Washington, DC: United States Department of the Interior.

Weeks, Kay D. 1986. *New exterior additions to historic buildings: Preservation concerns*. Preservation Brief No. 14. Washington, DC: United States Department of the Interior.

Yorke, Douglas A., Jr. 1981. Materials conservation for the twentieth century: The case for structural glass. *Bulletin of the Association for Preservation Technology*, 13(3): 18–29.

PART IV

Building Ornamentals and Finishes

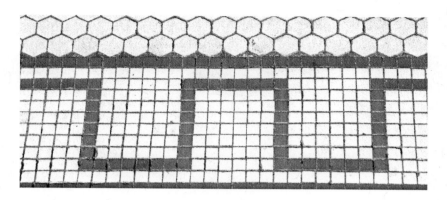

CHAPTER 13 Floors

The floor can be a significant character-defining feature of a space, particularly when it is composed of highly decorative finishes and materials. In many public spaces of commercial buildings, care was taken to install expensive flooring that conveyed a sense of wealth, power, sophisticated taste, and the owner's personal pride. Similarly, civic buildings included large, well-appointed public spaces that demonstrated these same traits at the community level.

This high level of ornamentation often carried over into the personal spaces of wealthy homeowners who could appreciate the detail and afford its cost. However, even in these homes, the level of ornamental detail, including floor finishes and treatments, was frequently modest in utility spaces and living spaces occupied or used primarily by servants.

For modest or less affluent businesses and homes, substantial ornamental detail was less common and, if present at all, may have occurred only in public areas. Even at that public level of display, there were cost-cutting measures that included using more expensive flooring only in areas left exposed, while less expensive flooring was used in locations expected to be covered with a floor cloth or area rug.

Flooring often was part of the design scheme of a space. Made from wood, stone, brick, or concrete, it could be polished or covered by floor cloths, linoleum, or a variety of tiling systems. The viewable surface was then treated with a protective finish composed of paint, clear finishes, stains, or other treatments to enhance its durability and appearance. In some instances, a decorative faux or "false" finish was painted on less expensive wood floors to simulate more expensive materials like oak or marble.

OVERVIEW

CONSTRUCTION AND FINISH TREATMENTS

The finished floor that is viewable within a space is the topmost layer of a construction sequence that begins with either wood skeletal framing (wood timbers or floor joists), masonry load-bearing construction, or a concrete floor slab. Once a smooth, level supporting surface has been constructed, the finished flooring can then be installed.

Dirt Floors and Early Finish Treatments

The early huts and shelters in the North American colonies had dirt floors that were compacted and mixed with a variety of materials to make them more stable. These materials included oxblood, plaster, lime, and ashes. Dirt floors may also have been covered with pine boughs, animal hides, or mats woven from local grasses. Although the exact date when wood floors were first widely used is unknown, the transition occurred throughout the seventeenth century as the colonists constructed permanent homes and other buildings (Von Rosenstiel 1978).

Wood

Early colonial wood floors consisted of planks or boards (some were 20 or more inches in width) secured in place with wooden pegs, hand-forged nails, or machine-cut nails (when they became locally available). Nails and pegs that were not concealed became more prominent as the wood flooring wore away through use. In the early colonies, sand was used as an abrasive cleanser on the wood, which also further eroded its surface. Floor boards had flush joints that shrank and swelled with seasonal moisture fluctuations. To fill these gaps, sand was swept into the joints. Although canvas floor cloths were available, the general poverty of the era prevented all but the wealthiest homeowners from obtaining them.

As the number of sawmills increased and the larger first-growth trees were used up, wood flooring became a two-layer system consisting of a subfloor and the finished floor. The subfloor was secured across the floor joists either perpendicularly or at a 45-degree angle to promote structural stability. The finished floor was secured to the subfloor. In masonry and concrete structural systems, wooden slats (also known as "sleepers") could be secured to the structure instead of a subfloor and the finished flooring was attached to the sleepers. The use of wood plank flooring, laid both as planks and as end-grain blocks, in mills and other industrial applications continued into the twentieth century, when wood flooring largely was replaced by reinforced concrete slabs that were plain or coated with a magnesite finish treatment.

By the late nineteenth century, the use of shiplap or tongue-and-groove flooring had been introduced. It used narrower strips of wood, typically 2 inches in width, which were secured using nails concealed from view. These nails were nailed into the lower flange of the shiplap or groove as each piece was secured to the subfloor. When the floor was completed, the exposed nails in the last strip were concealed by the baseboard trim installed along the perimeter of the floor. An alternate form of connection was to insert a wood spline into two facing grooves.

These strip flooring systems were generally laid in a uniformly parallel construction with a slight gap along the perimeter to allow for expansion and contraction due to changes in temperature and humidity (see Figure 13-1). This gap was typically concealed by baseboard or shoe molding. As design tastes evolved, geometric patterns and inlays of wood and other materials could be inserted in the floor to enhance their ornamental qualities. The first process created what is called a "parquet" floor and was made by cutting mitered joints into smaller wood pieces to form decorative borders. The second process was referred to as "marquetry" and used more

Figure 13-1 Wood flooring must be allowed to acclimate to on-site conditions before being installed. These floorboards have been brought to the job site to allow them to adjust to local humidity and temperature. Notice how they are stacked to allow air circulation.

curvilinear designs that were more elaborate. Marquetry had been developed as a decorative inlay on furniture of the period and was adapted as a decorative detail on flooring. These two floor treatments usually were found in the homes of the middle and upper classes.

While the construction practice incorporating the exposed finished floor system using narrow strips of tongue-and-groove-wood remains in use today, a significant modification occurred in subfloor construction: the use of plywood. First used as a subfloor material in the late 1930s (Jester 1995, 135), plywood had become a significant subfloor material by the end of the 1950s.

Concerns about the availability and cost of wood for conventional strip flooring systems in the late twentieth century led to the introduction of laminated flooring products. First developed in Europe, these systems came to the United States in the 1990s and now rival the use of conventional wood flooring. Laminated wood flooring typically consists of reconstituted wood faced with a veneer that has been painted to simulate various wood grains and finishes.

Masonry and Concrete

Stone or brick floors were constructed by installing a substrate consisting of a mortar bed or compacted sand base on the supporting masonry or subfloor structure. The use of these substrates created the necessary flat, level surface on which the stone or brick could be installed. Sand, mortar, or grout was used to fill the joints between the bricks or stones. Marble and slate are two of the more common stone flooring materials.

Concrete has been used for flooring in locations requiring an inexpensive yet durable material. It has been used as the finished floor surfaces as well as a subfloor. Concrete flooring was often left exposed in many manufacturing facilities, as well as in nonpublic service locations and storage spaces in commercial and residential buildings. In spaces that served a publicly accessible need, any number of finish treatments (e.g., staining, painting, embossing, and etching) or floor coverings could be used to enhance the overall appearance of the space.

Ceramic Tile

Tile floor construction methods date back to the earliest use of fired clay tile as a building product. The tiles themselves were crafted by cutting, extruding, or pressing clay into uniform shapes and sizes that were then fired in a kiln, much like terracotta and other fired clay products described previously. The traditional method for installing ceramic tile was to construct a flat, level surface on which mortar was placed to form a bedding layer for the tile. Early-twentieth-century methods included using hollow clay tile and tar paper as substrates when laying tile over a wooden floor to accommodate moisture issues in the mortar. Methods evolved further after World War II to enhance the compatibility of installation with reinforced concrete, expanded wire mesh, polyethylene, and waterproof plywood. Traditional ceramic tile floors typically were not sealed, and the only protective coating used on them was floor wax.

Prior to the start of decorative tile production in the United States in 1870, decorative tile was largely imported from other countries, such as England (Grimmer and Konrad 1996). Spurred on by the Arts and Crafts movement around the turn of the twentieth century, the American decorative tile industry flourished until the Depression and was curtailed by the Modernist movement, which rejected many of the ornamental features common in previous styles. While the tile industry continues to this day, it has a much smaller market share due to competition from the resilient flooring industry.

Most tiles came in standard sizes and shapes. Among the most common were square, rectangular, and hexagonal tiles that could be used with different colors to create patterns and decorative motifs and even to spell out names. Some tiles featured three-dimensional relief or applied decoration in glazed paint. Tiles could be installed in floors, on walls, and on ceilings.

Tile was classified as unglazed or glazed. Unglazed tiles were fired in a kiln without adding any materials to enhance their surface finish or color. They included quarry and encaustic tiles. Ceramic mosaic tiles were either unglazed or glazed, and most other floor tiles were glazed. Glazed tiles usually had a specific finish that could include one or more colors.

Glazes melted or were transformed by the heat of the kiln into a permanently bound surface finish or color. In addition to providing finish and color, glazes protected the tiles from moisture absorption and made them easier to clean. As the popularity of glazed tiles increased, so did the number of potteries that became noted for a specific type of glazing. As is often the case with trade secrets, the formulation or application technique for a particular glaze frequently became unavailable after the potter who invented it died.

Quarry Tile

Quarry tile was created as a less expensive replacement for floors that were otherwise made from a quarried stone, such as slate or flagstone. They were the simplest form of tile because they were basically flat clay slabs fired in a kiln.

Encaustic Tile

The majority of floor tile made in the United States before 1890 was known as "encaustic" tile. In this product, a base tile was formed in a mold that included relief carvings. When the clay was pressed into the mold, a depressed space was created by this relief carving. After the clay tile dried, clay slips (liquid clay mixed with pigment) of different colors were applied to the depression to form a specific pattern or shape. After each color dried, the next color was applied. This process was repeated until the face of the tile was completely built. This tile was then fired to create one solid piece. Later, less expensive tiles were made from cement or clay and painted, silk-screened, or decaled to imitate patterns found on encaustic tile.

Mosaic Tile

Small tiles of a single color were known as "mosaic" tiles. Building on the traditions of ancient European societies, mosaic tiles of various shapes and colors were installed to create figural images, geometric patterns, and other decorative details. Originally executed with pebbles and rock chips in the villas of Rome, the modern equivalent used systematically sized ceramic tiles to achieve a modern translation of this ancient craft (see Figure 13-2).

Terrazzo

Terrazzo was a formed-in-place flooring system with ancient roots. Essentially, the resulting floor was conceptually similar to the fine mosaic floors of Rome, Egypt, and Greece in that stone chips were embedded in a mortar or concrete substrate to form an imitation of a marble floor. Modern terrazzo flooring consisted of a mixture of marble (or other stone chips) mixed with a binder, usually concrete, and any desired pigments for coloration. The terrazzo was made in two ways. The first was as a precast panel known as "Art Marble." The second method was to cast the terrazzo in place directly on the substrate. Brass strips used to facilitate the casting were left in place to serve as an expansion joint (see Figure 13-3) to reduce cracking and minimize damage to adjoining sections of terrazzo if it became necessary to remove a damaged section. The surface was then ground down and polished to a lustrous sheen that mimicked marble (Plaisted 1939).

With the fall of the Roman Empire, the original traditions were carried forward by the western Christian and eastern Byzantine churches. Modern terrazzo was derived from practices used in Venice in the eighteenth century. While it was used in the United States before 1900, terrazzo did not gain wide acceptance until 1920, when cracking problems were solved. Terrazzo was a popular design element in the Art Deco and Art Moderne styles since the brass strips could be used to create

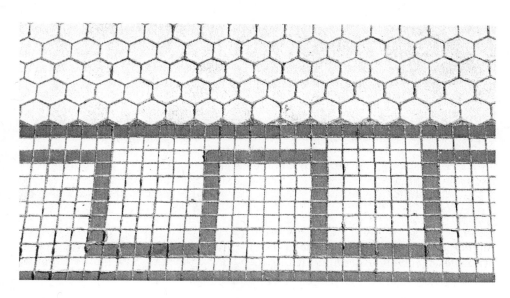

Figure 13-2 Tile flooring can produce a variety of ornamental patterns that become part of the character-defining features of a space.

Figure 13-3 Terrazzo could be used in a variety of patterns to create highly ornamental floors. In some cases, business names or logos were integrated into the floor.

curvilinear designs. Terrazzo competed with mosaic tile floor systems until it finally became more popular in the 1930s. Typical installations for terrazzo included hospitals, schools, airports, and commercial buildings (Johnson 1995).

RESILIENT FLOOR COVERINGS

Resilient floor coverings were used to cover otherwise plain floors. In some instances, these floor coverings were intended to serve a functional purpose, such as blocking drafts from joints between the floorboards, while in others they were intended to simulate a more expensive material at a fraction of the cost. Because of their construction, resilient floor coverings had some flexibility that cushioned the impact of an occupant's foot and to varying degrees softened the acoustical reflections in a room. Some resilient floor materials, such as linoleum and cork, were used as part of an acoustical treatment to reduce both sound reflection and transmission.

Floor Cloths

As the colonies became more firmly established and prosperous, those who could afford them imported floor cloths from Europe. Floor cloths (also known as "oil cloths" or "wax cloths") were typically made from canvas that had been waterproofed with linseed oil or shellac and then painted to simulate marble flooring or carpets. They were used as a substitute for rugs and carpets that were expensive or unavailable in the early colonies. Although a sign of status in early colonial homes, the floor cloth was actually a less expensive alternative to carpets and rugs. Although installed like the area rugs or carpets they simulated, floor cloths also can be seen as predecessors to linoleum and other resilient flooring.

Rubber Tile

Rubber tile flooring was used as early as 1853, when Charles Goodyear described a floor cloth composed of canvas, wallpaper, and rubber varnish. By 1894, interlocking 2 by 2-inch rubber tiles were in use for interior floors. In the early 1920s, Goodyear introduced sheet rubber flooring that could be used directly on a subfloor or cut into tiles of various sizes and shapes and mixed with tiles of other colors to make an assortment of patterns. The rubber tiles could also be made individually through a casting process (Park 1995).

Linoleum

Linoleum was invented by Frederick Walton in 1863. The manufacturing process consisted of a mixture of oxidized linseed oil and ground cork or wood flour that was calendered (pressed with heated rollers) into sheets with a burlap backing.

Linoleum could be delivered to the site in rolls and laid on a smooth surface. From the 1860s to the 1940s, linoleum was the preferred floor covering for residential and commercial purposes and was even used on countertops and backsplashes (see Figure 13-4).

There were two common methods for installing linoleum. The first was to coat the underside of the linoleum sheet with paste and lay the linoleum directly on the floor. The second method, often used with thinner linoleum products, was similar to the first except that a sheet of felt underlayment was glued to the floor first and then the linoleum was pasted on the underlayment. In both methods, the linoleum was then pressed with rollers to eliminate trapped air and create a good bond between the linoleum and the underlying material(s). A key issue when installing linoleum was the need to ensure that the flooring beneath it was flat, since any surface irregularities would eventually become evident in the linoleum. The outlines of uneven floorboards or irregular concrete surfaces eventually could be seen in the linoleum on many early installations that did not attempt to smooth the subfloor surface.

The asphalted felt–based and vinyl-based floor coverings introduced in the mid-nineteenth century competed with and eventually outsold linoleum. Linoleum was made in the United States until 1964 and has since remained available from manufacturers in Europe (Simpson 1997; Snyder 1995). Recognition of its value as a sustainable product (when compared with vinyl flooring) has caused a recent resurgence in its use nationwide.

Cork is a naturally occurring product taken from the outer bark of a cork oak tree. Originally, it was used as stoppers for bottles and jars. Experimentation in the 1890s in the United States and Germany led to the development of a heating process that transformed cork particles into a stable material that could be cut into tiles and sheet goods for flooring, insulations, and acoustical absorbers. Cork floor tiles were introduced in

Cork

Figure 13-4 Linoleum provided a durable, resilient surface and, besides flooring, was used for countertops and backsplashes (as shown here).

1899 but saw only limited use until the 1920s, when additives were provided that improved resistance to dirt, food, grease, ink, mild acids, and, most of all, moisture.

Cork flooring was available in various square and rectangular shapes and came with either a natural or prewaxed finish from the factory. Before installation of the flooring, the cork had to acclimate to the conditions at the site. Therefore, the cork was brought to the installation site and left for several days. Waterproof mastic was applied to the floor surface to serve as a bed for the cork. Once the cork was installed, headless brads were used to further secure it in place. After 1925, brads were eliminated and mastic or waterproof cements were used, Weights were placed on the cork while the adhesive cured to ensure a good bond (Grimmer 1995).

The use of cork tile after World War II declined when competition from other resilient floors, particularly vinyl, significantly diminished its popularity. Cork has recently enjoyed a resurgence since its recognition as a sustainable building product; the cork oak tree can regenerate the removed bark relatively quickly.

Magnesite

Magnesite (also known as "diato") flooring is a formed-in-place floor that was installed in a semiplastic state and screeded or spread to a self-leveling finish. When a seamless floor was desired, magnesite floors could be formed in one monolithic slab. Before hardening, the surface could be embossed with a pattern to simulate stone, brick, tile, or other materials. However, like terrazzo, sections of a magnesite floor could also be divided by brass strips that served as control and expansion joints. Magnesite was also made as individual tiles.

The durable surface of magnesite floors made them popular for industrial applications from the 1930s to the 1950s. They were commonly installed in buildings constructed in the Mediterranean Revival styles of this period as well. The ability to form them in place and then emboss them made them a low-cost alternative to the materials they imitated.

Magnesite floors contained magnesium oxide, and binders and fillers enhanced their resilience. These binders included sawdust, cork, and leather. Unfortunately, when these materials rotted away, pitting occurred and degraded the visual and structural integrity of the floor. The floor surface had to be sealed with floor wax or sealers and could not be used where oil, grease, or acids contacted them. Magnesite flooring typically contained asbestos (Plaisted 1939; Salter 1974).

Felt Base

In 1910, an alternative to linoleum was developed that was basically composed of roofing felt coated and painted to imitate the appearance of linoleum. Rag or paper felt was soaked with various forms of asphalt to harden and seal it. The wearing surface, consisting of a linseed oil composition, was added and then painted. First manufactured by roofing companies and later by linoleum companies, "fiber base looked like linoleum but it was not resilient, had no cork or burlap, and its thin coating of paint could easily wear through" (Simpson 1999b, 91). Due to its lower cost, felt base production surpassed that of linoleum in 1923 and continued to do so until the introduction of vinyl flooring after World War II (Simpson 1999b).

Asphalt Tile and Vinyl Asbestos Tile

Asphalt tile was developed in 1924 as a replacement for hand-troweled, jointless, formed-in-place asphalt flooring, which required greater skill to install correctly. Asphalt tiles could be installed by a competent floor layer. Originally available only in dark colors, these tiles were sold later in lighter colors that were popular in the 1950s.

In 1932, the addition of a resin known as "gilsonite" as an early substitute for previously used, more expensive vinyl resins was a first step toward making what became popularly known as "vinyl asbestos tile." Gilsonite was an asphalt material containing asbestos that increased the durability of the tile. In 1948, an adequate supply of vinyl resins and colorless plasticizers was developed and became available for use in the production of asbestos vinyl tile. While this product was originally produced in sheet form, it was later cut into tiles since the sheet was too brittle to be installed without breaking (Salter 1974).

Vinyl asbestos tile was extremely popular for its durability and was among the least expensive flooring products to purchase and install. As such, this tile was widely used in institutional, domestic, commercial, military, and industrial buildings (Berkeley 1968; Konrad and Kofoed 1995). It was composed of thermoplastic vinyl chloride polymers and copolymers mixed with plasticizers and stabilizers, short-fiber asbestos, crushed limestone, and mineral pigments. The tiles were variegated or marbled in appearance. Unlike other resilient floorings, vinyl asbestos tile did not typically curl up when in contact with moisture and was therefore used where damp conditions could occur (Salter 1974). The inclusion of the asbestos created a more scuff-resistant surface than that of early resilient flooring. Despite its name, asbestos was a nominal percentage of the composition. While the tiles were relatively harmless when whole, the asbestos fibers could become friable when the tiles were cut or broken. Vinyl asbestos tile must be removed following asbestos abatement guidelines. In many cases, it has been left in place and covered with other floor covering products.

Vinyl Tile and Sheet Goods

The post–World War II era saw the expanded use of synthetic materials nationwide. The thermoplastic polymer resins that became increasingly available in this period provided the ingredients to manufacture the durable resilient flooring that eventually captured this market sector by the late twentieth century. Thermoplastics soften when heated and harden (and may become brittle) when cooled. Technically, vinyl flooring could be classified as homogeneous vinyl or laminated vinyl. In the former, only the vinyl material was present throughout the tile; the latter had the same surface appearance as the former but was attached to a backing consisting of a variety of materials, including fiber-base compounds, solid vinyl sheets, resin-impregnated paper, asbestos fiber, and jute.

Homogeneous vinyl products were made in tile form, while laminated vinyl products were available as tiles or sheets. This flooring could be made in lighter and brighter colors than had been possible with other flooring. While the finish was smoother and more lustrous, the material was not as soft as rubber tile but it was softer than vinyl asbestos tile. By comparison to other floorings, vinyl had good wear resistance; good grease and oil resistance; superior resistance to many harsh chemicals; and could withstand exposure to moisture. The lustrous sheen of these products was vulnerable to scuffing and improper cleaning methods. Vinyl tile floors reacted badly to solvents, and cleaning them with products containing naphtha, turpentine, and petroleum distillates was to be avoided. Since the vinyl wearing surface on the laminated vinyl was much thinner than that of the homogeneous vinyl, the laminated version was less expensive. Both types were more expensive than other resilient floorings.

Vinyl flooring was intended to be installed on hard, smooth surfaces, such as concrete floors. Where wood floors were used, an asphalt underlayment was recommended. The vinyl was secured in place by using soft asphalt-based adhesives (Konrad and Kofoed 1995).

TYPICAL PROBLEMS

The broad range of floors and floor coverings has resulted in a variety of possible problems (see Figure 13-5). Beyond short-term losses through fire, vandalism, or natural disasters, these problems can be categorized as problems caused by moisture, soiling and abrasion, mechanical failure, and materials losses.

Moisture

As always, moisture is a common cause of floor decay. Not only can moisture cause rot, corrosion, and other forms of decay in the supporting structural system, it can also affect the flooring materials themselves. Wood flooring treated with protective coatings can withstand nominal exposure to water. However, with prolonged moisture exposure, untreated wood can absorb the moisture and swell. Conversely, wood can shrink due to the drying that occurs when moisture or humidity levels decline. In other flooring systems, moisture that penetrates into the subfloor can cause swelling and buckling of the substrate and can lift the edges and corners of resilient flooring, such as linoleum and cork tiles. When sufficient buckling has occurred, these tiles may break from being stepped on or from the stress caused by buckling.

In addition to the water used in kitchens and bathrooms, sources of moisture include deteriorated roof and exterior wall construction, leaking pipes, condensation accumulating below windows and radiators, snow and rain tracked in around entrances as people enter the space, pet urine, and poor maintenance practices. Pet urine is a source of moisture that affects flooring in a number a ways. Its salts and extremely high pH levels can mar the surface finish, and the residual smell can be an irritant to the occupants of the space. Moisture should not be allowed to remain on the flooring for prolonged periods to avoid deterioration. Along with the problems to finishes, moisture may adversely affect the adhesives holding the flooring in place. For example, the adhesives may be water-soluble and the moisture will dissolve their bonds with the flooring.

In floors directly exposed to the ground (such as basement floors and those over inadequately ventilated crawl spaces), condensation and groundwater can cause

Figure 13-5 Moisture, abrasion, mechanical failure, and surface losses are typical problems for floors and floor coverings.

problems. Floors in basements may be the eventual destination of water leaking from upper floors and the roof or they may be affected by hydrostatic pressure from fluctuating groundwater levels under the building itself. These problems are similar to the rising damp that affects load-bearing masonry construction above or below grade. In any situation, the source of the moisture should not be allowed to remain for extended periods of time or should be eliminated altogether to avoid the mold, rot, and insect problems that can contribute to the decay of the floor.

Wood floors and resilient floor coverings are susceptible to scrapes and damage caused by the users of a space as they move furniture or scuff the flooring while walking. Pets can scratch floor finishes, and the chemical reactions from their urine can discolor and degrade the finishes. Finishes can become cloudy or damaged by improper cleaning solutions or infrequent replacement of soiled water when cleaning the floor.

Soiling and Abrasion

Tile floors (especially those made of encaustic tile) can be worn away by foot traffic as well as floor cleaning machines. Areas of heavy foot traffic become evident by the soiling and wear that occurs on them. Softer tiles are prone to wear down faster. Resilient flooring has the widest range of resistance to soiling and abrasion, as noted above.

As materials age, they can become brittle. Seasonal moisture and temperature variations can cause materials to expand and contract, and if they are water permeable, they also swell and shrink. Sudden excessive moisture can produce extreme dimensional changes that can cause the flooring finish, the floor coverings, and the substrate to buckle and force connections to fail.

Mechanical Failure

As a building settles and shifts due to changes in structural equilibrium, resilient floorings and floor coverings can stretch and deform to some extent. However, when the shifting becomes too extreme, many flooring systems fail. Stone, tile, and terrazzo floors tolerate less movement and fail along joints of weak grout or mortar. The flooring materials themselves crack if the grout or mortar is significantly stronger than the floor materials or if no control joints or expansion joints are present. A control joint is an intentionally weaker joint included in the construction that allows the floor to fail in a controlled manner along the joint rather than across the material itself. An expansion joint compresses or expands as temperatures fluctuate.

Over time, even with appropriate care, whether through abrasion or mechanical failure, finished floor surfaces may wear down or individual pieces may become dislodged or broken. These losses can become pronounced when floor sanders are used for cleaning. This decay can lead to removal of the remnants of the worn-out elements or failing pieces. Left unchecked, this removal can cause material losses that detract from the overall appearance of the finished floor. This is particularly true for tiles that have become detached and are missing or have been either stored for later reinstallation or discarded.

Material Losses

Floor coverings can also suffer gradual losses from breakage, surface wear, or decay from poor maintenance. This is particularly true for materials originally secured by mastics or adhesives that have become brittle with age or have been compromised by moisture penetration. In some instances, the substrate on which the floor covering has been placed has reacted to the moisture and buckled or warped. This reaction, in turn, has caused the overlying finished floor or floor covering to shift and become uneven. The decay process is then accelerated by a combination of abrasion and mechanical failures.

Beyond the loss of flooring that can occur during a fire, the most common loss has occurred when the floor or a portion thereof was removed to access the floor cavity beneath it or in a bathroom or kitchen when the plumbing fixtures were updated. Unless specific care was taken in removing and saving the original finished flooring, the floor was more likely to have been patched (or entirely replaced). The patch may be of similar materials if they were available. More likely, the patch is made from another material that usually may only nominally, if at all, match the remaining materials. A number of these patching repairs have used a completely different material that matched neither the thickness nor the appearance of the original materials.

RECOMMENDED TREATMENTS

Identification of the source(s) of decay is an important aspect of treating the floor. When the source of any problems has been corrected and the damaged portions of the floor have been repaired and protected from additional decay, the methods for treating the problem can be evaluated. One treatment is to preserve the floor by conserving the existing materials and minimizing the effect of repairs, if any, that will occur. When the damage is extensive, such as significant water or fire damage, replacement of the damaged portions of the floor may be required if they cannot be readily repaired. When damaged materials and features must be removed, compatible infill and replacement materials should be used to replace them.

Preservation Strategies

The best process for preserving the floor is to review the operations and maintenance practices being used and determine whether they will allow the historic fabric to be preserved in place. The historic significance of a character-defining feature may be due to the decorative finish itself or the materials used to construct the floor. Harsh cleaning methods and improper use of floor cleaning equipment can irreversibly damage a decorative protective finish. If continued over a prolonged period, these actions may eventually destroy the floor itself. Check for the compatibility of the chemical cleansers, water impurities, and cleaning equipment. Modify their use to eliminate or minimize further damage.

Ultraviolet light will also cause finishes to change color or materials in the floor covering to fade or turn yellow. Humidity and temperature conditions can particularly affect the exposed surfaces of wood floors and the various floor coverings. Exposure to light should be reduced, and closer control of humidity and temperature should prolong the life of the flooring.

With wood, tile, and stone floorings, loss of surface finishes can be reversed or stabilized through the application of protective coatings or the reattachment of broken components. Prolonged exposure to moisture, ultraviolet light, and staining can cause flooring to deteriorate even faster. Resilient flooring is more problematic in that it may have begun to disintegrate and become physically unstable. If the flooring is to be preserved, these sources of decay must be removed or significantly reduced. Surfaces can be cleaned with mild cleansers and then sealed in accordance with manufacturers' guidelines to minimize further decay.

Repair Strategies

The opportunity to repair something implies that the damage is limited to small areas of the floor. Wood and stone floors can be repaired using methods discussed in previous chapters. Typically, stone flooring and pavers can be removed and reset so long as care is taken to restore the bedding substrates so that the finished floor level of the repaired area(s) matches the undisturbed portions of the floor. If wood flooring is to be repaired, care must be taken to use similar wood species and to match the finish surface treatment (e.g., paint, stain, or varnish) with the original floor materials adjoining the location(s) of the repair.

Tile floors present a specific problem: damaged or missing tiles may need to be replaced. Care must be taken to ensure that the new tiles match not only the characteristics of the surface of the original tile but the thickness of the original tile as well. If matching tiles are not available from a manufacturer, then tiles may be salvaged from less visible areas of the floor or from less public spaces that have the matching tiles. When removing the damaged tile, care must be taken not to disturb the surrounding undamaged tiles. Careful cutting of grout is required so that vibrations do not weaken the mortar bed or adjoining grout.

Stiff, formed-in-place floors without control joints or expansion joints, such as early forms of concrete, terrazzo, or magnesite flooring, often crack erratically across the entire surface of the floor. Cracking is especially problematic since it could be the result of structural issues or thermal expansion and contraction that may not be correctible without significant alteration of the floor. The presence and use of expansion joints, especially in terrazzo and magnesite, may limit damage to smaller sections of the floor. The extent and degree of cracking should determine the course of action. Smaller and shallower cracks may be left alone. As the depth and extent of the crack increase, it is necessary to remove the weakened or disturbed material adjoining the crack and fill in the void with a compatible grout or epoxy material. Patching using suitable dutchman techniques can be installed to repair small areas of damage. It is important to understand and relieve the source of the damage before repairs are made; otherwise, cracking may recur. Caution is also warranted to ensure that the physical and visual properties of the patching materials match those of the remaining floor.

For resilient flooring, the original manufacturer may still be in business and provide replacement materials, but the resulting patchwork may be less than satisfactory. Care must be taken to ensure that the patching materials match the thickness, color, pattern, and visual appearance of the original materials. Methods to remove the flooring include mechanical scraping that can be enhanced by using a heat gun (but not an open flame) or dry ice. Heat will soften thermoplastics, while dry ice will make them more brittle and easy to remove. Unfortunately, many of the adhesives, mastics, and some of the vinyl products themselves may contain asbestos, and appropriate asbestos testing and abatement practices must be followed in removing them. Like tiles, many resilient flooring products are no longer made or available. Unlike tile, it may not be possible to swap in compatible new flooring without compromising the visual appearance of the floor. The remaining resilient flooring products may show different wear characteristics or surface finish conditions due to heavy traffic or uneven exposure to ultraviolet radiation. Piecemeal repair may eventually reduce the overall visual appearance of the floor and lead to replacement of the entire floor.

Replacement Strategies

Replacement of missing features must be based on documented evidence and not simply conjecture. Replacement of extensively or irreparably damaged flooring can be based on the remaining floor. When original materials are damaged beyond repair, similar or compatible materials must be used as their replacement. Species of wood or stone must be matched. The dimensional thickness and colors of replacement tiles must be similar to those that remain.

As noted above, in some cases tiles may be obtained from a salvage company or from other, less public locations in the room when missing tiles are to be replaced. More likely, extensive damage will require complete replacement with new tiles. If so, adjustments to the mortar bedding may be needed to ensure that the final floor level matches the original.

While a number of the traditional resilient flooring products went into decline, if not outright demise, with the introduction of vinyl flooring products after World War II, the growing interest in sustainability in the late twentieth century led to the renewed availability of linoleum and cork flooring. Even though many of the later vinyl-based resilient flooring products were initially selected for their durability, some of these installations may have reached the end of their useful life. These vinyl products or compatible replacements may still be available from their original manufacturers.

References and Suggested Readings

Belle, John, John Ray Hoke, Jr., and Stephen A. Kliment, eds. 1994. *Traditional details for building restoration, renovation and rehabilitation from the 1932–1951 editions of Architectural Graphic Standards.* New York: John Wiley & Sons.

Berkeley. Bernard. 1968. *Floors: Selection and maintenance.* Chicago: Library Technical Program, American Library Association.

Calloway, Stephen, ed. 1996. *The elements of style: A practical encyclopedia of interior architectural details from 1485 to the present.* Rev. ed. New York: Simon & Schuster.

Carlisle, Alexander M. 1997. Historic linoleum: Analysis, cleaning systems, recommendations for preservation. *APT Bulletin,* 28(2/3): 37–43.

Carroll, Orville W. 1969. Linoleum used in restoration work. *Bulletin of the Association for Preservation Technology,* 1(3): 8–11.

Cummings, Abbott Lowell. 1979. *The framed houses of Massachusetts Bay 1625–1725.* Cambridge, MA: Harvard University Press.

Durbin, Lesley. 2005. *Architectural tiles: Conservation and restoration from the medieval period to the twentieth century.* Burlington, MA: Butterworth-Heinemann.

Fawcett, Jane, ed. 1998. *Historic floors: Their history and conservation.* Burlington, MA: Butterworth-Heinemann.

Fisher, Charles E. and Hugh C. Miller, eds. 1998. *Caring for your historic house: Preserving and maintaining: Structural systems, roofs, masonry, plaster, wallpapers, paint, mechanical and electrical systems, windows, woodwork, flooring, landscapes.* New York: Harry N. Abrams.

Grimmer, Anne E. 1995. Cork tile. In Jester, 1995, 228–233.

———and Kimberly A. Konrad. 1996. *Preserving historic ceramic tile floors.* Preservation Brief No. 40. Washington, DC: United States Department of the Interior.

Ierley, Merritt. 1999. *Open house: A guided tour of the American home 1637–present.* New York: Henry Holt and Company.

Jandl, H. Ward. 1998. *Rehabilitating interiors in historic buildings.* Preservation Brief No. 18. Washington, DC: United States Department of the Interior.

Jennings, Jan and Herbert Gottfried. 1988. *American vernacular interior architecture 1870–1940.* Ames: Iowa State University Press.

Jester, Thomas C., ed. 1995. *Twentieth century building materials: History and conservation.* New York: McGraw-Hill Book Company.

Johnson, Walker. 1995. Terrazzo. In Jester 1995. 234–239.

Konrad, Kimberly A. and Paul D. Kofoed. 1995. Vinyl tile. In Jester 1995, 240–245.

Labine, Clem and Carolyn Flaherty, eds. 1983. *The Old-House Journal compendium.* Woodstock, NY: Overlook Press.

Lavenberg, George N. and Sam Jaffe, eds. 1986. *Ceramic tile manual.* 2nd ed. Los Angeles: Ceramic Tile Institute.

Massey, James C. and Shirley Maxwell. 1991. Decorative tile: Art for the Victorian and Arts and Crafts home. *Old-House Journal,* XIX (March/April): 54–58.

———. 1995. The ceramic circus. *Old-House Journal,* XXIII (March/April): 46–51.

Moore, C. Eugene. 1997. *Inspiring 1950s interiors from Armstrong.* Atglen, PA: Schiffer Publishing.

Moss, Roger W. and Gail Caskey Winkler. 1986. *Victorian interior decoration: American interiors 1830–1900.* New York: Henry Holt and Company.

Park, Sharon C. 1995. Rubber tile. In Jester 1995, 222–227.

Parks, Bonnie Wehle. 1989. The history and technology of floorcloths. *APT Bulletin,* 21(3/4): 44–54.

Plaisted, Cornelia D. 1939. *Floors and floor coverings.* Library Equipment Studies Number Two. Chicago: American Library Association.

Salter, Walter L. 1974. *Floors and floor maintenance.* New York: John Wiley & Sons.

Simpson. Pamela H. 1997. Linoleum and lincrusta: The democratic coverings for floors and walls. *Perspectives in Vernacular Architecture,* 7: 281–292

———. 1999a. Comfortable, durable, and decorative: Linoleum's rise and fall from grace. *APT Bulletin,* 30(2/3): 17–24.

———. 1999b. Cheap, quick and easy: Imitative architectural materials, 1870–1930. Knoxville: University of Tennessee Press.

Snyder, Bonnie Wehle Parks. 1995. Linoleum. In Jester 1995, 214–221.

Stecich, Jack and Jerry G. Stockbridge. 1987. Turn-of-the-century floor construction. *APT Bulletin,* 19(3): 7–9.

Von Rosenstiel, Helene. 1978. *American rugs and carpets from the seventeenth century to modern times.* New York: William Morrow and Company.

——— and Gail Caskey Winkler. 1988. *Floor coverings for historic buildings.* Washington, DC: Preservation Press.

CHAPTER 14 Walls and Ceilings

Walls and ceilings are significant character-defining features of the spaces they enclose. Beyond the methods used to construct walls and ceilings, an important issue is how their surfaces are decoratively finished. This chapter introduces a number of wall and ceiling coverings and surface treatments. Further details on wood, plaster, and decorative and protective finishes are presented in later chapters. In the early North American colonies, wall and ceiling construction practices of Europe were adapted to local conditions. Although the early colonists were aware of the decorative finishes and ornamental details found in Europe, few were wealthy enough to immediately duplicate them.

By the early nineteenth century, prosperity, political upheaval, emerging trade networks, and the Industrial Revolution brought many opportunities to apply decorative treatments. Not only were decorative materials, such as paper and tile, being imported but they were also being made domestically. Changing technologies enabled expensive materials to be replicated or imitated more economically. Increased mechanization throughout the nineteenth century introduced new types of wallboard, decorative papers, embossed coverings, and premixed paints that eventually simplified construction or replaced many expensive materials that had been unavailable to the mass market.

Advances in the first half of the twentieth century introduced new products and processes that took advantage of emerging plastics industries. These advances included materials that were more affordable than the most recent products of the previous half-century. Many products in this era were sold to both commercial and

271

residential markets, but it was the residential market in the post–World War II era of housing mass production and do-it-yourself homeowners that embraced plastic laminates, vinyl coverings, and drywall systems.

WALL AND CEILING CONSTRUCTION

Based on the construction system used, the separation of interior spaces varied from the thickest constructions created by massive load-bearing masonry systems to the thinnest construction composed of a single thickness of plank. Once the structure was in place, the next step was to form the surfaces of the walls and ceilings. The intended use of the space and the wealth of the owner dictated the form of the walls and ceilings. In humbler buildings or utilitarian buildings, such as barns or warehouses, the structural system and the interior face of the exterior cladding were often left exposed. As the intended building use and the wealth of the occupant rose, so did the expectation that the walls and ceilings would consist of a continuous material (e.g., stone, wood, plaster) that formed the wall surface.

Protective coverings and decorative finishes were then applied. In some instances, due either to economic limitations, religious beliefs, or lack of suitable decorative treatments, the interior surface was originally left unfinished or only modestly decorated and later improved by the application of new finish treatments. As a result, many interior surfaces have multiple layers of finishes that provide a catalog of the building's chronology as a reflection of changing aesthetic tastes and building uses (see Figure 14-1).

In many American interiors, the surface of a wall was divided into three parts— dado (also known as "wainscot"), field, and cornice (Shivers 1990). The dado is the

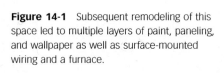

Figure 14-1 Subsequent remodeling of this space led to multiple layers of paint, paneling, and wallpaper as well as surface-mounted wiring and a furnace.

lowest 24- to 36-inch portion of the wall, which most likely had more contact with the occupants' activities. The field was the area above the dado, which had less physical contact with the occupants. This area was where decorative objects and finishes were most prominent. As economic conditions allowed, a chair rail was commonly installed to separate the dado and the field. The cornice was at the top of the field, where the wall and the ceiling intersected.

Various exposed structural materials and components served as the final interior surface, which was then decorated with paint or other coatings and left in view within the occupied space. Various surviving buildings (e.g., colonial-period timber-framed houses and churches, nineteenth-century stone or brick mills and factories, and early-twentieth-century revival and Arts and Crafts houses) "express" their structure in this manner. Additionally, the work of architects like Frank Lloyd Wright, the Greene brothers, and other early advocates of natural or organic architecture was noteworthy for the visual expression of structural forms that served as the final surface.

SURFACES

Stone, such as marble or granite, has been installed as a finished wall or ceiling featuring flat or carved relief surfaces. Following European traditions, stone was used in buildings of the wealthy and powerful. The use of stone became notable in the late nineteenth and early twentieth centuries as many early industrialists made their fortunes in banking, railroads, steel, automobile manufacturing, and other emerging industries. They commissioned architects to traverse Europe seeking examples of the finest craftsmanship in a variety of construction materials to enhance their designs. The buildings subsequently designed and built were a legacy to remind future generations of their fortunes. The use and treatment of stone in the interiors of these buildings is similar to that of structural stone described previously.

Stone

Although wood was used to cover interior walls as prosperity eventually allowed, a number of the original wall surfaces created in the first permanent but modest homes of the colonists were the exposed wattle and daub infill or other crude plastering systems. Early wood surfaces consisted of hand-sawn boards or plank walls used to separate rooms. In their most primitive form, wood boards were pegged or later nailed to the structural framing. On walls that had been previously plastered, baseboards (mopboards), chair rails, and other trim pieces were installed to protect plaster surfaces. Ceilings were frequently left with structural members exposed until plaster could be installed. Exposed timbers were embellished with carving, beaded or chamfered edges, and, in some instances, decorative paint.

Wood

As economic conditions improved, wood paneling became a status symbol. At first, the face of the fireplace wall was enclosed. Then, in varying stages, the dado was filled with wainscot and the field was covered with hand-crafted molding. By the mid-eighteenth century, in finer buildings, the eventual outcome was the complete enclosure of walls and some ceilings with hand-cut, stick-built wood paneling systems. The political events of the late eighteenth century brought about a decline in the popularity of this full-height paneling in favor of the more ornamentally detailed Neo-Classical and Federal period styles as a symbolic American architecture developed. Wood carvings were used as ornamental coverings, such as fireplace mantles, in the finest buildings. In the late nineteenth century, paneling saw a revival, as noted above when, emerging industrialists sought instant affirmation of their wealth and status by copying or acquiring entire rooms of paneling from cash-strapped European estates and manor houses.

In the meantime, the homes and businesses of less wealthy citizens were decorated less ornately using materials that imitated those found in more expensive buildings. Until the late nineteenth century, when transportation networks created the means to transport finer decorative materials throughout the country, wooden ornamentation was restricted to materials available from local mills, including simple baseboards and trim.

Plaster

Plaster was installed as a finished wall and ceiling treatment or as substrate for the outer layers of wall and ceiling coverings from the mid-seventeenth century well into the twentieth century, when it was eventually replaced by a variety of wallboard systems. Plaster had numerous decorative qualities, including surface texture and three-dimensional ornaments that emulated the carved stone and wood used in more expensive buildings.

Wallboard

Wallboard had numerous uses in construction throughout the nineteenth and twentieth centuries. Along with the various forms of wallboard (described in Chapter 9) used as exterior cladding and interior wall surfaces, perhaps the most well known is gypsum board, introduced by Augustine Sackett in 1894. The earliest forms of gypsum board were made from multiple layers of felt and gypsum plaster, but early-twentieth-century improvements led to the modern form, which is composed of one layer of gypsum plaster faced on both sides with paper.

Gypsum board was originally developed as a drywall alternative to wet plastering methods. However, due to its fire resistance, gypsum board was more often used initially as a plasterboard replacement for wood and metal lath. By the end of World War I, gypsum board began displacing both wood and metal lath. Manufacturers throughout the twentieth century experimented with enhancements to make gypsum

Figure 14-2 Drywall systems had replaced wet plaster systems by the mid-twentieth century. Drywall could be nailed or screwed to framing. The joints and fastening locations were then covered with joint compound that was sanded smooth after it dried. The finish treatment (e.g., paint, wallpaper) was then applied. This process took only a fraction of the time required by wet plaster systems.

board lighter, more decorative, and more fire-resistant. This led to its use without a plaster overlay during World War II, and subsequently it became the material of choice in the housing boom of the 1950s (Konrad and Tomlan 1995).

Drywall (as gypsum board is more commonly called now) has been used anywhere plaster and lath had been used traditionally. Drywall was secured to framing with nails or drywall screws. The joints were covered with flexible mesh or tape and sealed with joint compound (see Figure 14-2). The surface was then finished with a decorative paint or covering. Unfortunately, drywall could be readily punctured or broken by excessive force. Small punctures and cracks could be filled with joint compound or spackling. Repairing larger holes typically involved trimming the damaged drywall to a rectangular opening and fitting in a drywall patch of similar thickness. The joints were then taped and sealed with drywall compound. Drywall patches were one method used for repairing broken plaster as well.

DECORATIVE COVERINGS AND TREATMENTS

The decorative coverings and treatments used on the walls and ceilings of buildings include a wide variety of materials and finishes. The earliest, simplest, and crudest coverings emerged as humans attempted to enhance their comfort by using animal hides and woven fabrics to protect themselves from the cold. As civilizations became more sophisticated, the adaptation and use of wall coverings moved beyond simple comfort functions to include an increasingly high level of decorative art. Primitive coverings evolved into woven tapestries and tooled leather. As decorative traditions expanded with the rise of domestic and foreign trade, expensive and highly decorated coverings became synonymous with wealth and power. As wealth and power were amassed, expectations for the treatment of walls and ceilings changed as well. For those who could afford them, carved marble surfaces symbolized the highest level of sophistication. Throughout the Renaissance and subsequent design periods in Europe and North America, wall and ceiling materials became more ornate until the modernist movement of the early twentieth century rejected ornamentation.

The more common early surface treatments, such as stenciling and homespun fabrics in the North American colonies, were primitive by European standards. Even so, some houses did have a few pieces of ornamental or decorative coverings that had been brought from Europe. However, by the mid-eighteenth century, prosperity had increased the level of sophistication and expectations that led to the wider use of ornamental coverings and decorative treatments.

With the Industrial Revolution of the nineteenth century and the technological advances made throughout that century, various coverings, such as Lincrusta-Walton and Anaglypta, emerged that sought to imitate the visual qualities of more expensive treatments but make them more commonly available. As a result, less expensive methods to provide decorative coverings were introduced or filtered down to public and private buildings of the middle and working classes. This process continued as decorative coverings and treatments evolved and as completely new materials, such as vinyl wall coverings and plastic laminates, came into use by the mid-twentieth century.

Ceramic Tile

Tile was used on walls and ceilings and was subject to the same decay and remediation processes as floor tile. Tiles made specifically for use on walls and ceilings may have been thinner and lighter since they did not need to withstand the impact and wear of foot traffic. These tiles were often more decoratively painted and visually prominent due to three-dimensional surfaces on some ornamental tiles that

Figure 14-3 Decorative tile is an important character-defining feature that was especially popular in the Arts and Crafts and subsequent revival styles of the late nineteenth and early twentieth centuries.

complemented the flat surfaces of the field tile (see Figure 14-3). Tiles added dramatic flair to the visual quality of public and private spaces. In the late nineteenth century, the "hygienic movement" brought about the extensive use of glazed tile in bathrooms, kitchens, and other areas (e.g., subway stations, hospitals, laboratories) where sanitary conditions were required. Often these tiles were rather utilitarian; decorative qualities were the result of using combinations of colors, although white tiles were most common. Following the bathroom fixture manufacturers' lead in the early twentieth century, vibrant pinks, greens, yellows, and blues were added to product lines.

Textiles and Fabric

The use of textiles as wall coverings has extended from the Middle Ages to modern times. In Europe, woven fabrics hung to reduce drafts were subsequently enhanced and refined to form the tapestries serving that purpose in castles, churches, and the homes of the wealthiest citizens from the Renaissance period of the fourteenth to seventeenth centuries. Fabric coverings, such as velvets, taffetas, brocades, and damasks made from silk and other fibers, were used in the more finely appointed spaces and living quarters. These uses of printed linens and cottons and painted taffetas were the forebearers of decorative paper coverings (McClelland 1926). Textiles were secured along their perimeter or mounted in frames attached to the wall or ceiling.

While textiles were vulnerable to the many decay mechanisms affecting other wall and ceiling coverings, they were acutely sensitive to soiling, abrasion, insect, and fire damage. Cleaning such soiled fine fabric is problematic since the moisture used to clean it frequently carries impurities that further stain the fabric. Comprised essentially of organic or plant-based fibers, these fine fabric coverings were also prone to damage from mold, mildew, and insects when adequate moisture was present to sustain them. Similarly, the delicate finish of the cloth could be scuffed or abraded when something brushed across it. Because of the fragility of these fabrics, any work in preserving, conserving, cleaning, or repairing them should be done under the supervision of a textile conservator.

Decorative paper for walls was first made in the late fifteenth century in Germany, **Decorative Paper**
the Netherlands, and northern Italy. The earliest paper consisted of individual hand-
made sheets that were hand-painted and glued directly on the wall. Decorative paper was also applied to ceilings to simulate three-dimensional plaster molding and
other ornamental motifs.

Wallpaper was sold in Boston at the start of the eighteenth century and, by the
last quarter of that century, was available to the rich and the middle class. During
that period, wallpaper was imported from England. After the Revolutionary War,
wallpaper imports were regulated to allow American manufacturers to establish
themselves. Nevertheless, French wallpapers dominated the market for the better
part of the nineteenth century. In the final decades of that century, English wallpapers regained their dominance in the American market. Over time, displaying decorated wall and ceiling papers in prominent rooms was a sign of good taste.

Throughout the nineteenth century, manufacturers continuously changed their
production methods to meet the demands of the public. One of the last great periods of wallpaper fashion was the Arts and Crafts period of the late nineteenth century. In the early twentieth century, American architects and other advocates of good
taste began rejecting the use of wallpaper. With the advent of modernism in the
1920s, ornamental and decorative elements that recalled the past were summarily
dismissed as displays of unsophisticated taste. Without the incentive to consider wallpaper as a sophisticated art form, the mid-twentieth century embraced the use of
wallpapers that appealed to the cultural tastes and more modest budgets of the middle class. As a result, wallpaper was primarily used in working-class and modest middle-class homes for the rest of the century.

Wallpaper manufacturing has used several methods of production. The earliest
was hand-laid or handmade paper that was then hand-painted. Besides inks and pigments, other materials applied to the surface included flocking (shaved wool and
silk fibers), powdered mica, and metallic colors. In the early eighteenth century,
stenciling was applied using printed black ink outlines that were filled with thinned
watercolor paints. Block printing with distemper paints was introduced in the mid-
eighteenth century. Distemper paints were a form of whitewash that contained pigment, tinting colors, and chalk (calcium carbonate) mixed with water and glue size
or casein. Machine printing and production on rolls of paper came into wide use
by the mid-nineteenth century and continues to be used today. In 1851, the first
moisture-resistant and therefore washable "sanitary" paper was introduced. Sanitary
paper gained widespread acceptance by the 1890s. Cloth-based "ingrain" papers
made from cotton and wool rags were patented in 1877 and used into the 1920s.
Strippable lined wallpapers were introduced in 1904. Silk screening came into use
in the 1940s, and the first vinyl wallpaper was introduced in 1947 (Historic New
England 2007; Hoskins 1994; Lencek and Bosker 2004). Wall and ceiling papers
were hung by applying a wheat-based paste to the back of the paper and pressing it
to the surface. In some cases, a cloth or paper lining sheet was applied first to provide a smooth surface.

In its earliest uses in the homes of the wealthy, wallpaper was seen as an art form.
The imagery and patterns reflected what was considered the highest intellectual sophistication. As decorative arts changed, so did the way wallpaper was used. Various
application systems included dividing the wall into three sections—dado, filler, and
frieze (i.e., essentially nominal variations of the divisions mentioned earlier).
Sometimes, borders and paper pilasters divided the wall horizontally and outlined

door and window openings. Wallpaper patterns varied as aesthetic tastes and architectural styles changed. Single colors (intended to be accented with ornamental borders), stripes, and geometric or abstract shapes, a broad range of landscapes, pastoral scenes, ancient ruins, historic and national emblems, highly stylized flowers (e.g., the "cabbage rose"), and other figural and abstract representations of contemporary popular culture have all enjoyed varying cycles of popularity. For instance, when American fascination with ancient Greek culture emerged in the early nineteenth century, patterns that emphasized ancient Greek buildings and art were popular. Certain historic patterns have been so popular that a number of them are now available as reproductions from contemporary manufacturers.

Embossed and Preformed Materials

Early attempts to imitate plaster ornamentation on ceilings and walls led to the use of papier-mâché, which was composed of cellulose (e.g., recycled paper) and glue pressed into molds and sealed in linseed oil. Papier-mâché architectural ornaments first appeared in the seventeenth century and were manufactured in England, France, and Germany. Papier-mâché was used "extensively in the eighteenth century" (Simpson 1999, 121). Papier-mâché ornaments were imported to the United States in the eighteenth century. A prominent example of this ornamentation is the Miles Brewton House built in 1760 in Charleston, South Carolina.

In 1877, Frederick Walton invented the first embossed wall and ceiling covering of the Victorian era in the United States. Having previously invented linoleum, Walton had been seeking a comparable product for use on ceilings and walls that could imitate the three-dimensional details of plaster, tooled leather, or wood but at considerably less expense. He named the product Lincrusta-Walton. The manufacturing process was similar to that of linoleum, except that instead of ground cork dust, Lincrusta-Walton used wood pulp and paraffin wax as filler to create sheets with the three-dimensional forms and patterns. Intended for walls and ceilings, Lincrusta-Walton was thinner and more flexible than linoleum. In 1887, the use of a waterproof paper replaced the original canvas-based construction to make Lincrusta-Walton more flexible. Lincrusta-Walton was mounted to walls and ceilings using a glue-paste mixture applied to the back of the material. This covering was popular for its ease of hanging, durability, and washability. Lincrusta-Walton eventually was used in the houses and buildings of the middle and upper classes to imitate carved plaster or wood, embossed leather, metal, ceramic tile, and oak dado. The surface could be decoratively painted to simulate wood, marble, or other expensive materials.

A number of domestic and foreign products introduced later in the nineteenth century competed with Lincrusta-Walton in the embossed wall and ceiling coverings market. These included Anaglypta, Tynecastle Tapestry, Japanese paper, Cameoid, Subercorium, Calcorin, Corticine, Lignomur, and Salamander. The most successful was Anaglypta, which was patented by Thomas J. Palmer in 1886. Unlike Lincrusta-Walton, Anaglypta was a paper-based covering that was pressed between two rollers to impart a hollow relief pattern to the material. The hollowness of the paper made Anaglypta lighter than Lincrusta-Walton. Anaglypta could be decoratively finished with glazes, paint, and gilding in the same fashion as Lincrusta-Walton.

Due to the varied manufacturing processes, these products present unique challenges in their preservation and repair. For instance, paper-based materials, such as Anaglypta, are vulnerable to the same decay mechanisms as wood, while

Lincrusta-Walton has characteristics similar to linoleum. Therefore, any treatments for embossed and preformed wall and ceiling coverings should be performed under the direction of an experienced conservator.

Metal

Metal ceilings first appeared in the 1870s as corrugated metal panels used for structural fire protection. Henry Adler patented the first sheet metal ceiling in 1874, and Albert Northrup patented the first commercially viable metal ceiling system in 1884 (Simpson 1999). Commonly referred to as "tin" ceilings, these systems were actually made from stamped or rolled sheets of steel. They became widely popular in the early twentieth century as a replacement for decorative plaster (see Figure 14-4). This system was also adapted for use on walls.

Metal ceilings can be identified by the uniform, regular repetition of the decorative elements on them. In some instances, smaller panels were assembled into combinations that formed a larger pattern. Each section was separately attached to a furring strip or nailed to the building framing. Some metal ceilings were flat panels that were originally installed with a decoratively painted surface applied by the manufacturer that remains exposed to view to this day. Unfortunately, most original metal ceilings were later painted and now exist under multiple coats of paint and built-up layers of dirt and grime.

Decay and remediation strategies are similar to those for metals described earlier in this book. For metal ceilings and wall panels in particular, the two most common problems are corrosion due to moisture and damage caused by later remodeling. These ceilings and panels are vulnerable to punctures created when modern lighting, communication, and fire safety systems are mounted. Many metal ceilings were damaged when fasteners used to suspend lowered ceilings were installed or when larger spaces were divided into smaller ones. Careful assessment must be made to determine whether sufficient replacement sections are available. If replacement sections cannot be relocated from less public portions of the building or obtained

Figure 14-4 Metal ceilings were sold in a variety of patterns and provided an alternative to decorative plaster. Many metal ceilings were routinely abandoned in place when ceilings were lowered in the late twentieth century.

from a salvage company, custom-cast, fiber-reinforced plastic replacements can be made from a mold from a sample panel.

Acoustic Treatments

In 1895, Wallace Clement Sabine, a professor at Harvard University, identified reverberation (echo) as a primary cause of acoustical problems at the Fogg Museum at Harvard University. His research formed the basis for modern acoustical theory of enclosed spaces and led to his 1911 collaboration with the noted architect Raphael Guastavino. Together, they developed the Rumford tile, which became known as the first acoustical tile. Sabine and Guastavino also later developed a second tile known as Akoustolith (Pounds, Raichel, and Weaver 1999; Weber 1995). These tiles were used initially as part of a proprietary system. Tiles composed of clay and organic compounds were fired in a kiln that burned off the organic materials to create a porous tile that could be applied as a finish surface. Sound waves entered the holes in the surface and were trapped in the tile.

The 1920s saw numerous materials introduced to control reverberation in spaces. These materials included felt and membrane systems, various board and panel systems, and acoustical plaster. Gypsum plaster, mineral wool, wood fiber, portland cement, cork, flax, sugar cane, and perforated metal were used by various manufacturers. Many of these products also contained asbestos to enhance their strength and fireproofing characteristics. Early acoustical materials were limited to natural colors, but processes were developed that allowed the use of certain water-based paints and tints in their manufacture. However, simply painting the coarse surface of the acoustical tile reduced or eliminated its sound-absorbing qualities.

Eventually, felt and membrane systems were discontinued, and after World War II, acoustical plaster or tile became the preferred applications. Originally developed in the 1920s, acoustical plaster that included organic fibers to acoustically soften the surface and enhance sound absorption was applied by hand. After World War II, pneumatic application processes became available.

Acoustical tile is perhaps the most commonly visible acoustic treatment in buildings today. Derived from the various board and panel systems of the early 1920s, these products were typically glued to the ceiling or wall. In 1929, the Philadelphia Savings Fund Society building featured the first suspended acoustical ceiling. The suspension system was concealed using splines to attach the cast gypsum panels to the suspension system. Use of concealed suspension systems continues to this day. The now familiar exposed aluminum t-suspension system became widely popular in the 1950s and remains so today (Weber 1995).

Acoustical tiles and panels were vulnerable to the decay mechanisms that affected their constituent materials. They could be damaged by rough incidental contact or rough handling. In addition, they were prone to soiling from atmospheric pollutants or staining from water contact due to leaking pipes or building envelope failure. While typically not cleaned, they were sometimes painted to cover the soiling. When this was done, the fissures and holes that trapped sound may have become filled with paint and therefore were less sound absorbent. Only products that used standardized perforations could be painted so long as the perforations remained open. If the material became waterlogged, the tiles or panels could detach from the substrate or cause its suspension system to collapse.

Plastic Laminates

Plastic laminates have been used as decorative finishes for walls, doors, countertops, and casework. The first laminates (Walker, Konrad, and Stull 1995) were patented

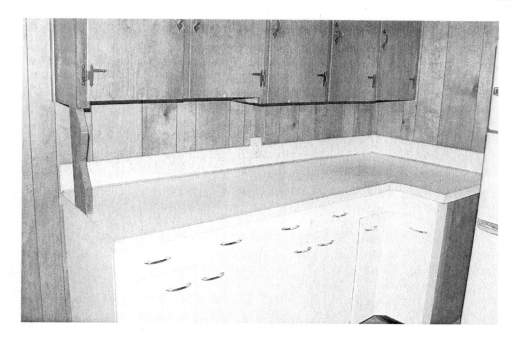

Figure 14-5 Plastic laminates became commonplace since they provided colorful and easily maintained surfaces for countertops (as shown here), walls, doors, and furniture.

by Leo Baekland in 1909 and consisted of fibrous sheets saturated with a phenol-formaldehyde resin. In 1913, Daniel O'Connor and Herbert Faber, while developing a substitute electrical insulator for mica, invented a plastic laminate sheet that became known as Formica. Originally, the dark color of the resins limited the material for use to black or dark brown. However, with the application of an opaque laminate covering made by Formica in 1927, lighter colors and lithographed wood grains could be used as the exposed surface layer. Throughout the next three decades, a variety of resin, polymer, or polyester-based laminates were introduced and gained wide acceptance by the late 1930s. By the 1950s, plastic laminates were being used in movie theaters, stores, diners, and kitchens where a colorful, decorative, durable, and easily maintained surface was desired (see Figure 14-5).

Plastic laminates could be installed using casein or resin glues or bonding cements; in some exterior locations, they were screwed in place. Despite their seemingly durable nature, plastic laminates did deteriorate. They did not react well to extremes of temperature and humidity and were vulnerable to staining from food and food preparation. Heat and moisture caused plastic laminates to change in dimension, blister, delaminate, and crack. Plastic laminates became brittle when cooled or dried out. Moisture penetration along cut edges was especially troublesome. Sunlight and ultraviolet light caused surfaces to yellow. The adhesive bond between the laminate and the substrate could fail due to improper installation methods. Although plastic laminates were usually resistant to chemicals, they have been known to react with hydrochloric acids, lye (sodium hydroxide), sulfuric acid, hydrogen peroxide, and acetone.

To extend the life of plastic laminates, care should be taken to limit and control exposure to light and extreme temperature and humidity fluctuations. The surface should be cleaned periodically with a nonabrasive, nonionic detergent and dried immediately. Although repair products for scratches and small surface defects are available, large infill and patching repairs are only nominally acceptable. When deterioration is more pronounced, replacement of an entire panel may be necessary. However, panels with matching patterns or colors may no longer be made or the original remaining panels may have changed color.

TYPICAL PROBLEMS

Due to the specialized and often fragile nature of the decorative materials used on or as the outermost layer of the wall, investigation of their condition generally requires consultation with a conservator who specializes in the specific material. Such consultation is particularly important when one or more layers of the decorative finish have historic significance that may be lost as the result of a specific planned treatment. For example, a wall covering from a later period may have to be removed to restore a room to an earlier period or when investigations of the subsurface portions of the wall warrant removal of the overlying layers to verify conditions (see Figure 14-6).

As with flooring described in the previous chapter, the broad range of coverings and surface treatments, as well as the underlying construction of the wall or ceiling itself, have produced a variety of possible problems. Beyond short-term losses through fire, vandalism, or natural disasters and longer-term losses from periodic updating, these problems include damage caused by moisture, soiling and abrasion, mechanical failure, and materials losses.

Moisture

Beyond the decay that it causes in the supporting structural system, moisture is a common cause of decay in walls and ceilings. Sources of moisture include deteriorated

Figure 14-6 Removal of overlying layers of paper and paint revealed this pattern. This is one full repeat of the pattern (notice how the blossom and leaf pattern begun at the top of the image starts to repeat along the bottom of the image).

roof and exterior wall construction, leaking pipes, condensation, leaking windows, and poor maintenance practices.

Materials that have been treated with protective coatings may withstand nominal exposure to moisture when properly cleaned, but when the exposure becomes prolonged or occurs where the water can penetrate into the material, the moisture can initiate decay mechanisms corresponding to the material (e.g., wood, stone, masonry, or metal) affected. Water may also discolor and stain the material with oxidized minerals that were captured as the moisture migrated through the assembly from the original point of entry. Additionally, various glues and binders react to the moisture and lose their strength, such as wheat-based glues used on wallpaper.

Soiling and Abrasion

Walls and sometimes ceilings are susceptible to scrapes and damage caused by incidental contact. The use of picture rails in the Victorian era was an attempt to eliminate damage caused by hanging pictures on the wall. Pets can scratch and chew on finishes within their reach. Finishes can become cloudy or damaged by improper cleaning solutions or infrequent replacement of soiled water when cleaning the surfaces.

Smoke and soot from fireplaces and candles will accumulate on any exposed surfaces, including walls and ceilings. Soiling also occurs as smoke and vapors from cooking appliances come into contact with the wall and ceiling. Tobacco smoke will deposit a brownish nicotine residue that will noticeably accumulate over time.

Material Failure

Seasonal variations in temperature can cause materials to expand and contract; if they are water permeable, they will also swell and shrink. In dry conditions, many wall and ceiling materials can become brittle. In humid conditions, they or their adhesives may lose strength.

As a building settles and shifts due to changes in structural equilibrium or failure of structural support components, wall and ceiling materials may withstand the forces exerted to some extent. However, when these forces exceed the strength of the substrate and structure, walls and ceilings begin to fall apart. This mechanical failure may at first be subtle, especially when concealed by a protective or ornamental covering, such as Lincrusta-Walton, paneling, or multiple layers of paper. This failure can remain undetected until bulges in the finish material are noticed or the surface cracks or breaks apart. As the building settles, doors and windows may also begin to stick while being opened or closed.

Material Losses

Walls and ceilings can deteriorate gradually from breakage, surface wear, or decay due to poor maintenance. This decay is particularly true for materials and the adhesives that originally secured them that have become brittle with age or compromised by moisture. In some instances, surface moisture acts directly on the finish; in others, the substrate upon which the finish or covering has been placed reacts to moisture within the wall and gradually begins to fail over the long term. This failure causes the overlying ornamental finish or covering to shift and become uneven, accelerating material losses through a combination of abrasion and mechanical failures. Sudden extensive loss can also occur as a by-product of water damage from a burst pipe or water used to fight a fire elsewhere in the building.

Insensitive remodeling and accidental or intentional damage by occupants can cause disruptions in the surface and finishes. Removing or installing wall- and ceiling-mounted fixtures, such as attaching fixtures or removing a portion of the wall or ceiling to install wiring, may damage the wall or ceiling. Damage can also occur when coverings are removed to update a space to reflect current aesthetic tastes.

Over time, even with appropriate care, whether through abrasion or mechanical failure, finish surfaces may wear down, break apart, or become dislodged. This can lead to removal of the remnants of the worn-out elements or failing pieces. Left unchecked, this decay can result in material losses that detract from the overall appearance of the building.

RECOMMENDED TREATMENTS

The treatments described below are an overview for the materials described in this chapter. Further information is given in other chapters for tile, millwork, plaster, and decorative coatings. Once the problem source has been identified and the damaged portions have been protected from additional decay, the methods for treating the problem can be evaluated. One treatment consists of preserving the materials by conserving the existing portions and minimizing the adverse effect of any contemplated repairs. When damage is extensive, replacement of the damaged portions may be required.

When the removal of damaged material becomes necessary, compatible infill and replacement materials should be used. Since asbestos was a common component in many wall and ceiling treatments and their adhesives, testing to verify its presence should be completed before disturbing the materials. When the project requires the disturbance or removal of multiple layers of wall and ceiling finishes (e.g., paint or paper), each layer should be documented (e.g., photographs or measured drawings) to retain a record of the historic chronological evidence of the changes of the finishes over time. This process extends the original sampling done at the start of a project to define the construction chronology.

Preservation Strategies

Preservation of a wall or ceiling decorative finish may be best accomplished by limiting its exposure to ultraviolet light and reviewing the maintenance practices being used. Changes in hygrothermal conditions (i.e., the combination of temperature and humidity) can particularly affect unprotected surfaces. Limiting exposure to light and closely controlling humidity and temperature should prolong the life of the wall and ceiling surface materials.

Before electric lighting was used, light fixtures produced soot from an open flame. Accordingly, decorative papers and other moisture-permeable wall coverings were cleaned periodically using an assortment of putty-type cleaners and erasers. The moderately adhesive nature and nominally abrasive action of the putty pulled the soiling from the surface. Any loose residue remaining on the surface after cleaning was then lightly brushed off. For moisture-resistant surfaces, mild soap and water was used or the surface was simply covered with a new finish treatment (e.g., new paint or paper).

To preserve the surfaces of decorative ornamental features, aggressive cleaning or harsh chemicals must be avoided. Careful verification of modern cleaning products and their compatibility with the surfaces being cleaned is necessary to ensure long-term retention of the historic fabric. As the historic significance and fragility of the surface being cleaned increases, conservators specializing in the materials in question should be consulted. Restoration work on any historic decorative covering or finish should be done only by a specialized conservator.

Repair Strategies

Damage in limited areas can often be repaired. The repair materials must be physically and visually compatible with the existing materials adjoining the repair. Since many early installations may not conform to the standardized thicknesses of modern materials, repair materials and installation procedures may need to be adapted to produce an appropriate finished repair. Dimensional thicknesses of wall and ceiling coverings can vary by manufacturer or type of product. Over time, manufacturers developed ways to increase production volumes or reduce costs by changing either the thicknesses of their products or the way they were installed. For example, a gypsum board drywall patch can be installed to repair a hole in a three-coat plaster wall, and then the surface can be built up to the desired thickness with joint compound. Regardless of the repair method, the finished surface should match the profile of the existing adjoining surfaces.

Small individual infill repairs may often be undetectable when properly done. However, when the repairs become too numerous, they eventually reduce the overall visual appearance of the space. The decision to make a multitude of small repairs that may compromise the integrity of the wall or ceiling in question should be weighed against the historic significance of that wall or ceiling. Failing wall and ceiling materials may be temporarily stabilized by securing protective coverings (e.g., mesh or rigid plastic sheets) in place. These temporary protections should be secured in a way that minimizes surface damage but allows air flow and prevents moisture accumulation.

When retaining portions of the wall or ceiling is dangerous and limiting access to the space is not possible in the long term, the entire wall or ceiling covering, its substrate, or both may need to be replaced in order to continue using the space safely.

Replacement Strategies

Replacement of missing features must be based on documented evidence and not simply conjecture. Replacement of extensively damaged or irreparable wall and ceiling features can be based on the remaining existing features. When original materials are damaged beyond repair, similar or compatible materials must be used as their replacement. As noted above, in some cases appropriate replacement materials (e.g., wallpaper) may be obtained by having them reproduced by a custom fabricator.

References and Suggested Readings

Belle, John, John Ray Hoke, Jr., and Stephen A. Kliment, eds. 1994. *Traditional details for building restoration, renovation and rehabilitation from the 1932–1951 editions of Architectural Graphic Standards.* New York: John Wiley & Sons.

Bradbury, Bruce. 1984. A laymen's guide to historic wallpaper reproduction (an overview of historic and modern production techniques, some jargon unraveled, and some tips on dating historic patterns). *Bulletin of the Association for Preservation Technology,* 16(1): 57–58.

Calloway, Stephen, ed. 1996. *The elements of style: A practical encyclopedia of interior architectural details from 1485 to the present.* Rev. ed. New York: Simon & Schuster.

Chase, Sara B. 1992. *Painting historic interiors.* Preservation Brief No. 28. Washington, DC: United States Department of the Interior.

Craft, Meg Loew and M. Nicole Miller. 2000. Controlling daylight in historic structures: A focus on interior methods. *APT Bulletin,* 31(1): 53–59.

Cummings, Abbott Lowell. 1979. *The framed houses of Massachusetts Bay 1625–1725.* Cambridge, MA: Harvard University Press.

Dierick, Mary. 1975. Metal ceilings in the U.S. *Bulletin of the Association for Preservation Technology*, 7(2): 83–98.

Durbin, Lesley. 2005. *Architectural tiles: Conservation and restoration from the medieval period to the twentieth century.* Burlington, MA: Butterworth-Heinemann.

Fisher, Charles E. and Hugh C. Miller, eds. 1998. *Caring for your historic house: Preserving and maintaining: Structural systems, roofs, masonry, plaster, wallpapers, paint, mechanical and electrical systems, windows, woodwork, flooring, landscapes.* New York: Harry N. Abrams.

Frangiamore, Catherine Lynn. 1977. *Wallpapers in historic preservation.* Washington, DC: United States Department of the Interior.

Gorman, J. R., Sam Jaffe, Walter F. Pruter, and James J. Rose. 1988. *Plaster and drywall systems manual.* 3rd ed. New York: McGraw-Hill Book Company.

Gould, Carol. 1997. Masonite: Versatile modern material for baths, basements, bus stations, and beyond. *APT Bulletin*, 28(2/3): 64–70.

———, Kimberly A. Konrad, Kathleen Catalano Milley, and Rebecca Gallagher. 1995. Fiberboard. In Jester 1995, 120–125.

Historic New England. 2007. Wallpaper in New England

Horie, Velson, ed. 1999. *The conservation of decorative arts.* London: Archetype Publications and the United Kingdom Institute for Conservation and Historic and Artistic Works.

Hoskins, Lesley. 1994. *The papered wall: History pattern technique.* New York: Henry N. Abrams.

Hull, Brent. 2003. *Historic millwork: A guide to restoring and re-creating doors, windows, and moldings of the late-nineteenth through mid-twentieth centuries.* New York: John Wiley & Sons.

Jandl, H. Ward. 1998. *Rehabilitating interiors in historic buildings.* Preservation Brief No.18. Washington, DC: United States Department of the Interior.

Jennings, Jan and Herbert Gottfried. 1988. *American vernacular interior architecture 1870–1940.* Ames: Iowa State University Press.

Jester, Thomas C., ed. 1995. *Twentieth century building materials: History and conservation.* New York: McGraw-Hill Book Company.

Katzenbach, Lois and William Katzenbach. 1951. *The practical book of American wallpaper.* Philadelphia: J. B. Lippincott Company.

Konrad, Kimberly A. and Michael A. Tomlan. 1995. Gypsum board. In Jester 1995, 268–271.

Lavenberg, George N. and Sam Jaffe, eds. 1986. *Ceramic tile manual.* 2nd ed. Los Angeles: Ceramic Tile Institute.

Lencek, Lena and Gideon Bosker. 2004. *Off the wall: Wonderful wall coverings of the twentieth century.* San Francisco: Chronicle Books.

Lynn, Catherine. 1980. *Wallpaper in America: From the seventeenth century to World War I.* New York: W. W. Norton & Company.

McLelland, Nancy. 1926. *The practical book of decorative wall-treatments.* Philadelphia: J. B. Lippincott Company.

Merritt, Frederick S., ed. 1958. *Building construction handbook.* New York: McGraw-Hill Book Company.

Milley, Kathleen Catalano. 1997. Homasote: The "greatest advance in 300 years of building construction." *APT Bulletin*, 28(2/3): 58–63.

Moss, Roger W. and Gail Caskey Winkler. 1986. *Victorian interior decoration: American interiors 1830–1900.* New York: Henry Holt and Company.

Nylander, Richard C. 1992. *Wall papers for historic buildings.* 2nd ed. Washington, DC: Preservation Press.

Phillips, Morgan W. and Andrew L. Ladygo. 1980. A method for reproducing Lincrusta papers by hand. *Bulletin of the Association for Preservation Technology*, 12(2): 64–79.

Poore, Patricia, ed. 1992. *The Old House Journal guide to restoration.* New York: Dutton.

Pounds, Richard, Daniel Raichel, and Martin Weaver. 1999. The unseen world of Guastavino acoustical tile construction: History, development, production. *APT Bulletin*, 30(4): 33–39.

Shivers, Natalie. 1990. *Walls and moldings: How to care for old and historic wood and plaster*. Washington, DC: Preservation Press.

Simmons, H. Leslie. 1990. *Repairing and extending finishes (Part 1)*. New York: Van Nostrand Reinhold.

Simpson, Pamela H. 1995. Quick, cheap and easy, part II: Pressed metal ceilings, 1880–1930. *Perspectives in Vernacular Architecture*, 5: 152–163.

———. 1999. *Quick, cheap and easy: Imitative architectural materials, 1870–1930*. Knoxville: University of Tennessee Press.

Van den Branden, F. and Thomas Hartsell. 1984. *Plastering skills*. Homewood, IL: American Technical Publishers.

Walker, Anthony J. T., Kimberly A. Konrad, and Nicole L. Stull. 1995. Decorative plastic laminates. In Jester 1995, 126–131.

Weaver, Shelby. 1997. Beaver board and Upson board: History and conservation of early wallboard. *APT Bulletin*, 28(2/3): 71–78.

Weber, Anne. 1995. Acoustical materials. In Jester 1995, 262–267.

CHAPTER 15 Art and
Stained Glass

OVERVIEW

While windows play an important role in the visual appearance of historic buildings, art and stained glass have a special niche in the aesthetic expression of a building. Early usage of art and stained glass in buildings conveyed a sense of power and wealth, since only the most powerful or richest building owners could afford to use them. As time progressed, the popularity of art and stained glass rose and fell in response to political and economic upheavals. Along the way, high aesthetic taste and cultural sophistication became associated with the use of art and stained glass while, ironically, manufacturing processes made them available to a larger market than ever before. In the development of this medium, many specific fabrication and conservation processes have evolved to meet the demands of the owners of a variety of building types, from religious and political institutions to commercial buildings to individual homes. This chapter will discuss the use of art and stained glass and how to maintain and conserve it in the built environment.

Historic Background

With a rich history that mirrors the changing political, economic, and aesthetic climates of the past two millennia, art and stained glass has seen several periods of widespread use followed by sharp declines. This type of glass provided a medium for the expression of religious beliefs, political authority, and aesthetic tastes. Starting out as a simple mosaic, art and stained glass has been raised to an extremely high art form, and has been decried and despised by conflicting schools of artistic thought.

Rise and Fall

Art or stained glass in windows can be traced back to Roman and earlier times. While the window of today is a product of the past three centuries, earlier windows included colored glass set into lead cames to form panels that filled window openings. Augsberg Cathedral in Germany has the oldest stained glass windows in continuous use, documented to 1065 AD (Davison 2003). Much information about early stained glass comes from the writings of Theophilus, a Germanic monk who described stained glass making in *De Diversis Artibus* (Theophilus 1979, 72):

> [T]o assemble simple windows, first mark out the dimensions of their length and breadth on a wooden board, then draw scroll work . . . and select colors that are to be put in. Cut the glass and fit the pieces together with the grozing iron. Enclose them with lead cames . . . and solder on both sides.

Over time, glassmaking evolved into such a prized art that some countries excused glassmakers from military duty or forbade them to leave the country under penalty of death. Glassmaking centers developed in Italy, England, France, Germany, and Belgium.

Stained glass was often transformed by additives that imparted to glass vibrant hues of red or blue. Since additives were expensive and the trade secrets of the glassmakers were closely held, only powerful monarchs or religious institutions commissioned the earliest stained glass windows. Vitreous enamel paints made from ground colored glass fired onto a piece of glass were developed and used as well. One method of decoration, known as "grisaille," used varying densities of a color to form objects, textures, and shadow patterns on an otherwise clear piece of glass.

The popularity of stained glass in Europe revolved around building technology and political conflict. The development of the pointed arch and the flying buttress caused a shift in architectural styles from the Romanesque buildings that flourished at the end of the first millennium to the flamboyant Gothic churches of the Middle Ages and the Renaissance. The arch and the flying buttress allowed walls to open up and display wealth and power through the use of immense stained glass windows. Building heights soared, and the large openings were filled with a multitude of decorative glass designs.

The twelfth through sixteenth centuries were the original golden age of stained glass. Beginning with the Gothic style, this period saw tremendous use of stained glass (Brisac 1986; Reyntiens 1990). This age abruptly ended due to political upheavals of the sixteenth and seventeenth centuries in which, for example, church reformation led to the decree that outlawed the use of stained glass to display representations of religious icons and events in all English churches. Many stained glass windows in England were then destroyed or replaced with clear glass, grisaille glass, or heraldic symbols. Similar fates befell stained glass across Europe as political and religious dictates forbade the use of stained glass. The demand for stained glass windows waned significantly.

The lack of stained glass in early North American colonies can be traced not only to the low status of stained glass in Europe at the time and the declining number of skilled stained glass artisans but also to the poverty of the colonists. Political turmoil in Europe, coupled with emerging preferences of the Renaissance period, severely reduced the demand for stained glass artisans working in the earlier medieval traditions. Throughout Europe, the decline continued to the point where many earlier skills were considered lost. As the glass historian H. Weber Wilson

(Wilson 1986, 7) notes: "Significantly, this also meant that there was basically no transfer of decorative glass traditions to the architectural heritage in the New World."

Revival

By the nineteenth century, political and architectural trends had again changed. With the European introduction of the Gothic Revival, followed by the Aesthetic, Art Nouveau, and Arts and Crafts movements, came a reinvigorated stained glass industry worldwide. Eugene Emmanuel Viollet-le-Duc, Augustus Pugin, Charles Winston, Charles Eastlake, and William Morris created much debate as these styles were translated into stained glass (Sloan 1993; Wilson 1986). They extensively employed stained glass and wrote numerous books on the subject. Their work had a pronounced impact on the stylistic tastes of Europe and the United States. The first Gothic Revival period introduced stained glass to the United States. Richard Upjohn's Trinity Church (1846) in Manhattan and Minard Lefever's Holy Trinity Church (1848) in Brooklyn are recognized as containing some of the earliest major works of stained glass in the United States (Sloan 1993) and denote the growing use of stained glass. The styles that emerged throughout the nineteenth century, promoted by architects and architectural pattern books, fostered the use of stained glass in nearly every architectural venue.

By the late nineteenth century, America had become a leader in decorative glass through the methods used by John La Farge and Louis Comfort Tiffany. Each had patented fabrication methods focused on plating techniques using several separate layers of opalescent glass in one came to create colors and patterns that were unavailable in one sheet of glass. Opalescent glass is a translucent milky glass of one or more colors, often in a swirled or streaked pattern (see Figure 15-1). Opalescent glass was a uniquely American product and by the twentieth century dominated the American stained glass market (Raguin 2002).

At this time, Frank Lloyd Wright promoted the use of geometric forms in an early type of abstraction that reflected the regulating lines of his Prairie Style architecture of the early twentieth century. Meanwhile, the second Gothic Revival period emerged and renewed the use of Gothic designs in ecclesiastical buildings. This style was championed by such noted artisans and studios as Charles Connick, Wilbur Herbert Burnham, Joseph, Francis & Rohnstock, Willett Stained Glass Studios, and Emil Frei.

Artistic Shifts

Shortages during World War I reduced the availability of lead for cames and the creation of new stained glass windows. In the 1920s, artistic thought led to the abstract, modernist, and avant-garde movements in France, the Netherlands, and Germany, and, later, albeit more slowly, in the United States. The Art Deco, de Stijl, and Bauhaus movements caused a radical shift toward abstract expressionism. World War II put a hold on new window installations, but the postwar period saw new explorations in stained glass. A significant development in this midcentury period was the resurrection of the medieval method known as "dalle de verre" or faceted glass. Unlike the medieval practice of setting glass in clay or stone, the twentieth-century fabrication of windows included the use of colored slab glass configured in a mosaic of abstract or simplified figural forms and set in mortar or epoxy resins. Some notable examples in this period were the First Presbyterian Church in Stamford, Connecticut (see Figure 15-2), designed by Gabriel Loire (Raguin 2002)

Figure 15-1 The introduction of opalescent glass in the late nineteenth century provided new opportunities for visual expression, as demonstrated by the outside of this house in Galveston, Texas.

Figure 15-2 Dalle de verre, literally meaning "slab of glass," re-emerged in use just prior to World War II and became popular in the United States in the 1950s. Unlike stained glass windows with lead or zinc cames, dalle de verre glass used concrete or epoxies to hold the slabs of glass in place.

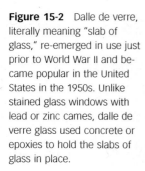

and the United States Air Force Academy Cadet Chapel in Colorado Springs, Colorado, designed by Walter Netsch, Jr.

A stained glass window consists of glazing held together by cames or, more recently, by mortars and epoxies that form a panel inserted into a frame of stone, wood, or metal. The colors, figural elements, heraldic symbols, geometric patterns, and abstract forms that make up the images found on a stained glass window have varied through time based on architectural style, political dictates, and the use of the building for which the window was intended.

Components and Their Fabrication

Cartoons

All stained glass windows started with a "cartoon" or template. The artisan could develop a window by creating an original cartoon using one from a plan book or by tracing an existing window. A marking pen or paintbrush was used to outline all the pieces as they would appear on the finished panel. The lines and voids showed where glazing, cames, mortar, or epoxy were located. One copy of the cartoon was secured or traced onto the work table. A second copy was then cut apart using pattern scissors to create templates for each opening. The templates were then traced onto the assorted pieces of glass intended for the finished window, and the glass was cut accordingly.

For the method using cames, the outlines formed the pattern for the cames themselves. Once the cartoon was in place, two edge cames at one corner of the panel were affixed to the work surface using farrier's nails secured to the work surface on the side opposite the one where the glass would be inserted. The stained glass was cut to match the outline of the cartoon and inserted into the corner. Next, the interior came(s) were secured along the remaining exposed edges of glass. This process was repeated and the window was then built outward from this starting point, much like assembling a jigsaw puzzle. As the work approached the corners, the remaining edge cames were secured along the remaining perimeters of the panel. For the dalle de verre method, a cartoon may have been used. Other methods included laying out a "full-scale gouache or powder colour painting" (Reyntiens 1967, 155) on the work table.

Glazing

In each opening formed by the cartoon outlines, the intended glazing may include a single piece of colored glass; multiple pieces of similarly shaped glass to build up colors and depth; an enameled glass insert with figural qualities, such as a head, face, hands, or feet; or a piece of dalle de verre glass.

In various parts of the world, the methods of stained glass production have survived continuously from ancient times, were lost and rediscovered, or were newly developed in the industrial era. While source materials varied over time, glass today is a mixture of three basic ingredients: silica, soda (an alkali), and lime (an alkaline earth). Silica, obtained from sand, forms the vitreous network of glass and requires very high heat to melt, so a network modifier is introduced in the form of an alkali to allow glass to be created at a lower temperature and make it more workable. Two alkalis used for glassmaking have been potash and soda. Potash was used extensively, but by the mid-nineteenth century it was known that soda formed more stable glass. A network stabilizer, usually lime, was added to rebuild the network and produce a stable glass (Raguin 2003).

The term "stained glass" originally referred to the earliest pieces of colored glass that were made by adding metal to a silica-potash-lime recipe to create colored glass. This early pot-metal glass was used in small pieces in a mosaic technique. Various metallic oxides were used to color the glass. In glassmaking, success in generating color was the result of maintaining either a reducing or an oxidizing atmosphere in the furnace. An atmosphere that does not admit oxygen is a reducing atmosphere, while one that does admit oxygen is an oxidizing atmosphere (Freun 2002). For example, adding copper oxide to clear glass produces bluish green glass in an oxidizing atmosphere and red glass in a reducing atmosphere (Davison 2003).

Glass sheets could be made from crown and cylinder glass or produced by slabs of rolled, embossed, or cast glass (also known as "cathedral glass"). The fuel source for melting glass ranged from wood to coal to natural gas with an accompanying shift in fabrication methods to account for the specific uses of each fuel. Flashed glass, in which a layer of clear glass and a thin layer of colored glass were fused together to form colored glass, could be carved or etched to allow varying degrees of color gradients. The most common color was ruby red. The overall result was that colored glasses could be attained using flashed glass at a fraction of the cost of using a comparable thickness of colored glass while allowing greater artistic freedom. Another development in stained glass was the use of vitreous paints on the glass. These fired-glass paints were grisailles, stains, or enamels that were fused to the glass using a kiln. Grisailles (black, dark brown, or dark red vitreous paints) were used to create shadows and contours by stippling or crosshatching the paint onto the figural objects depicted in the glass (see Figure 15-3). In the fourteenth century, a silver stain was developed using silver oxide that turned yellow in the kiln. This stain acted like a translucent paint that produced gold details on clear glass. A sanguine or Jean Cousin stain that created red hues was made from gold from the Renaissance through the nineteenth century. Enamels were translucent paints made from crushed glass. These methods permitted better depiction of faces, hands, and feet. After the

Figure 15-3 Stains and dark enamel paints were commonly used for shadows, lettering, and other details. This window from the Chapel of the Transfiguration in the Grand Tetons National Park shows the effects created by this process.

fifteenth century, many recipes and methods were lost until the industrial era, which saw the reuse or new introduction of pot-metal glasses and the invention of opalescent glass in the late nineteenth century and the resurrection of dalle de verre in the mid-twentieth century. Cold paint is sometimes found in American stained glass and is simply paint applied to glass without firing in a kiln.

A number of ornamental glass forms can be found in stained glass windows, including rondels, jewels, and beveled inserts. Rondels were round pieces of glass created by spinning or machine pressing. Jewels were relatively small pieces of glass that were cast, spun, pressed, carved, sculpted, or hammered to form a specific geometric shape or pattern. Beveled inserts were flat pieces of glass formed in geometric and irregular shapes that had a bevel along each edge. These elements were used to create borders and patterns that added visual interest to the stained glass panel.

Cutting

Until the diamond glass cutter was invented in the sixteenth century (Sloan 1993), a heated iron rod was used to make a rough cut of the piece of glass from its larger slab. When the heated iron rod was placed on a "wet" line (created by wetting the glass with water, wine, saliva, or other liquids), the thermal stress created caused the glass to break away in thin sheets from the slab along the line of the rod. A grozing tool was then used to chip away the remaining glass to achieve the desired shape. True medieval glasses display this chipped edge. In the sixteenth century, diamond glass cutters made this method obsolete since diamond cutting allowed more delicately defined shapes to be cut more easily. The diamond scored a straight or curved line along the surface of the glass, creating a weakness in the glass. The glass was then tapped lightly beneath this line, causing the glass to break along it. While grozing practices continued after the introduction of the diamond cutter, the diamond cutter remains in more common use today, sometimes supplemented by an electric-powered grinding wheel. The thicker dalle de verre glazing requires the use of an anvil and hammer to break the glass slabs along the scribing lines. The cut sections of glass could also have been knapped (chipped at the edges) to produce a faceted appearance.

Cames

Stained glass windows consisted of numerous pieces of glass set into a mosaic format to create figural shapes. Each piece of glazing was cut and placed on the work surface in its corresponding location on the cartoon and then secured in place by a piece of came shaped to conform to the exposed edge of the glass. A came was the channel made of lead and cast into an H or U shape that was used to hold the glass (see Figure 15-4). When all pieces had been inserted and trimmed with cames, the joints created by the cames were soldered to hold the assembly together. Putty was then rubbed onto both faces of the panel and forced into the joint between the glass and the cames to make them weatherproof. The panel was then ready for insertion into its window frame.

Zinc, copper, and brass cames also have been used. Zinc cames, introduced in 1893, were stronger and had an expansion coefficient that was more compatible with glass than lead cames. Brass and copper cames were used only for a brief period between 1890 and 1920 (Vogel and Achilles 1993). An alternative to cames was the copper foil method used by Louis Comfort Tiffany, in which a strip of copper foil was used to separate the pieces of glass and then soldered along its length to form the final connection.

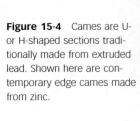

Figure 15-4 Cames are U- or H-shaped sections traditionally made from extruded lead. Shown here are contemporary edge cames made from zinc.

In the dalle de verre window of the mid-twentieth century, there were no cames. The mortar and, subsequently, epoxy poured between the glass pieces were allowed to cure, thus providing structural integrity to help support the glass in place. These windows were supplemented with armatures and reinforcing materials cast within the mortar or epoxy.

Supports and Frames

Once fully assembled, the glazing panel was inserted into a frame that was then fit into the window opening. Framing materials included wood, stone, architectural metals, cast stone, and terra-cotta. Zinc cames were stronger than lead cames, but both needed reinforcement to prevent sagging of the panel. Thus, a variety of flat, round, and diamond-shaped bars were used in various eras to allow the panel to stand vertically without slumping out of the frame. These bars were attached by wires to the window or soldered directly to the window. The structural support for earlier windows was enhanced by using bars and rods mounted at various intervals across the panel and secured to the cames and the frame. Round bars were used from the Middle Ages to 1920, and flat bars were introduced in 1890 (Vogel and Achilles 1993).

Dalle de verre windows were constructed in sections and secured in place with integral structural supports within the window frames. These windows, using concrete mortar as the filler between the glazed pieces, were fitted with internal armatures buried within the mortar. Similarly, metal or fiberglass reinforcements may have been embedded in other sections of the casting locations. The mortar or epoxy was then poured between the design elements.

TYPICAL PROBLEMS

Glass is a fragile material, and a number of factors can be introduced during its manufacture, installation, and maintenance that contribute to its success or failure in the built environment. These factors include the composition of the glass itself, the way it was installed and subsequently supported in window openings, and the methods used, both appropriate and inappropriate, to maintain or repair it.

Glass failure can be due either to its internal composition or to external factors that act upon it. In the former case, little can be done to correct the problems caused by poor manufacturing techniques beyond either accepting the problem or replacing the glass. On the other hand, the external factors may be correctible without necessarily altering or replacing the glass.

Glass Failure

Composition and Clarity

In early houses in North America, the use of potash and inappropriate amounts of lime caused glass to be unstable and lose clarity. The advent of silica-soda-lime formulas in the United States in the mid-nineteenth century largely eliminated this problem (Sloan 1993). Still, dirt accumulation makes glass less translucent, and may react with moisture in the atmosphere to form crusts or pitting and permit biological growths to occur. Crizzling (a network of cracks in the glass) can form on the glass due to the use of improper formulas to make the glass. Devitrification (internal formation of crystal) occurs when the glass has been exposed to high heat. In a process known as "solarization," manganese in the glass will interact with sunlight to create a purple hue in the glass (Davison 2003).

In some instances, paints appear to have faded, but it is more likely that the paint is exfoliating (flaking) from the surface. Fading can occur due to improper paint manufacture or application, inadequate firing techniques, weathering, or corrosive effects of material in contact with the paint (Sloan 1993). Likewise, modern cleaning products are sometimes too harsh and can remove the paint that has fused with dirt on the glass surface.

Abrasion and Breakage

The primary causes of decay in glass are abrasion and breakage (see Figure 15-5). Abrasion occurs through friction or the use of harsh cleaning chemicals. Breakage can be caused by striking the glazing; stresses due to failure of support systems; and poor glass fabrication. In the first case, the window is hit directly. In the second, thermal expansion of the cames or shifts in the frame cause stress that breaks the glass. And finally, the glass itself may contain impurities or may have been improperly annealed (heated and slowly cooled) in the lehr (cooling oven). Methods used to create opalescent glass make that glass vulnerable to crizzling and devitrification

Figure 15-5 Vandalism has significantly damaged this stained glass window at the Soldiers and Sailors Monument in Cleveland, Ohio.

that can cause the glass to become opaque, form an iridescent shimmer, or exfoliate from the window. There are no methods to solve these problems, and finding matching replacement glass is difficult (Sloan 1993; Vogel and Achilles 1993).

Structural Failure

Besides the glass, the structural support system can also fail. Traditional stained glass windows using cames have demonstrated many of the following failure mechanisms. The use of metal cames allows continuous expansion and contraction of the metal daily and seasonally. When combined with dirt or other accretions, the stress created can fracture the glass and cause the material to succumb to metal fatigue. This stress may cause soldered joints to fail, admit moisture, and accelerate overall decay. This decay can be noted by the presence of open, fractured cames and solder joints as well as moisture penetration and corrosion products (salts) on the came and the glass.

Along with cames, stained glass windows were supported by horizontal saddle bars connected by twisted copper wires soldered to the glazing panel or attached directly to the glazing panel. Corrosion or galvanic action can cause these connections to fail. The window will then sag and eventually slump out of its opening.

Early dalle de verre windows of the twentieth century suffered from failure of the concrete used to hold the glass dalles in place. Later experimentation led to the use of integrated reinforcement with epoxies and mortars that reduced this problem.

Other structural failure can occur in the window frame itself. Materials prone to rot or corrosion (e.g., wood and many metals) can lose structural integrity as they fail and introduce stresses into the window assembly. In some instances, the building structure may fail due to settlement or seismic activity.

Improper Replacements and Modifications

In Europe, many nineteenth-century restorations that followed the philosophy of Viollet-le-Duc, who felt at liberty to improve a restoration based on conjecture as to what the original owner or builder really wanted, caused the improper replacement of original glass with then contemporary interpretations of something considered more appropriate. While this widespread practice did not impact American windows as significantly at the time (most likely because the American stained glass traditions were not as old), subsequent restorations of the twentieth century did have an impact. Whether through insensitive methods or inexperienced personnel, a number of attempts to clean, repair, or restore art and stained glass windows resulted in loss of historic materials. As preservation practices matured in the late twentieth century, the available information, guidelines, and number of restoration artisans have grown and this problem has declined. However, this possible problem remains if there is an insufficient budget or lack of historic appreciation to do the work properly.

A second issue has been alterations due to security, vandalism, and energy conservation concerns (see Figure 15-6). Common approaches to solving these problems have been to enclose the window with a sheet of polycarbonate or acrylic plastic or other glazing, install storm windows on the exterior, or install metal grills. When this is done, the visual integrity of the window is often compromised since these solutions do not necessarily conform to the original lines of the existing window. On the interior of the building, the entering light is altered by the possible discoloration of the plastic or the shadow lines of the new windows or grills. In any event, additional problems posed by these solutions include concerns about the allowance of venting to relieve heat and moist air trapped between the installation and the original window and access to the exterior of the original window for

Figure 15-6 Attempts to reduce energy costs in the 1970s led to a number of poor treatments. In this case, a contemporary storm window has visually compromised the appearance of the stained glass window of this church in Mt. Carroll, Illinois.

ongoing maintenance. In addition, the installation of these modifications could damage the existing historic materials around the original windows.

A number of common cleaning products and practices may lead to accelerated deterioration. The appropriate cleaning, maintenance, and repair of historic art and stained glass windows often requires glass conservators and skilled artisans. The decay mechanisms are complex and often repaired in a piecemeal manner. A systematic assessment of the conditions and factors contributing to the decay is required before attempting significant repairs or replacement of deteriorated historic art and stained glass windows.

RECOMMENDED TREATMENTS

Overview

Cleaning windows can cause damage if done using harsh or abrasive cleansers. Built-up dirt reduces light transmission and conceals actual conditions. Care should be taken when determining the extent of cleaning and what processes are used. Removing dirt may expose otherwise protected metal or glazing to further decay. Abrasive or acidic cleansers should never be used. For unpainted glazing, cleaning media depend on the significance of the glass. Soft water should be used initially and then supplemented with nonionic detergents. Some glazing may need solvents, alcohol, or mineral spirits to remove yellowed shellac, lacquer, varnish, or stubborn grime. Enamel painted glass must not be cleaned before the stability of the paint is known.

Before doing any repair, the existing conditions should be documented and assessed. Hire a stained glass consultant who can do a condition survey, list the extent of needed repairs, suggest a plan for repairing or restoring the windows, and do the actual repairs. The initial documentation is necessary to ascertain the level of repairs needed and then to assess the appropriateness of the completed repairs.

Economic considerations may make repairing damage in situ seem attractive, but this strategy can accelerate deterioration. Extensive damage may warrant removing the windows and completing repairs in a studio where the window is disassembled and reconstructed incorporating the repairs. Photograph window before removal. Then, in the studio, make a full-sized rubbing of the window by placing a continuous sheet of acid-free vellum over the entire window and rubbing it with a wax rubbing stone. The rubbing is used to assign reference numbers to each opening and the glazing within it. Each piece of glass or glass fragment is given a reference number that is used to track the work. Then the original documentation is used to reassemble the window.

Strategies for Small Repairs

The decision to repair in situ must not be made lightly. Small repairs can be done to stabilize conditions, but long-term retention of the windows must be considered. Small repairs should be performed by an expert.

Stabilizing Loose Glazing

Loose glazing may occur from damaged cames or the loss of material holding it in place. The former problem can be remedied by applying blackened window putty to the gaps between the glazing and cames. For the latter, a single came may be repaired by using repair leads, replacing the damaged flange, or replacing the came using the copper foil method. However, if numerous cames are damaged, then consideration must be given to releading the entire window.

Cracked Glass

For cracked glass, repair may be made most readily using the copper foil technique. Modern epoxies have also been used successfully; the edges of the crack are secured together using the epoxy instead of soldered copper foil. These epoxies have the advantage of saving the visual integrity of the window. A disadvantage is that some epoxies are sensitive to ultraviolet discoloration and may require a shielding panel on the entire exterior of the window (Brown and Strobl 2002).

Broken or Missing Glass

Glazing that has broken free from the window can be inserted in the opening by peeling back the came and reassembling the pieces using either the copper foil or epoxy repair method. If the piece is lost, then the missing piece may be replaced in kind if possible.

Damaged Cames

Cames become brittle with age or metal fatigue and can be further damaged if the window is bowing. Small damaged sections that have opened can be gently pressed closed and resoldered as needed. Came flanges can be replaced or resoldered using a variation of the copper foil method. Any bowing should be assessed to determine if the window can be safely repaired in situ, as the window may be further damaged by simply pushing the bowing back in place.

Saddle Bars

Bowed windows may be caused by a loose or disconnected saddle bar. Due to corrosion or metal fatigue, the bars may have come out of their sockets in the window frames or the twisted wire or solder connection may have broken. As with damaged cames and bowing windows, assess the condition of the saddle bar and the connection. Confirm whether this is a single incident or the result of a larger, more extensive problem.

When the number of small individual repairs becomes excessive, large portions of the window assemblage may need to be removed so that repairs can be made more safely and efficiently in a studio. Removal, repair, and reinstallation should be done under the supervision of a stained glass conservator.

Releading

Much like mortar in brickwork, lead cames were intended to be a sacrificial material that was expendable for the sake of retaining the glazing. If there is damage to the cames over an extensive portion of the window, then the best course of action may be to replace them with comparably sized and profiled cames. In essence, the window is disassembled and the rubbing and other documentation are used as a guide for its later reassembly. Attention should be paid to the profile and dimension of the came flanges and core since even a slight difference can change the overall panel size and necessitate alteration of the framed opening. Selecting a similar profile is also important to match the original appearance of the window.

Paint Stabilization

There is no method for slowing deterioration or reattaching paint short of removing the window and placing it in a stable environment. Attempting to infill or replace the missing paint on the glass is not recommended since this method is largely irreversible. Unless specific documentation is available showing the missing detail, there may be no way to legitimately restore the missing paint. In certain instances, it may be possible to find physical evidence of shadow lines of the original paint on the glass, but modern paints do not typically match the expansion characteristics of the original paint and can further damage the original paint. Missing details can be re-created on a separate piece of clear glass plated into the same opening as the deteriorated piece. The paint can either be fired glass paint or cold paint.

Consolidation of fragile paint may be desired, but this presents a difficult challenge specific to the situation. Because of concerns about a variety of physical properties, consolidation is best left to a stained glass conservator (Sloan 1993).

Energy Conservation

Energy conservation in the 1970s resulted in alterations to stained glass windows that significantly affected their visual integrity. The most common one was the addition of storm windows or plastic panels. As a rule, these products should not be added since the drawbacks outweigh the benefits (Vogel and Achilles 1993). Aside from causing loss of visual integrity, many of these alterations did not include moisture and heat relief in the space created by the enclosing panel or subsequent maintenance practices allowed the vents to become plugged. In the former case, the trapped heat and moisture accelerated deterioration of the glazing, the cames, and the frames. To correct this problem, the modification should be altered to enable heat and moisture relief. In the latter case, cleaning plugged vents may be all that is needed. In either case, these steps do not mitigate the loss of visual integrity (Sloan 1993).

Two alternatives exist. First, a storm panel custom fit to the window opening may be mounted to the interior side of the window, with air gaps to relieve heat and moisture. Alternatively, using a more invasive procedure devised in Europe, protective glazing is installed in the original location of the stained glass panel and the stained glass panel is moved to the interior side, with venting for moisture and heat relief (Brown and Strobl 2002). These methods significantly alter the appearance of the window.

References and Suggested Readings

Brisac, Catherine. 1986. *A thousand years of stained glass*. Garden City, NY: Doubleday and Company.

Brown, Sarah and Sebastian Strobl. 2002. *A fragile inheritance: The care of stained glass and historic glazing*. London: Church House.

Calloway, Stephen, gen. ed. 1996. *The elements of style: A practical encyclopedia of interior architectural details from 1485 to the present*. Rev. ed. New York: Simon & Schuster.

Caviness, Madeleine H. 1996. *Stained glass windows*. No. 76 in Typologie des Sources du Moyen Age Occidental. Turnhout, Belgium: Brepols.

Davison, Sandra. 2003. *Conservation and restoration of glass*. 2nd ed. Burlington, MA: Butterworth-Heineman.

Fisher, Charles. E., Deborah Slaton, and Rebecca A. Shiffer. 1997. *Window rehabilitation guide for historic buildings*. Washington, DC: Historic Preservation Education Foundation.

Freun, Lois. 2002. Ancient glass. (accompanies *The Real World of Chemistry*, 6th. ed. Dubuque, IA: Kendall/Hunt Publishing). www.realscience.breckschool.org/upper/fruen/files/Enrichmentarticles/start.html.

Isenberg, Anita and Seymour Isenberg. 1998. *How to work in stained glass*. 3rd ed. Iola, WI: Krause Publications.

Jackson, Albert and David Day. 1992. *The complete home restoration manual: An authoritative, do-it-yourself guide to restoring and maintaining the older house*. London: Simon & Schuster.

Lee, Lawrence, George Seddon, and Francis Stephens. 1976. *Stained glass*. London: Mitchell Beazley.

Raguin, Virginia C. 2002. *Reflections on glass: Twentieth century stained glass in American art and architecture*. New York: American Bible Society.

────── 2003. *Stained glass: From its origin to the present*. New York: Harry N. Abrams.

Rambusch, Viggo B. A.1981. Preservation and restoration of leaded glass windows. *Bulletin of the Association for Preservation Technology*,13(3): 11–17.

Reyntiens, Patrick. 1967. *The techniques of stained glass*. New York: Watson-Guptill Publications.

────── 1990. *The beauty of stained glass*. Boston: Bulfinch Press.

Sloan, Julie. L. 1993. *Conservation of stained glass in America*. Wilmington, DE: Art in Architecture Press.

Smith, Baird. 1978. *Conserving energy in historic buildings*. Preservation Brief No. 3. Washington, DC: United States Department of the Interior.

Theophilus. 1979. *On divers arts: The foremost medieval treatise on painting, glassmaking and metalwork*. Translated by John G. Hawthorne and Cyril Stanley Smith. New York: Dover Publications.

Vogel, Neal and Rolf Achilles. 1993. *The preservation and repair of historic stained and leaded glass*. Preservation Brief No. 33. Washington, DC: United States Department of the Interior.

Wilson, H. Weber. 1986. *Great glass in American architecture: Decorative windows and doors before 1920*. New York: E. P. Dutton.

CHAPTER 16 Wood Carving
and Millwork

In the earliest periods of each colony, survival was the driving force and construction activities focused on providing shelter and security for the colonists. Whatever ornament existed was mostly an indirect result of being an integral part of the woodwork that had come directly from the hands of the master builders. Surfaces were plain and often severe by modern standards. Ornamental woodwork was not a priority. As the seventeenth century proceeded, when prosperity did occur and security was assured, ornamental objects became more common features in building construction. As time, technology, and economics allowed, decorative embellishments became part of the craft of creating interior surfaces. With long-term prosperity, carvers (another occupation-based surname) became available who could create more elaborately shaped objects and decorative pieces to adorn otherwise plain buildings.

Architectural details varied with contemporary opinions based on the political, aesthetic, intellectual, and economic disposition of the owner. Using the latest construction methods and materials was often viewed as the means of demonstrating these qualities. The improvement of cutting-edge technologies in the eighteenth century enabled a more sophisticated set of woodworking tools to be created. As a result, the Federal style, popular at the turn of the nineteenth century, was noted for more finely cut handcrafted wood details (see Figure 16-1).

The early stages of the Industrial Revolution radically transformed manufacturing systems to increase production capacity, including the nineteenth-century introduction of steam-powered saws. While a dispersed population of woodworkers still

303

Figure 16-1 The Federal architectural style featured the pinnacle of hand-carved ornamental woodwork that can be seen here in the McIntyre Historic District in Salem, Massachusetts.

existed, the profitability of concentrating workers in a central location to fabricate such things as doors and windows was recognized. This became especially true as transportation networks expanded.

As a result, woodworking became a nominal component in a larger production process to maximize production output and eliminate inefficiencies. The former separate craft operations were organized into a network of sequenced activities that fed the mill raw materials and enabled the greater production of goods that could be sold less expensively than the products of smaller independent operations. Skilled workers using a combination of power and hand tools could produce significantly more board feet of lumber that was milled into trim and moldings or subsequently crafted into separate window and door units. While not quite the assembly line later devised by Henry Ford in building cars in the early twentieth century, this process established the mill as a powerful economic presence.

Independent shop operations that could not compete with the lower-priced products coming from mills failed, and their owners often eventually went to work for the mills. The construction of wood buildings and their components was then completed by rough and finish carpenters, and the subtleties of the early crafts were lost. The products formerly made by the independent trades were then collectively called "millwork."

By the mid-nineteenth century, carvers were crafting the models for casting molds used to create objects of cast iron, plaster, and composition (consisting of chalk, glue, linseed oil, and resin) details to reproduce ornamental details originally crafted in wood. Creating the molds used for these operations required carvers who could make wood models that were used to make molds. Thus, carvers too became part of the process of creating millwork.

The nineteenth century produced many architectural styles that all relied heavily on wood construction (see Figure 16-2). Among these were Gothic Revival and its variant Carpenter Gothic, Italianate, Second Empire, Queen Anne, Stick, Eastlake, and Shingle style. All brought an increasingly higher demand for low-cost wood products. By the late nineteenth century, an array of catalogs, displaying a multitude of building products from mills, were readily available. The railroad and the postal service made it possible to ship and receive an assortment of goods from across the country.

In this period, ornament became larger but in many cases less finely detailed. Mechanization could speed up production but was constrained in producing three-dimensional objects with finely crafted details. Carvers created individual components, such as mantles, fireplace surrounds, brackets, and decorative infill pieces, for larger pieces of assembled millwork.

The Arts and Crafts movement of the late nineteenth century became popular in part due to the recognition that mechanization had eliminated much of the human craft in building ornament. This movement promoted the idea of individual craftspeople producing more finely crafted details by hand. Despite this movement,

Figure 16-2 Millwork became widely used in the nineteenth century for exterior decoration. Shown here is the William Mason House (1845) in Thompson, Connecticut, as constructed from a design from A. J. Downing's *Cottage Residences*, published in 1842.

milling operations continued in the same direction taken since the early Industrial Revolution. Cutting and shaping profile moldings and ornamental details were based on cost reductions. Intricately carved details were still available from artisans but were more expensive than the standard details produced at a mill.

In the late twentieth century, advances in computerized cutting technologies were developed that theoretically allowed more intricate detailing to be produced. However, many of the most intricate carvings are still done by hand. The use of plastics and injection foam systems designed for new construction has also been suggested for historic preservation work.

CONSTRUCTION

The ability to shape and craft objects and buildings from wood was long known and well practiced in Europe. The long-standing system of master and apprentice had produced buildings that ranged from simple and utilitarian to those that were finely appointed with the highest level of ornamental craft.

Materials

Natural wood obtained directly from trees was the essential component of woodworking for centuries. Later, other materials were developed to assist in producing wood ornament more easily. A group of early plastics known as "composition" was devised to help create highly elaborate but often repeated details. In the twentieth century, sustainable design efforts led to the introduction of medium-density fiberboard (MDF) and various resins and plastics to reproduce the visual aspects of historic woodwork.

Wood

Wood was used because it was typically readily available and could be cut and shaped. The use of exotic woods, such as mahogany, was limited to the wealthiest building owners. In many instances, less expensive woods were used in the private and service spaces and more expensive species in the public spaces. When expensive woods were not available, faux finishes that imitated their color and graining patterns were used instead. For example, pine and fir could be painted to look like the more expensive oak, maple, or mahogany.

When hardwoods, such as oak, were readily available, they were often the choice for durable surfaces needed for flooring and stair treads. Softwoods, such as pine and fir, were used as subflooring and trim that was less likely to be exposed to wear or in place of hardwoods in less expensive construction. The placement and use of these two different types of wood often indicated the prosperity of the building owner and the community.

Composition

Composition was an early form of thermoset plastic (i.e., it hardens when it cools or softens when heated) that was used to duplicate numerous repetitive details. Rather than carving every detail separately, a composition (or "compo," as it is also known) element was cast. To make the detail, a wood model of the detail was carved and then a mold was made from it. Composition was a mixture of chalk, glue, linseed oil, and resin that was heated and poured into the mold and allowed to set. Once set, the piece was removed from the mold and glued to what otherwise would have been a simple flat surface. This process saved the time needed to repeatedly carve multiple copies of the same detail. Compo was typically a medium to dark brownish-orange and could be painted, glazed, or stained.

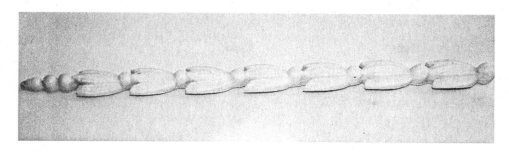

Figure 16-3 Modern casting resins can serve the same function as composition, as shown in this ornamental casting.

Modern Substitutes

Efforts to reduce costs led to the use of wood and plastic laminate veneers on hollow-cored doors or pressed wood substrates. Similarly, stamped metal doors were used where stronger or more fire-resistant materials were warranted. When a health or life safety code requires their use in historic buildings, metal doors can be decoratively painted to match historic door finishes.

Two modern materials that have found growing acceptance are embossed MDF and resin castings of detailed three-dimensional ornaments (see Figure 16-3). Use of these substitute products must be approved by any oversight body if the project is undergoing design review that follows the *Standards*.

Tools

Originally, hand tools were used manually to cut wood into various shapes and sizes. Cutting tools were made from iron sheets and bar stock cut and shaped by the blacksmith. The cutting edges were honed by hand on a grindstone or with a file.

Saws were designed in one of two ways. A rip saw was intended to cut along the length of the wood with the grain, and a cross-cut saw was intended to cut across the grain. Large saws were used for rougher cuts. Saw blades became increasingly small, with finer teeth, as required cuts became more delicate.

Hand-made molding planes allowed specific shapes to be created. The edge of the metal blade was cut and filed to the desired shape and beveled for easier cutting. Later developments in steel led to the use of steel secured along the blade to improve the durability of the edge. When steel became more commonly available in the nineteenth century, the entire blade was made from steel.

Other cutting tools included augers, braces and bits, chisels, gouges, rasps, and scrapers that were used to shape the wood. Chisels and gouges came in a wide assortment and range of shapes (flat, curved, and V-shaped). Wooden mallets were used to strike the chisels and gouges to chip away at the raw surface (hence, the term "chip carving" is used to describe carving). Mallets were also used to force joints together. A variety of clamps, straps, bench dogs, and vises evolved to hold the assembly in place on the bench while the assembly was being worked, glued, or secured together (e.g., pegged, nailed, or screwed).

Measuring and laying out work was the most critical aspect; thus, marking and measuring tools were developed for a specific range of uses. They included gauges, rulers, squares, levels, compasses, calipers, and assorted scribing devices.

The use of hand tools was a slow, careful craft process, but the increased demand for woodwork goods in the early nineteenth century brought the opportunity to mechanize these tools. Previously limited by human endurance, woodworking tasks were completed in a fraction of the previous time by machines that included the cutting edges of hand tools. The high cost of these tools limited their immediate use to entrepreneurs who were backed by investors or family money. The nineteenth century saw the introduction of a variety of cutting devices including band saws, joiners (planer), drill presses, lathes, and other power tools that employed water, steam, and eventually electric power.

Although seemingly immune to mechanization, by the late nineteenth century carving was being replicated by embossing machines that formed ornamental detail using a stamping die. In essence, a metal die was crafted to press a design into a piece of wood. In some cases, bevels on the dies carved out portions of the raw wood; in others, the wood fibers were simply crushed into a specific shape that formed the detail.

Terminology

Joinery is the craft of joining pieces of wood together. To accomplish this, several standard joints came into common use. These included:

- Butt: This joint was formed when two flat edges were placed next to one another, either in parallel or at a 90-degree angle to one another. This joint was used in early casework, flooring, and some vertical paneling but had a tendency to reveal gaps as wood shrank in response to temperature and humidity changes.

- Dadoes and grooves: These slots were respectively cut across and along the grain of a board to receive a shelf or a panel.

- Dovetail: This joint featured interlocking triangular, flared grooves cut along the ends of two pieces of wood commonly used to secure cabinet carcasses or drawers together.

- Half-lap: This joint, commonly used in window muntins and cabinet carcasses, was formed when complementary corresponding portions of the framing were removed so that when the framing was secured together, the material formed a full thickness at the joint itself.

- Miter: This joint was formed by cutting a 45-degree angle along the length or ends of the wood to form a corner that was commonly used in paneling and muntins.

- Mortise and tenon: This joint was commonly used in making doors, windows, and paneling. A protruding tenon was shaped on the rail (horizontal piece) and a corresponding mortise was cut into the stile (vertical, receiving piece). A "stopped" or "blind" tenon was one that was concealed completely within the mortise. When the mortise extended completely through the stile, it was called a "through" tenon (see Figure 16-4).

- Shiplap: This joint was formed when alternating rabbeted edges were placed next to one another to form a continuous plane. It was commonly used in flooring and in horizontal and vertical paneling to accommodate shrinkage without allowing a gap to show.

- Spline: This joint was formed when two grooved edges were connected by filling in the gap with a continuous thin strip of wood (spline). A variant of this was a biscuit joint, where thin discrete pieces of wood (biscuits) were fit into two adjoining slots. These were used in flooring and to build up smaller widths of wood into a panel.

a   b

Figure 16-4 Mortise-and-tenon joints: (a) through tenon and (b) stopped or blind tenon.

- Tongue and groove: This joint was formed by cutting a continuous protruding tongue along the edge of one piece of wood that mated into a matching groove cut along the edge of the adjoining or receiving piece of wood.

Early mechanical fastening systems included animal hide glue and nails (see Chapter 4 for a review of nail technology). The increasing mass production of screws, hinges, and other fastening mechanisms enabled their widespread use for many millwork assemblies in the late nineteenth century.

In its earliest form, decorative trim was an embellishment of the material used for commonly constructed objects, such as doors, windows, trim, paneling, and other ornamental objects. Cutting and shaping with hand tools was a slow, laborious process. In some instances, this process could be speeded up using a reciprocating treadle that provided the motive force to turn a lathe or other shaping device.

Methods

The introduction of water-powered sawmills began the process of introducing cutting and shaping methods that fostered the use of millwork to create and sell decorative enhancements in wood on a broader scale. Trees could be cut more readily into timbers, planks, and, most importantly, the boards and stick work that could be used to create the subcomponents used to assemble doors, window sashes, paneling, and decorative assemblies.

The subsequent introduction of steam power and eventually electric power increased production capacity. By the late nineteenth century, the vast majority of decorative woodwork was being mass-produced in mills and factories around the country. This woodwork became increasingly evident in the Carpenter Gothic style and culminated in the Queen Anne style at the end of the nineteenth century. The complete abandonment of manual production was stalled by the Arts and Crafts movement that emerged at the turn of the twentieth century. Early-twentieth-century revival styles revived the use of various ornamental elements, originally done manually, updated by the use of mass production.

Today, computer-controlled milling machines can be combined with digital scanning technology to create identical reproductions of historic ornamentation. The twentieth century also introduced plastics, laminates, and other substitute materials that can provide ornamental detail for a fraction of the cost of a manually produced decorative element.

Decorative Elements

Many decorative elements consisted of smaller components crafted and assembled in the mill and made in specific widths and lengths with a uniform profile or integrated details that were formed using a router or drill. These details were promoted in a variety of trade catalogs and were shipped nationwide to remote job sites. This process promoted the nationalization of decorative taste and further moved millwork into the public eye as it became more readily and economically available.

Exterior Elements

Throughout the historic period described in this book, numerous forms of exterior wood ornament were used. In some instances, the ornament was an embellishment on a structural member; in others, it was simply a nonstructural aspect of the exterior cladding or porch. Assorted shutters, trim moldings, brackets, vergeboards (also known as "bargeboards"), pendants, cutwork, fretwork, spindles, railings, steps, and other elements were used when a particular architectural style called for it (see Figure 16-5).

Figure 16-5 High Victorian ornamentation was constructed from wood using a variety of power tools that could produce a number of repeated details.

Interior Trim

Interior trim included baseboards, chair rails, picture rails, and window and door casework. Typically, humbler buildings had less elaborate examples of these pieces, if any at all. As trim work became more available and as building owners prospered, trim became more extensively used and elaborately detailed. Still, many wealthy building owners used more decorative forms in public spaces and more modest or utilitarian versions in private and service spaces.

Baseboards or mopboards served several purposes. When the use of plaster surfaces was anticipated, the baseboard was installed to serve as a base to gauge the thickness of the plaster above it. (Note: early baseboards were attached directly to the wall framing. Later systems applied the plaster over the full height of the wall and then attached the baseboard.) The baseboard also provided a means to conceal the joint in the flooring along the wall that allowed for expansion of the flooring and to conceal irregularities in the length of the flooring materials. The harder surface of the baseboard allowed a mop to strike it without fear of damaging the plaster wall surface. As with other ornamental elements, what often started as a utilitarian object was eventually decorated to improve its appearance. The basic form was simply a board attached to a wall. Many more finely decorative baseboards consisted of a board with a decorative profile planed into it. Further embellishments included decorative cap trim along the top and a quarter-round shoe molding along the base. These moldings increased the baseboard's profile and height. The appearance of this molding varied and, in more public portions of the building, was likely to be taller and more ornamental. Today, a historic profile can be re-created by combining several separate stock moldings in current production.

Chair rails were used to define the top of the dado (the lower 2 to 3 feet of the wall) and to protect the wall from damage due to the top of a chair rubbing against it. The profile of the molding ran horizontally along the face. Chair rails were sometimes used even if there was no wainscoting in the dado below it.

Picture rails were located at the top of the field above the dado. In some styles, a frieze was located above the picture rail; in others, the field was terminated at the cornice. Nails and other pinning elements were attached to the picture rail. Cords were then used to suspend pictures and other decorative pieces along the field. In this manner, damage from pounding nails into the wall was avoided. Picture rails could have a linear profile along their face or could be carved or embossed with a repeating pattern (e.g., egg and dart molding).

Casework around windows and doors was used to cover the edges of the rough construction beneath them. The simplest casework was a flat board, but it could become quite elaborate when budget and availability allowed. In less ornate styles, the casework met as a butt joint or possibly a mitered joint. As the woodwork became more ornate, corner blocks and intermediate rosettes were used to add decorative touches.

Wainscoting

Wainscoting was a vertically mounted tongue-and-groove board mounted on the dado portion of the wall. Wainscoting served both utilitarian and decorative functions. Its utilitarian value was in protecting the plaster subsurface behind it from damage. Decorative embellishments included incorporating a V-groove or an incised bead along the edge (and sometimes repeated it across the board). In some cases, raised paneling was used in place of wainscoting.

Paneling

Paneling came into use by the late seventeenth century in the North American colonies. The simplest form consisted of boards mounted vertically or horizontally to the face of a wall. The boards could be fastened to the framing using wood pegs or, if available, hand-wrought nails. In the eighteenth century, raised and inset paneling came into use with the Georgian period of architecture

The traditional form of raised paneling was created using a system of vertical stiles and horizontal rails. The stiles were mortised to receive the tenon of the horizontal rail. A groove was cut along the interior edge of the stile and rails (and the intervening muntins). Early forms of this groove included a decorative molding that was cut by a molding plane as part of the groove. After the Industrial Revolution and the advent of mills, this molding profile became a separate piece. The inset panels were then cut to fit into the grooves and fill the opening created by the assembly of the stile and rails. The inset panel typically had a beveled edge that gave the appearance of raised relief to the remainder of the panel. In highly ornamental paneling, these inset panels could also be carved with an assortment of sculptural forms. These inset panels were not physically fastened to the stiles and rails. Instead, they were allowed to simply rest in the surrounding grooves along their edges. In this manner, as they shrank or expanded in reaction to moisture and temperature, they were free to move. Over time, as they were painted to suit aesthetic tastes, these panels often became fixed in place by the paint. When this occurred, they became vulnerable to splitting along the grain.

The use and location of paneling varied with architectural styles and the prosperity of the owner. Paneling was typically associated with wealth and good taste and consequently was found in more expensive and finely appointed buildings. The use of raised paneling peaked in the late Victorian styles of the nineteenth century and the revival styles of the early twentieth century but, like other ornamental finishes, fell out of favor with the rise of modernism.

As with wallpaper and other ornamental coverings, the working class and modest middle class wanted to use paneling as a decorative wall covering. Their demand for wood paneling led to the development of veneer-faced plywood paneling and panel-faced gypsum board by the late 1950s and, subsequently, the use of plastic laminates.

Doors

Among the most simple door types was the board and batten door. It was constructed of vertically aligned planks or boards joined together horizontally by two or more boards. These doors were common in the earliest colonial period, not only for their ease of construction but also for the security they provided.

As prosperity and security came, doors (and windows) were built using mortise-and-tenon construction similar to that of raised paneling. Exterior doors were typically constructed of heavier and more durable materials but over time became elaborately decorated with wood carvings, inset windows, and ornate hardware. Fanlights and transoms above the doors and sidelights flanking the doors became popular in the late colonial period.

Interior doors are often character-defining features that can be used to identify and date historic construction and remodeling periods of the building. It is important to remember that doors were often relocated to other parts of the building (or to another building altogether) when later remodeling occurred. Sometimes, these doors were stored on site in a basement, attic, or outbuilding. Like exterior doors, interior doors became increasingly elaborate as construction budgets expanded.

Stairways

Stairways were typically composed of risers and treads. When one or more sides of the stairway were not directly attached to a wall or were open, a combination of a newel post, banister, and any number of balusters was used to prevent accidental falls. The earliest colonial houses that had stairways often used only minimal detailing. They had flat square-edged treads and plain risers. Balusters were often small square posts used to support a plain banister.

As houses became larger, the stairway located in the main entrance hall often gained stature, including ornamental newel posts and turned balusters that supported a contoured banister. The risers and other exposed woodwork were decorated with hand-carved wood or cast compo ornament. The treads had contoured edges and sometimes a small piece of molding to conceal the joint between the tread and the riser. A flying staircase, where neither side of the stairs was attached to the wall, was considered the single most dramatic form of woodwork and was found in only the finest and most expensive buildings. These stairs were produced by the finest stair builders that money could afford.

Stair design and detailing varied with architectural styles and reached its pinnacle in the revival styles of the early twentieth century. In larger houses of the wealthy in the early nineteenth century, there was often a service stair that allowed servants to access the upper floors without entering the public space. This stair was often far less ornamental, smaller, and narrower than the stairway used by the family and their guests.

Mantles, Fireplace Surrounds, and Overmantles

As with stairways, as households prospered, a variety of ornamental treatments occurred around the fireplace, including paneling and ornamental carving. Carvers and woodworkers were commissioned to craft exquisitely detailed ornamental features. The original bare mantle became enhanced with molded profiles along its edges and was supported by a variety of columns and decorative posts recalling classic orders of architecture. As styles dictated and money allowed, these features became increasingly ornamental and spread upward to cover the entire facing of the chimney. Shelves, mirrors, and other ornamental flourishes, such as carved heads, were extremely popular in the late Victorian period. At the fireplace hearth, decorative tile and stone were often an integral part of the overall assembly.

In most homes, mantles were modest and simply decorated. Whatever ornament existed usually consisted of shallow molding or minimal detailing. Millwork companies and their retail counterparts, such as Sears and Roebuck, however, capitalized on the image of the elaborate mantles, fireplace surrounds, and overmantles found in finer houses and offered scaled-down versions that featured some compo, carved or embossed details, or decorative pieces that were glued or inset into the otherwise flat face of vertical surfaces.

Cabinets and Cupboards

By the late seventeenth century, cupboards and open shelves formed the majority of storage space. Niches and built-in corner cabinets, often a later addition to the original building, were used to store and display dishes. Throughout the eighteenth century, food and food service preparation activities led to the use of counters as work surfaces. Shelving was installed above and below the counter to store the foodstuffs and tableware needed for meal preparation. Valued items were locked in boxes and food safes. By the late eighteenth and early nineteenth centuries, butler's pantries and other food service preparation areas adjoining the kitchen had been created and contained either open shelving or lockable cabinets. In more affluent homes, these rooms featured glass doors to allow the contents to be viewed.

The nineteenth century saw the expanded use of both built-in and freestanding cabinets and shelving. In stores, display cases and shelving became common for a variety of dry goods. These units could be built in place or fabricated elsewhere and then installed on site. For businesses, storage systems were made by numerous manufacturers nationwide. Furniture companies saw the opportunity to expand into domestic kitchens. Metal and wood cabinets for storage and iceboxes were sold by a number of companies, including the Hoosier Manufacturing Company and Kitchen-Maid Incorporated. The early twentieth century saw the introduction of standardized cabinet sizes known as the "unit system" (Beecher 2001). This standardization simplified the planning and layout of increasingly small kitchens built during the mid-twentieth century. In addition, manufacturers gained an advantage in both pricing and marketing of factory-built metal and wood cabinet systems. This advantage became particularly great in the construction boom of the 1950s.

The introduction of plastic laminates and pressed wood substrate during the twentieth century, coupled with the do-it-yourself ethos of this period, further decreased the custom-built cabinet market as modularity and low prices found favor with kitchen designers and homeowners.

TYPICAL PROBLEMS AND RECOMMENDED TREATMENTS

The problems associated with wood carving and millwork have been outlined and discussed in earlier chapters. The treatments discussed earlier in this book for wood may all be applied here as well. A brief overview of issues directly related to millwork and wood carving is given below.

Conservation Strategies

Wood can be readily conserved in place. If the integrity of the ornament is at risk after the source of decay has been removed, it may be necessary to use modern materials, such as epoxies and consolidants, when they can be concealed by paint or will not be directly viewable. Severely damaged original materials may need to be removed and stored in an appropriate way to conserve them for future reference. In this instance, a replacement object would be fabricated using the original as a model.

Repair Strategies

Wood can be readily worked and shaped using modern or traditional tools and methods. The principal repair issues in many cases are whether a specific species of wood is available and whether the wood will remain exposed to view when the repairs have been completed. The use of old-growth wood in many historic buildings raises the possibility that a matching grain pattern may not be possible with newer wood when clear finishes or tinted stains are used. Although salvaged materials may be available, matching the grain patterns exactly may prove difficult. In this case, there are two approaches. First, the new wood used for the repair can be left clear finished or painted to differentiate it from the remaining original wood. Second, the wood can be decoratively painted to match the grain pattern of the earlier construction. Some early historic wood constructions were decoratively painted to resemble a more expensive wood, so this concealment is not unknown.

Small portions of the damaged or missing millwork may be newly milled using cutting tools that use profiles taken from existing millwork. Mills typically have setup charges for creating a molding profile that is not part of their standard stock run. When only a short run of molding is needed, it is possible to craft scraping tools to hand-make short runs of molding (see Figure 16-6). When larger quantities are needed, having the millwork cut at a mill may be more economical.

Replacement of a severely deteriorated or damaged wood ornament or piece of millwork is acceptable when physical and visual evidence that supports the replacement is available. One form of replacement is salvaged pieces of millwork, such as doors, that match the missing or damaged pieces. Locating salvaged matching woodwork, although difficult, is not impossible. When salvaged resources are not readily available, the replacement must be delayed until they are found or budgets must be devised to allow reproduction of the replacement woodwork.

With the advent of modern technology, digital scanning and numerically controlled milling machines can replicate three-dimensional wood carvings. Corner blocks, moldings, and three-dimensional shapes can be electronically measured and re-created as needed to replace missing pieces (see Figure 16-7). Some large carvings can, to some extent, be crafted using this same process. However, certain highly

Replacement Strategies

Figure 16-6 When small amounts of millwork are needed, metal cutting blades (as shown here) can be crafted to replicate the profile and then make new molding.

a   b

Figure 16-7   This original piece of millwork (a) was cleaned and digitally scanned to re-create and replace other similar missing or damaged pieces using a computer numerically controlled (CNC) milling machine. A replicated piece (b) is shown after installation.

detailed objects may still be made only by hand carving them. Master wood carvers can still be found to re-create large, detailed carvings that are beyond the scope of modern mills.

References and Suggested Readings

Beecher, Mary Anne. 2001. Promoting the "unit idea": Manufactured kitchen cabinets (1900–1950). *APT Bulletin*, 32(2/3): 27–37.

Calloway, Stephen, ed. 1996. *The elements of style: A practical encyclopedia of interior architectural details from 1485 to the present*. Rev. ed. New York: Simon & Schuster.

Clark, Clifford, Jr. 1986. *The American family home: 1860–1960*. Chapel Hill: University of North Carolina Press.

Comstock, William T. 1881 (reprinted 1987). *Victorian domestic architectural plans and details: 734 scale drawings of doorways, windows, staircases, moldings, cornices, and other elements*. New York: Dover Publications.

Cummings, Abbott Lowell. 1979. *The framed houses of Massachusetts Bay 1625–1725*. Cambridge, MA: Harvard University Press.

Davies, Mike. 1997. *Woodcarving techniques and designs*. Cincinnati: North Light Books.

Englund, John H. 1978. An outline of the development of wood moulding machinery. *Bulletin of the Association for Preservation Technology*, 10(4): 20–46.

Fisher, Charles E. and Hugh C. Miller, eds. 1998. *Caring for your historic house: Preserving and maintaining: Structural systems, roofs, masonry, plaster, wallpapers, paint, mechanical and electrical systems, windows, woodwork, flooring, landscapes*. New York: Harry N. Abrams.

Hull, Brent. 2005. *Historic millwork: A guide to restoring and re-creating doors, windows, and moldings of the late nineteenth through mid-twentieth centuries*. New York: John Wiley & Sons.

Ierley, Merritt. 1999. *Open house: A guided tour of the American home 1637–present*. New York: Henry Holt and Company.

Jandl, H. Ward. 1998. *Rehabilitating interiors in historic buildings*. Preservation Brief No. 18. Washington, DC: United States Department of the Interior.

Jennings, Jan and Herbert Gottfried. 1988. *American vernacular interior architecture 1870–1940*. Ames: Iowa State University Press.

Jester, Thomas. 1995. *Twentieth century building materials: History and conservation*. New York: McGraw-Hill Book Company.

Johnson, Ed. 1983. *Old house woodwork restoration: How to restore doors, windows, walls, stairs and decorative trim to their original beauty*. Englewood Cliffs, NJ: Prentice Hall.

Jones, Bernard E. 1998. *The complete woodworker for the craftsman or the beginner*. Berkeley, CA: Ten Speed Press.

Lee Valley Tools, Ltd. 1984. *The Victorian design book: A complete guide to Victorian house trim*. Ottawa, Ontario, Canada: Lee Valley Tools.

Leeke, John. 1993a. *Columns*. Portland, ME: Historic Homeworks.

———. 1993b. *Exterior woodwork details*. Portland, ME: Historic Homeworks.

———. 1993c. *Mouldings*. Portland, ME: Historic Homeworks.

———. 1993d. *Wood gutters*. Portland, ME: Historic Homeworks.

———. 2004. *Wood-epoxy repairs*. Portland, ME: Historic Homeworks.

Moss, Roger W. and Gail Caskey Winkler. 1986. *Victorian interior decoration: American interiors 1830–1900*. New York: Henry Holt and Company.

Palliser, Palliser & Co. 1878 (reprinted 1990). *American Victorian cottage homes*. New York: Dover Publications.

Rodriguez, Mario. 1998. *Traditional woodwork*. Newtown, CT: Taunton Press.

Shivers, Natalie. 1990. *Walls and moldings: How to care for old and historic wood and plaster*. Washington, DC: Preservation Press.

Shoppell, R. W., et al. 1890 (reprinted 1984). *Turn-of-the-century houses, cottages, and villas*. New York: Dover Publications.

Thornton, Jonathan and William Adair. 1994. *Applied decoration for historic interiors preserving composition ornament*. Preservation Brief No. 34. Washington, DC: United States Department of the Interior.

Walker, Anthony J. T., Kimberly A. Konrad, and Nicole L. Stull. 1995. Decorative plastic laminates. In Jester 1995, 126–131.

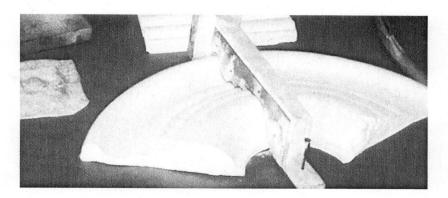

CHAPTER 17 Decorative and
Flat Plaster

The use and application of mud, clay, or other cementitious materials to protect surfaces predate written history. This process has been called by various names, including "daubing," "rendering," "plastering," and "stuccoing." In modern practice, only the last two are in use. The former describes the process when performed indoors and the latter describes it when done outdoors.

Plastering methods today use many of the same processes, tools, and basic ingredients used in ancient Egypt, Greece, and Rome. Over time, two separate areas of expertise in plastering have emerged—flat and ornamental; both require substantial training. In the late nineteenth century, adaptations of the construction process led to the introduction of what eventually became known as "drywall," the common gypsum board substitute for plaster. Use of this material dramatically reduced the use of plaster to the point where, by the end of the twentieth century, the plaster trade had become a small specialty segment of the construction industry.

Flat continuous surfaces were made by building up layers of plaster and then tooling them to a smooth finish. The most common application of flat plaster was for ceiling and walls. Various forms of plastering were used in the earliest colonies and continue in use today. The earliest plastering method used to infill spaces between timber framing members consisted of daubing clays directly onto a network of thin sticks, reeds, or branches known as "wattle." Hence; this process was also referred to as "wattle-and-daub construction."

OVERVIEW

FLAT PLASTER

As economic conditions improved in the mid-seventeenth century, interior surfaces of new buildings featured plastered walls. In finer buildings, ceilings may have been plastered, while in lesser buildings they may have been left bare or plastered only between the framing members. By the early eighteenth century, plaster had become a common substrate or finish material for both walls and ceilings and remained so well into the twentieth century. Plaster surfaces in the early colonies could be left plain or were later stenciled, adorned with hand-painted murals, covered with wallpaper, or painted. With the advent of cheaper drywall systems in the 1950s, the traditional multicoat plaster systems were viewed as expensive and became obsolete.

Construction Materials and Methods

Early plaster was made from lime putty, aggregate (sand), and water. Lime putty was made by grinding and then calcining (roasting) limestone or oyster shells. The calcining process removed the carbon dioxide, and the remaining calcium oxide, also known as "quicklime," formed the basis of the lime putty. Quicklime was then slaked (mixed with water) in sand-lined pits, where it could be stored indefinitely. The resulting lime putty was then mixed with sand and water (and other additives) to make plaster.

Early plasterers used multiple coats of plaster applied directly to a masonry surface or to a system of wood laths attached to wood framing. Most common was a three-coat system consisting of an initial scratch coat to establish the base for the succeeding brown coat. These two coats were composed of coarser aggregate and sometimes had animal hair or shredded fibers from rope or plants to serve as a strengthening agent. Atop the brown coat was the finish coat, which had a higher percentage of lime putty that formed a harder and more durable finish than the coarser coats beneath it (Shivers 1990). In some cases, a two-coat system was used that combined the first two coats of a three-coat system while retaining the finish coat.

To hold plaster in place, wood lath was attached to the structural framing with a small gap between each lath. When plaster was troweled onto the lath, it was forced through the gap and sagged due to gravity to form a key (see Figure 17-1). When the plaster hardened, this key held the plaster in place. The earliest lath was hand-split and was characterized by ragged edges. The next generation was cut using sawmills and featured a regular straight edge and a consistent thickness. At the turn of the twentieth century, a new lath was developed by cutting slits into a sheet of metal and stretching the metal to expand the slots into a diamond-shaped opening. This process created what is known as "expanded metal lath." Because of its resistance to insects, fire, rot, and other decay mechanisms associated with wood, expanded metal lath became the industry standard during the first half of the twentieth century.

Prior to the mid-nineteenth century, plaster was applied to the wall after the baseboards were installed. The wooden plaster grounds introduced at that time were used to assist the plasterer in creating a plumb or level surface across the flat plaster field. After this introduction, wood molding, such as baseboards and picture rails, were installed by nailing them to the plaster grounds.

The standard formula for plaster changed in the early twentieth century with the use of gypsum instead of lime as the binding material, and in some instances perlite (a lightweight silica-based volcanic stone) was used instead of sand. Gypsum plaster was stronger and did not require reinforcement materials such as hair or fibers. More importantly, gypsum plaster set up (hardened) more quickly and took

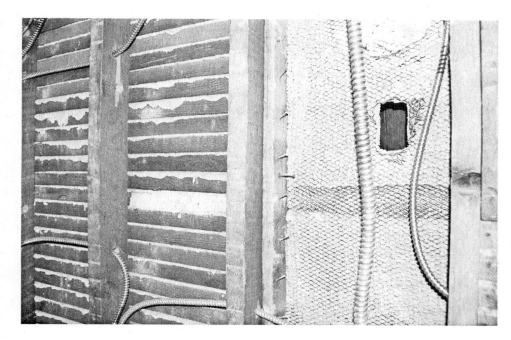

Figure 17-1 The formation of plaster keys is critical in the construction of flat plaster walls and ceilings. Wood lath is shown in the left-hand portion of the photograph and expanded metal lath is shown in the right-hand portion. The keys form when the plaster slumps after being pushed through openings in the material.

significantly less time to fully cure than lime plaster. Inadequately cured plaster could not be painted or covered with wallpaper; it contained too much moisture. Gypsum plaster cured much more rapidly than lime plaster. With the changeover to gypsum plaster, what had been a several-month to year-long curing process was shortened to a few weeks (MacDonald 1989).

Although lime plaster formulations are available or can be re-created, gypsum plaster is typically used today. The term "gauging plaster" refers to gypsum and additives to control the setting rate of the plaster that is used for the finish coat. Prior to the late nineteenth century, gauging plaster had been limited to decorative plaster ornaments because it set so rapidly (Van den Brandson and Hartsell 1984). Care must be taken when repairing plaster walls to use a compatible formulation using gypsum or lime. This decision is often left to the experienced plasterer.

Derivatives and Alternatives

The evolution of lath over the past 100 years contributed to the declining use of plaster. In the early twentieth century, following the introduction of expanded metal lath, gypsum board was introduced. It was first used as lath strips and subsequently as a replacement for the scratch coat of the three-coat system. The addition of the brown coat and the finish coat to this substrate forms what is known as "veneer plaster." If desired, the finish coat could be mixed with additives to re-create the finish traditionally found in early colonial plaster surfaces.

As the use of gypsum board expanded, holes were cut into the board to assist in forming keys and created what became known as "rock lath" (see Figure 17-2). The development of adhesive compounds and tape to seal the joints between the separate panels subsequently led to the adoption of gypsum board alone as a "drywall" alternative to wet plaster. This drywall system gained considerable popularity in the 1950s and remains in use today.

As mentioned earlier, the late nineteenth and early twentieth centuries saw the introduction of a number of panel and sheet products that provided the hard, flat,

Figure 17-2 The use of gypsum board pierced to form rock lath led to the eventual adaptation of gypsum board as a drywall alternative to wet plaster systems.

smooth surface desired for ceilings and walls. Pressed wood, hardboard, plywood, paneling, and asbestos-cement board have all found their way into the building to be used as a substitute for plaster.

DECORATIVE PLASTER

The dictates of the Federal style introduced in the late eighteenth century and the Greek Revival style in the early nineteenth century prompted the expanded use of architectural ornaments, such as brackets, modillions, cornices, and other three-dimensional elements. For the wealthy, this ornament was often crafted from stone. For others, it was made of plaster or wood. On the exterior of finer buildings, stucco was applied or parged onto brick columns to simulate more expensive stone. As with many other indicators, a number of less affluent households featured decorative plaster only in the public spaces of the house, while the private spaces were less adorned.

Inspired by the philosophical ideals of ancient Greece being revived in Europe and the classical orders of architecture being revealed in archaeological investigations in Greece at the time, Americans embraced this form of ornamentation. By the late nineteenth century, plaster studios offered numerous elements that could be created, shipped to a building site, and installed directly. Beyond the direct visual qualities that ornamental plaster provided, these elements were gilded or glazed to enhance their three-dimensional light-reflecting qualities, as well as being faux-finished to imitate more expensive carvings made from granite, marble, or wood. With the advent of the modernist styles of the 1920s, ornamental plaster fell out of favor.

Construction Methods

Ornamental plaster could be cast as a single piece on a work bench in a studio, constructed on site, or both (Flaharty 1990; Gorman, Jaffe, Pruter, and Rose 1988; Sawyer 1951; Van den Branden and Hartsell 1984).

Bench Casting

The first step in creating a piece was to create a positive model. This was done by carving wood or plaster into the desired shape. In a restoration project, this model may be obtained by demounting an existing element that needs to be reproduced and may require a composite reassembly of several damaged elements to reconstitute the desired single element. Once the model was cleaned and all defects were corrected, a mold was made. While historic methods varied, modern technology has generated a number of elastomeric and resin-based materials for making molds (see Figure 17-3).

a

b

c

d

Figure 17-3   The flood molding process can be used to create a mold from an original positive model of the object to be replicated: (a) object to be replicated, (b) flood mold with object, (c) pouring in molding polymer, and (d) mold and replicated casting.

For larger, complicated elements, the model was constructed in several components that were later assembled to make the final product. The cleaned model was coated with mold release (i.e., liquid soap) and was placed inside a flood mold (an open-ended container fixed to the bench). The liquid urethane rubber for the mold was then poured, slowly to minimize or prevent air bubbles, into the container, where it was left to cure. The hardened mold was then removed, taking care not to stretch and tear it.

The mold was then used to receive the casting material for the desired element. Traditionally, plaster of Paris and specially prepared molding and casting gypsum plasters were used (Van den Brandon and Hartsell 1984). In modern construction, various resins and epoxies have been used. The material was left to set (harden) inside the mold. After it set, the casting was removed from the mold, cleaned to remove any surface defects, and ready to be installed at the job site. The mold itself could be reused. The number of reuse times depended on the durability of the molding material and the complexity of the object being cast. Molds made from more durable materials lasted longer but were stiffer. Stiffer molds could be used for simpler elements. More flexible molds could be used for more complicated elements but could tear or stretch out of shape after just a few castings.

On-Site Construction

Like millwork for wood ornament, certain linear shapes could be created by running a profile along the material. The first step in running a cornice or another linear element was to create the desired profile. The outline of the profile was drawn on a piece of paper that served as a template. This template was used to trace the profile onto a piece of sheet metal that was cut to the exact profile shape. The profile was also transferred to a piece of wood that was likewise cut to the exact profile shape. The sheet metal and the block were aligned, secured together, and installed on a horse, a hand-held wooden frame that was run along the surface of a wall or ceiling to create the profile. To create the cornice (see Figure 17-4), a wood guide strip was secured to the wall to guide the horse. Plaster was applied to the wall or ceiling. This process required multiple passes of the horse. As more plaster was built out, the metal edge removed any material that exceeded the profile with each successive pass along the material surface. Eventually, the cornice was fully built out and finished to a smooth, continuous surface. Cornices with large cross sections often required wood blocking to be installed to reduce the quantity and weight of the plaster being installed (Flaharty 1990; Sawyer 1951).

Circular elements, such as modillions, can be run in place by attaching an armature to the horse and pinning the assembly to the ceiling (or the bench). The single pin connection allows the armature to be rotated 360 degrees to complete the element.

Combined Bench and On-Site Construction

For larger elements with more complex surface details, a combination of bench and on-site construction methods was used. By incorporating flat faces in the profile, space could be left to insert a more ornamentally detailed component (e.g., angel, dentil, leaf, egg, or dart). A more detailed and decorative piece could be attached using thinned plaster or animal hide glue as an adhesive in combination with small nails and hooks implanted in the back of the piece or the site-constructed element.

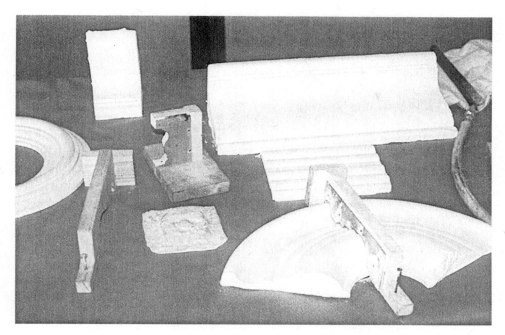

Figure 17-4 A horse can be used as a running mold to create a linear element, such as a cornice, or can be pinned and rotated to create circular ornamental elements.

In 1856, fibrous plaster was patented. It incorporated canvas as reinforcement and enabled large panels to be precast off site with ornamental details already included and then nailed in place on site (Calloway 1991).

In this manner, the majority of the ornamental element could be constructed less expensively on site, while the decorative bas-relief details could be made off site. As a result, a great variety of plaster ornament eventually came to be made elsewhere and was promoted and distributed via catalog services. Later development of catalog sales methods prompted the use of suspended constructions, delivered to the job site in sections, and attached to the concealed substructure with stiff wire hangers. This method allowed ornamental compositions, such as coffered ceilings and domes, to be bench-made while reducing the weight (and shipping costs) and ultimately lowering on-site labor costs.

Derivatives and Alternatives

Scagliola was a highly specialized form of decorative plaster that simulated natural stones, such as marble. This effect was achieved by mixing granite or alabaster rock chips with pigments and plaster of Paris. The mixture was then pressed into a mold to form a variety of architectural elements. The final product was then polished to a smooth, glossy finish that closely resembled polished stone but cost much less to construct.

Throughout the nineteenth century, products and construction practices were invented to cheaply imitate more costly materials and reduce the expense of construction. As noted previously, these products included wall and ceiling coverings like stamped metals, Lincrusta-Walton, and Anaglypta, as well as molded composition and papier-mâché. In the late twentieth century, this practice was repeated when fiberglass, plastics, and an assortment of expanded foam and cast-resin products came into use in new construction. Their use in rehabilitation and restoration work has been highly debated.

TYPICAL PROBLEMS

Among the first signs of problems due to weak or damaged plaster are cracks that first appear as hairline cracks in the surface finish coat and then, as damage increases, begin to move into the underlying layers of the brown and scratch coats. Some hairline cracks can be expected in new construction as the plaster cures. These cracks may even vary slightly due to seasonal changes in temperature. They are usually only a cosmetic problem. As the house reaches equilibrium over time, certain cracks will become inactive and stop growing. Changes in groundwater levels, remodeling, or vibration from traffic or nearby construction can activate a crack or initiate new ones.

Next is the occurrence of irregularities in the surface. Exposed plaster may begin to show signs of pitting as the surface breaks down and materials sift out of cracks. Since most plaster has been painted or covered numerous times, a crack may be concealed until the plaster keys fail and a section of plaster breaks loose. Prior to failure, this problem may only appear as a bulging surface and, in some instances, the multiple layers of wall coverings may be the only thing holding the plaster in place.

Often, without nondestructive testing methods to recognize impending collapse, the problem is recognized only after plaster fails. Testing methods include simply tapping on the plaster to assess its soundness, sophisticated thermographic imagery, and long-term computer monitoring. The testing method selected should be based on the level of historic significance of the plaster. Testing and monitoring can be performed to determine the extent and activity of cracks and other problems. In some cases, when time permits, the activity level of visible cracks may be measured using a crack gauge temporarily fixed across the crack and left in place for several days or longer.

In general, there are three major problems (MacDonald 1989) that cause plaster to fail: poor construction methods, moisture, and mechanical failure. Like other exposed finish materials, plaster surfaces are also vulnerable to the accumulated effects of occupation, such as soiling, abrasion, and assorted decorative surface treatments (e.g., multiple layers of paint and wall coverings).

Poor Construction Methods

Economic factors, the desire to speed up the process, and poor craftsmanship all affect the way plaster is installed. A variety of defects can occur due to poor construction practices. These defects include improperly mixing materials or using inferior materials, installing plaster in extreme temperatures, skimping on materials, or using incompatible finish and base coats.

The plasterer may have mixed too much or too little aggregate (sand or perlite), binders, and retarding agents into the plaster mix. The net result is defective plaster. Too much sand will make the surface weak and crumble easily and perhaps impossible to repair. Too much retardant will slow the setting process and cause the surface to be weak and powdery.

To cure properly, plaster needs to be installed between 55 and 70°F. Beyond this range, several problems can occur. When it is too hot, the plasterer can spray the wall with alum water to allow proper setting. Unfortunately, when plaster is installed in cold temperatures, especially in freezing conditions, or where air circulation is poor, the plaster will not set properly and will have to be redone, literally from scratch.

Improper application of the base coat or the finish coat can result in cracks in the finish coat. When a finish coat is applied to an overly dry base coat, is improperly troweled, or is made from an insufficient amount of gauging plaster, the surface will crumble into a series of chip cracks resembling an alligatored paint surface. When a finish coat is applied to a base coat that is too thin or has too much sand, random or map cracking (i.e., long, irregularly shaped cracks resembling map boundaries) will occur (Diehl 1960; Van den Branden and Hartsell 1984).

Moisture

Moisture can be detrimental to plaster in two ways. First, moisture reacts with gypsum plaster by dissolving the captured gypsum in the plaster as it saturates the plaster. As the moisture recedes, the gypsum forms crystals that expand and create pressure that can weaken the overall structure of the plaster. This process, referred to as "sugaring," is similar to efflorescence in masonry. With repeated exposure to moisture, the plaster eventually disintegrates (see Figure 17-5). In flat plaster, the keys and the separate coats can all be damaged. In decorative plaster, the surface can become pitted and delicate details may fall apart. In lime plaster, the moisture is simply absorbed into or pushes against the plaster. This increased force can break the keys or the flat plaster itself and cause the plaster to break away from the lath.

Moisture can also affect other aspects of plaster construction. Wood lath can rot. Nails and metal lath can corrode. Inadequately treated paper coatings in rock lath can absorb water and fail as well. Within the wall cavity, the moisture can also attract mold and insects.

Common sources of moisture are fairly obvious and include leaking pipes, roof, or exterior cladding. Other sources are less obvious since they occur in either short- or long-term cycles, such as condensation and rain penetration around windows and doors, rising damp, and groundwater fluctuations (affecting basements and crawl spaces). Properly installed and maintained gutters, downspouts, French drains, and lawn sprinklers that prevent water from splashing on the building and landscaping practices that do not trap moisture (snow, rainwater, or groundwater) against the building are important to reduce moisture migration into the building. These prac-

Figure 17-5 Gypsum plaster will form crystals when drying after being exposed to moisture. Repeated cycles will cause the crystals to disintegrate the plaster, a process referred to as "sugaring," as seen along the top of the baseboard at the lower right. The vinyl wall covering had concealed this problem.

tices are particularly important in maintaining masonry walls. Many older buildings were not constructed with insulation or vapor barriers. With little or no insulation, the interior surface temperature of the wall is lower than that of an insulated wall, which can cause condensation either on the surface or within the wall cavity if water vapor permeates the materials. Often a result of the high humidity levels found in kitchens and bathrooms, this condensation can cause deterioration seemingly from within the wall itself even when other typical internal or external sources of moisture are present.

Mechanical Failure

Like other masonry-based products, plaster has relatively low tensile strength. The most common causes of mechanical failure in flat plaster are cracks in the plaster, breakage of the plaster keys holding the plaster in place, and inadequate bonding between layers of plaster. Failure can occur as (1) a building moves due to settling or expansive clay soils beneath it; (2) excessive loads are introduced by remodeling (especially in ceilings from the floor above) or vibration; and (3) moisture or fire weakens the structural bonds of the plaster itself.

Cracks that run diagonally often are indications of settlement, structural deformation, or excessive vibration. These cracks often run from corners and near openings, such as doors and windows. Sticking doors and windows can further indicate a structural deformation or settlement problem. Horizontal cracks may occur when the underlying lath has broken free.

Occupant-Related Damage

Although early economic hardships and construction practices may have resulted in a bare plaster wall or ceiling in the first few years after its installation, flat plaster was never left as an exposed finish over the long term as periodic updating to current aesthetic tastes occurred. However, this recurring updating frequently led to multiple layers of paint and wallpaper being applied over the original finish surface of flat plaster. Like flat plaster, some decorative plaster was painted in later redecorating efforts that often obscured surface details due to the accumulation of paint. Abrasion damage and accumulated soiling (commonly due to smoke, soot, and nicotine residue) were often simply covered with a new coat of paint and other coverings. Plaster could be further damaged indirectly by vibrations caused by doors slamming or directly by the stresses created when nails were hammered into the wall for hanging pictures (hence the use of picture moldings above the field in some houses with plaster walls). Two final causes of damage were accidental breakage and intentional vandalism.

RECOMMENDED TREATMENTS

The repair or replacement of plaster is complicated by the fact that it normally has had a decorative finish applied or that it is often a substrate for other decorative treatments, such as wallpaper or paint. While the historical significance of the plaster itself may be nominal, the decorative treatment adorning it is often a significant character-defining feature of a space. Accordingly, the repair and replacement alternatives should be evaluated carefully to minimize the loss of not only the plaster but the finishes on the plaster as well.

When it is necessary to remove large sections of wall coverings to assess the condition of the underlying plaster or when replacement of all of the plaster may be the chosen treatment, the layers of existing wallpaper must be documented prior to removal. This documentation includes exposing each layer and photographing it. A graphic scale indicator should be used in each photograph to enable accurate re-

Figure 17-6 Graphic scale indicators can aid in determining the size of the pattern and the repeating figures within it. This image shows three variations at the upper left, upper right, and right center.

production if one is restoring to a specific period of the building's history (see Figure 17-6). Due to the delicate, fragile nature of older wallpapers, this task may require the services of a professional wallpaper conservator or conservation laboratory to appropriately expose each layer.

Before attempting to repair or replace damaged plaster, the source of the damage must be identified. If the source cannot be immediately corrected, the damaged plaster should be stabilized until appropriate remediation treatment can be performed. For example, loose sections of plaster should be held in place using plywood and bracing.

While some damage may be only cosmetic, an overall assessment must be done to determine the full cost of these repairs. In many cases, individual repairs can be readily made; in others, the overall decay of the plaster may indicate the need for replacement (Flaharty 1990; MacDonald 1989; Simmons 1990; Van den Brandon and Hartsell 1984).

Repair Strategies

Small surface cracks and small areas of damage can be repaired using patching compounds, such as high-gauge lime putty or an all-purpose drywall joint compound. The putty is worked into the defect and smoothed with a trowel. Larger cracks that are not caused by structural problems can be repaired by removing plaster along both sides of the crack in an inverted V profile. The gap created as a result of this process is then filled with patching material. For repaired cracks that recur, the process is repeated and drywall tape is secured across the crack and covered with a quick-setting drywall compound. After this first coat dries, a second coat of compound is added and sanded smooth.

Cracks caused by structural problems should be repaired only after the structural defects have been corrected. These cracks are cleaned back down to the lath and to a width of 6 inches. Metal lath is inserted into this gap, and a three-coat system of compatible plaster is constructed.

In some cases, one or more layers may have delaminated or broken away. If the damage has not exposed the lath, a plaster bonding agent can be applied and the surface repaired using a lime finish coat. Where missing plaster has exposed the lath, the area should be cleaned and the lath reattached to the framing. Wood lath should be moistened with a light spray of water for a better bond with the new plaster and to prevent warping. To strengthen the patch, expanded metal lath should be secured to the existing lath. At this point, one or more base coats are applied, depending on the depth of the hole, to build out the surface. Finally, a finish coat is applied that matches the surface profile and texture of the existing plaster around the patch.

The repaired surface may still have other undesirable irregularities (e.g., missing sections of wallpaper). If a smooth, regular surface is required, there are coverings that can be applied directly to the surface to smooth it. Applied as a canvas or fabric, this material can then be used as the substrate for applying paint or other desired wall coverings.

Repairing flat plaster ceilings is more difficult than repairing plaster walls since ceiling plaster pulls away from wood lath as the keys break. When the back of the lath is accessible, usually from the attic or by selectively removing floorboards in the space above, verify that the lath is attached to the framing and can support the added weight of the reinforcing materials described below. If the lath is not secure, then repairs are needed. If the lath is in good condition, carefully push the plaster against the lath and prop the plaster in place. After the lath is wetted, a bonding agent can be applied and worked into the gaps between the lath and the plaster. Before the bonding agent dries, plaster-saturated jute scrim is firmly pressed against the bonding agent.

One specialized method used to conserve historically significant wall coverings or decorative surfaces is to inject acrylic adhesives through small holes drilled into the face of the plaster or the back of the lath. The loose plaster is propped in place until the adhesive sets. This should be done by a skilled plaster conservator.

For repairing ornamental elements, updated variations of traditional casting methods exist so that it is possible to make molds from existing elements of existing features. Note, however, that demounting an intact ornament for use as a model for other similar but damaged ornaments may be required to get a proper sample to serve as a model for new castings. Individual pieces of a larger composite element (or several if none can be removed without damage) can be demounted and a mold created using methods described in the section on construction. For features that cannot be demounted, other methods are used. For example, the profile of an existing molding can be obtained by either cutting a slot into the cornice, inserting a piece of thin sheet metal, and tracing the profile directly; removing a complete cross section of the molding from the wall and tracing the profile at a bench; or applying a thixotropic casting material (e.g., a gel that eventually will harden into the shape against which it is pressed) to create a mold of the ornamental element in place and obtaining the cross section from the section cast from that mold. At this point, once the metal strip is cut and affixed to the running mold, the traditional method of running the molding can be used.

For other components, an intact ornament may need to be sacrificed to obtain specimens for what are called "sinkages." Sinkages are the strips or blocks of three-dimensional ornament that were sunk (placed) into the receiving trench or slot formed in a running profile. Since sinkages were typically secured by animal hide glue, removing a sinkage for modeling a replacement will most likely result in the destruction of a section of otherwise intact molding (which, of course, will need to be replaced). The newly cast pieces may then be inserted into or attached to the larger decorative element and reattached at the location of the missing or removed piece.

For decorative plaster whose details are obscured by paint, a decision must be made on how to treat the paint. Although breakage losses may occur, careful removal of the paint can reveal the underlying details. This process creates a dilemma. Moldings made from a cleaned specimen will display sharper features than the existing moldings in the space. Conversely, moldings made using a specimen with the built-up paint intact may match the existing molding, but if a later project includes removing the paint, these repaired sections would then need to be recast to match.

Replacement Strategies

A simple misunderstanding has led to the misperception that substantial replacement of plaster construction is necessary to meet modern fire codes. Unfortunately, many people on the project team and their building code enforcement counterparts who have typically worked only on new construction may not realize that the fire ratings of plaster construction are defined in the *IEBC*. A sad result has been the unnecessary removal of sound historic plaster walls and ceilings (and their finishes) and their replacement with modern drywall construction. Allowances for archaic materials, including plaster, granted in the *IBC* and the *IEBC* should be confirmed before proceeding with plans to demolish plaster walls and ceilings. If demolition is done, OSHA-compliant precautions should be taken to protect the workers from the accumulated soot, lead paint dust, and asbestos fibers released during the demolition process.

Replacement of flat plaster can be done in three ways: apply new plaster to existing wood lath or remove the lath and install either new expanded metal lath or rock lath. Replastering existing wood lath requires that the lath be wetted to enhance bonding and prevent warping. Although the lath can be wetted in place, it can also be removed, soaked in a tub, and then reattached to the framing. However, once wood lath is demounted, installing metal lath or rock lath may be more cost effective. Whichever course is chosen, the new plaster can be applied using traditional methods and finished to match the original plaster surfaces.

For ceiling replacements, a new ceiling can be installed below the original one using new lath and a three-coat plaster system secured to furring strips attached through the original ceiling to the underlying structural framing. A contemporary alternative to any of these approaches for either walls or ceilings is to install a veneer plaster system consisting of gypsum board panels on furring strips over the existing plaster and then finish with two thin coats of plaster that can be troweled to match the original surface appearance. Caution is needed to ensure that neither of these two new plaster surface types compromises the visual appearance or integrity of any remaining ornamental plaster or woodwork.

For decorative plaster, the replacement process for severely damaged or missing features is comparable to the repair casting methods discussed previously. Replacement of missing features must be based on documented evidence or the remaining ornamentation and not simply conjecture. When original materials are damaged beyond repair, similar or compatible materials must be used as their replacement.

References and Suggested Readings

Calloway, Stephen, ed. 1996. *The elements of style: A practical encyclopedia of interior architectural details from 1485 to the present*. Rev. ed. New York: Simon & Schuster.

Cummings, Abbott Lowell. 1979. *The framed houses of Massachusetts Bay 1625–1725*. Cambridge, MA: Harvard University Press.

Diehl, John R. 1960. *Manual of lathing and plastering*. New York: MAC Publishers Association.

Eckel, Edwin C. 1928 (reprinted 2005). *Cements, limes, and plasters*. Dorset, UK: Donhead Publishing.

Fisher, Charles E. and Hugh C. Miller, eds. 1998. *Caring for your historic house: Preserving and maintaining: Structural systems, roofs, masonry, plaster, wallpapers, paint, mechanical and electrical systems, windows, woodwork, flooring, landscapes*. New York: Harry N. Abrams.

Flaharty, David. 1990. *Preserving historic ornamental plaster*. Preservation Brief No. 23. Washington, DC: United States Department of the Interior.

Gorman. J. R., Sam Jaffe, Walter F. Pruter, and James J. Rose. 1988. *Plaster and drywall systems manual*. 3rd ed. New York: McGraw-Hill Book Company.

Jandl, H. Ward. 1998. *Rehabilitating interiors in historic buildings*. Preservation Brief No.18. Washington, DC: United States Department of the Interior.

Jennings, Jan and Herbert Gottfried. 1988. *American vernacular interior architecture 1870–1940*. Ames: Iowa State University Press.

Jester, Thomas C., ed. 1995. *Twentieth century building materials: History and conservation*. New York: McGraw Hill Book Company.

Joway, Hubert F. 1981. Reattachment of loose plaster. *Bulletin of the Association for Preservation Technology*, 13(1): 40–41.

Konrad, Kimberly A. and Michael A. Tomlan. 1995. Gypsum board. In Jester 1995, 268–271.

Labine, Clem and Carolyn Flaherty, eds. 1983. *The Old-House Journal compendium*. Woodstock, NY: Overlook Press.

MacDonald, Marylee. 1989. *Restoring historic flat plaster*. Preservation Brief No. 21. Washington, DC: United States Department of the Interior.

Moss, Roger W. and Gail Caskey Winkler. 1986. *Victorian interior decoration: American interiors 1830–1900*. New York: Henry Holt and Company.

Phillips, Morgan W. 1980. Adhesives for the reattachment of loose plaster. *Bulletin of the Association for Preservation Technology*, 12(2): 37–63.

——. 1995. Aqueous acrylic/epoxy consolidants. *APT Bulletin*, 26(2/3): 68–75.

Sawyer, J. T. 1951 (reprinted 2007). *Plastering*. Dorset, UK: Donhead Publishing.

Selwitz, Charles. 2002. New consolidants for historic plaster: Studies at the Knickerbocker mansion in upstate New York. *APT Bulletin*, 33(4): 13–17.

Shivers, Natalie. 1990. *Walls and moldings: How to care for old and historic wood and plaster*. Washington, DC: Preservation Press.

Simmons, H. Leslie. 1990. *Repairing and extending finishes (Part 1)*. New York: Van Nostrand Reinhold.

Simpson, Pamela H. 1999. *Cheap, quick and easy: Imitative architectural materials, 1870–1930*. Knoxville: University of Tennessee Press.

Staehli, Alfred M. 1984. Scagliola: Restoration of an antique plaster finish in the Portland City Hall, Oregon. *Bulletin of the Association for Preservation Technology*, 16(2): 44–50.

Van den Branden, F. and Thomas L. Hartsell. 1984. *Plastering skills*. Homewood, IL: American Technical Publishers.

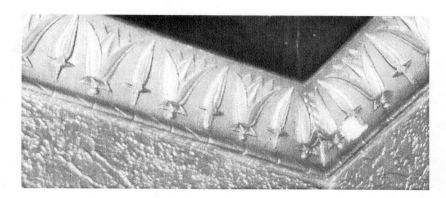

CHAPTER 18 Protective and Decorative Finishes

Unprotected and exposed materials that are vulnerable to moisture and sunlight will decay over time. As a precautionary measure, protective coatings were applied to stop this decay. At the same time, as civilization progressed, decoration was applied to common objects to make them more pleasant to look at or use. The expanded use of decoration, like the expanded use of metal, became a hallmark of the society's sophistication and power.

Over time, many protective and decorative coatings became one and the same thing. As such, numerous types of coatings and finishes developed as technology and aesthetic tastes changed throughout the nineteenth and twentieth centuries.

Coatings and finishes were used to protect and decorate objects, highlight architectural elements, or add visual interest to otherwise plain surfaces. For the purposes of this book, the classification of a coating or finish is based on how readily the original surface can be seen through the applied coating. These classifications include paint, stain, and clear finishes.

Paint has traditionally consisted of a liquid vehicle and pigments. The liquid could be oil (e.g., linseed oil, fish oil, or other oils) or water that both carried the pigments and acted as a binder as it dried. Pigments fell into two categories—concealing and tinting. Concealing pigments distinguished paint from stain and clear finishes as they concealed the surface (e.g., wood grain, patterns, or colors) on which the paint was applied. Concealing pigments included clay, whiting (ground-up chalk), various types

333

of lead, and titanium oxide, typically in assorted shades of white. Tinting pigments gave paint its coloration. Early tinting pigments were drawn from locally available umbers and ochers (e.g., clays and soils that were ground to a fine powder) that contained a variety of minerals and oxides that produced a yellow, reddish brown, or other earth tone. Lampblack (soot) was used to darken colors. In the twentieth century, additives that enhanced hardness, mildew protection, and other qualities were added.

Stains were made with a liquid vehicle and tinting pigments but had less or no concealing pigment. Due to its lower viscosity (resistance to flowing), stain penetrated the surface of the raw wood to which it was applied. Shingles and other exterior cladding and decorative elements were stained to protect them from the weather and possibly to provide some coloration.

Clear finishes were composed of various mixtures similar to the liquid component of paint and included specific plant-or-animal based ingredients that sealed the surface. The finish was clear or had some tonal coloration due to its specific ingredients. For example, shellac comes from the secretions of the lac beetle that impart a range of golden-brown colors, depending on the concentration. Clear finishes were also applied in the form of waxes, oils, and polishes that were hand-rubbed into the surface of the object being finished. Beeswax, linseed oil, walnut oil, and poppy seed oil are examples of these wood-finishing products.

As the coating dried, a chemical reaction occurred that caused it to harden. Water and solvents evaporated, while oil oxidized. The result was a surface that had some degree of durability. Early experimentation with various oils led to the recognition that some oils dried faster than others; hence, linseed oil became more popular than fish oil.

Composition

Due to the varying degree of locally available or affordable materials used to make protective and decorative coatings, a range of products has evolved. While the modern formulas may be significantly different from those of the eighteenth and nineteenth centuries, they remain largely known and categorized by their early origins as oil-, water-, or solvent-based media.

Oil-based coatings contained varying amounts of pigments, linseed oil, and white or red lead. These were more expensive than their water-based counterparts but lasted longer. Water-based coatings contained water, pigment, and a binder (e.g., animal hide glue, natural glue or gum). Solvent-based coatings contained solvents, tinting, and concealing pigments. The following list describes the composition of each type of coating:

- Acrylic paint: Often referred to as "latex paint," this paint uses acrylic resin, pigment, and water. Latex paint has become the leading paint used in interiors today.

- Casein paint: This paint is made from slaked lime, pigment, casein (a component of milk), and oil and is the modern enhancement of milk paints. Prior to the early nineteenth century, milk paints were made from skimmed milk and derived their color from earth pigments or tinting pigments made from the stems, leaves, and berries of various plants. Casein paint manufacturers use only the casein component of milk.

- Distemper: This water-based paint included water, glue, whiting, and coloring pigments. It was also called "calcimine" or "kalsomine paint" and was used for interior applications.

- Enamel paint: Traditionally, varnish was added to oil-based paint to enhance its sheen and was called "enamel paint." In discussing the paint used on stained glass windows in Chapter 15, the term "vitreous enamel paint" referred to the use of ground glass that formed the translucent tinting pigment of the paint.

- Fuming: Fuming is a process rather than a coating. When woods with a high tannic acid content (e.g., white oak) were exposed to ammonia fumes, the wood darkened. Longer exposures meant darker finishes.

- Glaze: Glaze consisted of linseed oil and varnish that sometimes had tinting pigments added. Used on interior applications, glazes added tonal depth to the decorative surfaces of flat walls and three-dimensional ornament made from plaster or composition.

- Gouache: This was a water-based paint that consisted of water, whiting, pigment, and gum arabic. It was used for decorative applications.

- Lacquer: This was a solvent-based coating that contained resin and solvents that evaporated to leave a hard finish. It originated in Asia and became popular in the late nineteenth century for furniture and built-in fixtures, such as fireplace mantels. The solvent used today contains a combination of naphtha, xylene, toluene, and acetone.

- Sand paint: This was a paint to which sand was added. It was applied to surfaces to give the object the appearance of stone.

- Shellac: Also known as a "spirit varnish," shellac derived its name from resins produced by the lac beetle that were mixed with various concentrations of denatured alcohol. Although other resins were used, "shellac" became the generic name for the product. Because shellac was made with alcohol, cleaning products containing alcohol should be avoided on shellac finishes, as they will dissolve the finish.

- Tempera: This was another water-based paint that included water, egg (yolk or white), and pigment. It was used for decorative applications.

- Varnish: This was made by combining a drying oil, such as linseed, walnut, or poppy seed oil, with resins, such as elemi, balsam, benzoin, rosin, sandarac, and mastic. In the 1700s, the most common varnishes were made with boiled linseed oil and rosin with lesser amounts of other resins. Varnish provided a harder finish than spirit varnishes, such as shellac, but by modern standards it is considered soft (see Figure 18-1). Modern synthetic varnishes are clear by comparison with natural varnishes (Allen 1994).

- Whitewash: This was a combination of water, slaked lime, and salt. Some mixtures included whiting (ground-up chalk), flour, glue, tallow, or soap. Whitewash was a nonstable finish that deteriorates rapidly and was used primarily for interiors. It must be removed before applying other types of paint, as they will not bind with it.

Manufacturing Process

As noted above, early paints were made by hand by combining the liquid binder with the pigments. Linseed oil was pressed from flaxseed and then boiled to remove impurities. Unfortunately, linseed oil tended to impart a yellow tone to an otherwise white paint and continued to yellow with age. Although a variety of other oils,

Figure 18-1 The varnish on this door has been marred by abrasion from keys.

such as tung and fish oil, were tried, linseed oil was most widely used. Until adequate lime and other pigment sources were identified and developed in the early colonies, these components were imported from England

The formulations were mixed by hand by painters who contracted with the building owner to make the paint, prepare the surface, and apply the paint. The limited availability of paint and painters made house painting an expensive proposition. Although by 1750 new houses were being painted in prosperous communities, many earlier houses of less wealthy owners were not painted until the nineteenth century (Little 1989). Some still have not been painted.

In the early to mid-nineteenth century, paint was still largely made by hand. By the 1860s, premixed paint was being manufactured and packaged for sale. For the first time, paint was more readily available, which made painting buildings less expensive. Throughout the late nineteenth century, production processes were improved and paint manufacturers began to offer designer palettes so that homeowners could understand how to select paints that complemented or contrasted with one another. As paint became more available, combinations of two or more colors were illustrated on images showing buildings with the colors applied to them.

The twentieth century saw a number of improvements that enhanced the performance of the coating as it was applied and extended the life of the coating. Among these was the introduction of acrylic-emulsion resins. Acrylic latex paint was introduced after World War II. Similarly, changes introduced by the mid-1960s (Carden 1991) to oil-based paints used alcohol and acid enhancers and led to the description of modern oil-based paints as "alkyd paints." The term "alkyd" was derived from the combination of *al*-cohol with the odd phonetic spelling of a-*cid* that was pronounced as "kyd."

Stylistic and Color Preference

The early North American colonists brought with them an awareness of and appreciation for decorative arts that included painting. While decoration was not their immediate concern in the first few decades of the seventeenth century, recognition of the value of protective finishes led to the increasing use of paints, varnishes, and shellac by the end of the century. While a number of houses built in the first century of settlement were never completely painted, trim pieces that were vulnerable to weathering were being finished in some manner by the late seventeenth century. The earliest coatings were typically whitewashes used on interior walls and exposed structural framing. Oil-based paints, if available at all, were used on exterior trim. The earliest documented use of distemper paint in New England occurred at the Eleazar Gedney house in Salem, Massachusetts, in ca. 1664 (Cummings 1979; Moss 1994). Distemper paints were used for indoor applications.

Since the early eighteenth century, particular color palettes were popular with specific architectural styles. In limited use in the early eighteenth century, paint coloration was restricted to browns, reds, and yellows that were made using pigments developed from locally available materials. As chemical formulations changed and various pigments were invented or imported, the color palette changed. For example, Prussian blue was introduced in the 1720s. Colors for both interior and exterior applications expanded into medium tones of greens, blues, reds, and yellowish whites throughout the Federal and Neo-Classical periods that ended in the early nineteenth century. In the mid-nineteenth century, with the Gothic Revival movement led by Andrew Jackson Downing and Alexander Jackson Davis, this palette shifted back much more to earth tones.

The aesthetic tastes of the Victorian period called for multiple colors for exteriors (see Figure 18-2) and clear finishes on dark woods, such as mahogany, or faux finishes on lighter woods to simulate the darker woods. Exploration and the opening of trade with Japan and China led to the use of japanning (black lacquer finishes) applied to a variety of furnishings and such features as fireplace mantels.

The early twentieth century was a period of numerous revival styles that also revived interest in the color palettes of earlier eras. This ultimately led to numerous

Figure 18-2 In the Victorian period, several paint colors were used to decorate the exterior of houses. Their purpose was to highlight specific details, such as window sashes, muntins, shutters, and other elements. Many of these details were painted in one color (often white) in the early twentieth century.

attempts to confirm what these earlier palettes may have actually been. Unfortunately, some researchers failed to account for the fact that linseed oil yellows over time. As a result, a number of the restorations in this period featured a bias toward more yellow tones that later researchers found to be inaccurate.

As the twentieth century progressed, competing architectural styles created an even wider array of paint schemes. Art Deco and Art Moderne brought forth combinations of lighter pastel colors, while the Prairie School embraced the earth tones. Commercial building technology shifted from masonry to curtain walls made of glass and metal that could be coated with porcelain enamel panels of virtually any color desired. By the end of the century, there was largely a bias toward lighter earth tones and pastel shades that appeared markedly different from many of the color palettes of the previous two centuries, especially those of the Victorian era.

APPLICATIONS

Protective coatings extend the useful lives of objects, especially those made from moisture-absorbent or moisture-sensitive materials. Paint, stains, and clear finishes provide the basic protection necessary to accomplish this. However, when the protective coating is enhanced by the use of color and pattern, the result extends the concept of protection to include decoration.

Decorative Painting

From the prehistoric cave paintings at Lascaux, France, to the murals and wallpapers of the late Renaissance, humans have been decorating the walls, ceilings, and floors of their living spaces with painted imagery for thousands of years.

The practice of enhancing the exterior appearance and perceived value of a building or object by making it appear to be constructed of more expensive materials has been common throughout history. Many of the earliest decorative painters were originally from France and Italy. The terminology used to describe decorative painting and finishes thus has French or Italian origins. Decorative painting terms like "faux finish," "trompe l'oeil," and "fresco" were taken directly from their country of origin to describe the process.

Decorative painters performed a variety of applications. Some were skilled artists who could paint murals much as a painter would paint a still life or landscape. Others created patterns that were painted freehand or employed a system of stencils. The various periods of decorative painting gave rise to notable examples of each technique. Rufus Porter was an acclaimed stencil painter whose work along the eastern seaboard is now highly prized. Muralists, such as John LaFarge, decorated many buildings across the country.

Stencils

Stencils were hand-painted on virtually any surface on the interior of the building. Stenciling was promoted as a sanitary alternative to wallpaper, since it did not attract insects and other vermin and was less expensive than wallpaper. Stencils were applied to the entire wall, the field above the chair rail, the frieze just below the ceiling, or as a border around window and door openings. Stenciling was also done on floors and ceilings, especially along the perimeter near the wall. While stenciling had been used in the eighteenth century, there were two periods of heightened popularity in the nineteenth century. The first occurred between 1810 and 1840 and the second during the Victorian period of the late nineteenth century. The popularity of stenciling declined significantly in the early twentieth century.

A stencil could be composed of one or more colors. Some simple patterns were considered naïve and could have been done by a less skilled painter. Many, however, were sophisticated and took into account color relationships, the proportion of elements, and the pattern proportion relative to the size of the room. A stencil was created by drawing the outline of the decorative subject on a thin piece of tracing paper. The drawing was then secured to a thin but sturdy piece of paper that served as the stencil template for applying a particular color. Pinholes were punched along the outline of the decorative pattern. A stencil was created by removing the paper from the area(s) outlined by the pinholes in the stencil paper. In multiple color applications, stencils were also pierced with registration marks to align each succeeding stencil.

Once the sequence of stencils was established, the first one was secured to the wall and its color was painted into the openings created by the holes in the stencil paper. That stencil was then removed, and when the paint dried, the next stencil was attached to the wall, aligning the registration marks. This process was repeated until all the colors had been applied and the stenciled pattern was complete. The process was then repeated along the surface until the stenciling formed a continuous repeating pattern. In some instances, the colors were applied solidly to fill the opening in the stencil paper; in others, artistic or painterly effects were added to each color as it was applied.

Stencils were either repeating classical or folk patterns. Classical patterns popular during the Federal and Greek Revival periods favored geometric patterns including Greek keys, alternating stripes, swags, urns, and simulated stone carving details, while folk patterns included vines with leaves, berries, flowers, stars, hearts, birds, stripes, and curves. Stencils from the late Victorian period were typically emboldened with brighter colors and larger elements, such as crests, fleur-de-lis, and other robust symbols of that period.

Trompe L'Oeil

Trompe l'oeil, literally meaning "trick the eye," was devised as a means of concealing an otherwise plain or unsightly surface and adding visual delight (see Figure 18-3). Originally, trompe l'oeil was used to re-create the effects of shadow and light produced by a three-dimensional object. One common method was the grisaille technique that took advantage of varying shades of gray to simulate the lighting effects found on an actual plaster molding.

On interiors, in addition to simulated three-dimensional ornament, flat plaster walls were decorated with trompe l'oeil that began to create such things as striped tent panels, furniture, pastoral views through a nonexistent window, and many other whimsical or serious images. On exterior walls, trompe l'oeil has often been used to create advertisements or depict a more interesting facade than what may have originally existed.

Murals

In the eighteenth century, murals or frescoes were painted on walls. In some cases, the murals were painted on sheets of paper secured to the ceiling or wall; in others, the painting was done directly on the plaster ceiling or wall surface. Murals were also painted on wood paneling over fireplaces as well as on doors. Murals were less expensive than the decorative papers imported for use on ceilings or walls and were created by freehand brush stroke painting methods. They were popular throughout

Figure 18-3 Trompe l'oeil was used to trick the eye into seeing something other than a plain surface. In this case, the image of a former post office in Portland, Maine, was painted on the wall of this building that is adjacent to the site of the post office, which was razed in the late twentieth century.

Figure 18-4 Murals were an important decorative feature used on interior walls instead of wallpaper and on the exterior of buildings to advertise services or commemorate events. This mural is on the exterior wall of a violin shop in Salt Lake City, Utah.

the eighteenth century and the first half of the nineteenth century and had a revival in the early twentieth century in commercial and civic buildings (see Figure 18-4).

While murals declined in popularity in private dwellings, large civic buildings (e.g., state capitols, city halls, museums, and libraries) constructed in the past 150 years feature a number of murals that celebrate aspects of local historic and cultural events. During the Depression, the Works Progress Administration (WPA) commissioned unemployed painters to paint murals in post offices throughout the country.

Faux Finishes

False or faux finishes (see Figure 18-5) were the most common means of giving less expensive materials the appearance of more expensive ones. These techniques, called "faux bois" and "faux marbre," were a form of trompe l'oeil that transformed lesser materials into imitation wood and marble (Brown 2003). Pine and fir were painted to look like oak, maple, mahogany, and other exotic woods in the faux bois process known more commonly as "graining." The faux marbre process was used to simulate marble and is also called "marbling." Graining consisted of matching both the color and pattern of the figure (grain) and ground (background) of a specific species of wood as it would appear under a clear finish. An assortment of brushes, feathers, rags, sponges, shaped pasteboard (later succeeded by rubber and plastic), and other implements were used to re-create the wavy patterns revealed when wood was cut along the grain. Knots, burls, and other features could be created if so desired. Similarly, in marbling, veins and intrusions could be simulated to create the appearance of marble and other expensive stones.

Architectural Metals

Architectural metals provided an opportunity to decoratively finish a variety of materials. Wood and metal objects were covered in gold, silver, or copper. Lesser metals were plated to give them a more decorative finish. In addition to adding

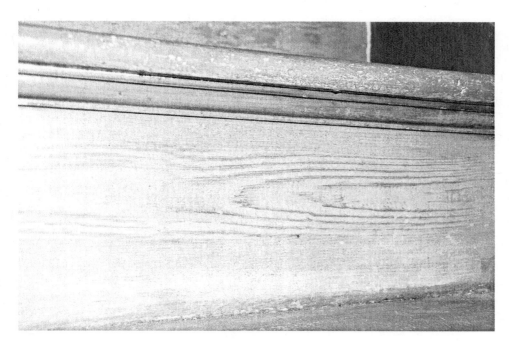

Figure 18-5 Faux finishes were a common way to make a less expensive material look like a more expensive one. This fir stair riser was grained to look like oak.

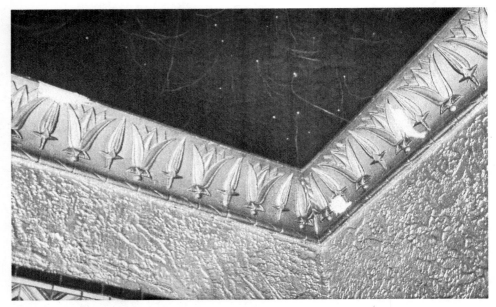

Figure 18-6 Dutch metal can be used in place of more expensive gold leaf. Shown here is a plaster cornice finished with Dutch metal. The white spots are recent repairs awaiting a new Dutch metal finish.

decorative metal to surfaces, a number of methods used chemical washes to create a decorative patina on them. In a more general conservation strategy, protective coatings had long been used to protect metals from decay.

One form of decorative surface finishing dating back several millennia involved gilding metal onto nonmetal surfaces by using sheets of metal hammered to an extremely thin, almost tissue-like, dimension referred to as a "leaf." The metals originally used were gold, silver, and copper. The surface being leafed was treated with gesso or sizing, and then the leaf was applied. A coating, usually shellac or lacquer, was applied to seal and protect the leaf. Leafing could be done to interior or exterior elements of the building. In the nineteenth century, to reduce costs, tissue-thin sheets of Dutch metal (a brass consisting of 84 percent copper and 16 percent zinc) were applied, with little or no visible difference from the more expensive gold leaf (see Figure 18-6).

Electroplating was used to secure an ornamental metal to the surface of a less valuable metal. For example, a thin surface layer of nickel or German silver could be secured to the surface of lesser metals. In this process, the object (e.g., a brass faucet) that was to be plated with a nickel finish was attached to the negative lead of a battery and placed in a salt solution containing nickel ions. On the positive lead, a nickel rod was attached and placed in the solution. As the electrical current flowed from the positive to the negative conductor, the nickel ions were attracted to the plating object and secured through an oxidation process. Meanwhile, ions from the nickel rod replaced the ions removed to the plated surface. When a sufficient deposit was made on the plated surface, the object was removed, dried, and polished to the desired finish.

A variety of methods have evolved to create patinas on a metal surface for ornamental purposes. These generally involve mixtures of acids or salts that react with the metal surface to either oxidize or color it in some way. In this manner, brass can be darkened or copper and bronze can be aged in appearance. The degree to which the patina forms is a function of chemical strength, metal reactivity, and dwell time (i.e., the time that the metal remains in contact with the solution). In many modern rehabilitation projects, brand-new materials are treated in this fashion to match existing original fixtures.

Since exposed ferrous metals corroded easily, applying a lead-tin mixture (terne) or zinc oxide paint became a common means of inhibiting corrosion on exposed ferrous metal sheet stock surfaces. Historically, these coatings were applied either by dipping the metal into a molten bath (i.e., hot dip galvanizing) of the coating material or by electroplating. A variety of paint coatings and sealants were also devised and used to inhibit rust and corrosion.

Typical Problems

Like other historic fabrics exposed to air, light, and moisture, paints and clear finishes can decay. Many of the same decay processes from pollution, moisture, and abrasion that affect other materials can also affect these finishes. Often it is the failure of the surface finish that signals the presence of decay, especially when an opaque coating, such as paint, is involved. Beyond the mechanical failure of the surface, other problems include overpainting and soiling. These last two may actually be the result of poor maintenance practices. Surfaces have not been adequately maintained and become dirty. This leads to new paint being applied or improper cleaning methods being used, which instead damage the surface finish. Lastly, lead was a common component in paint for many centuries. Its presence has now been linked to health problems in young children.

Surface Failure

Paint and finish failures (see Figure 18-7) are caused by a variety of factors. Without proper surface preparation, such as sanding and scraping, when repainting, the eventual accumulations of paint may begin to break apart. This failure first appears as light surface cracks known as "crazing," but then the surface continues to break

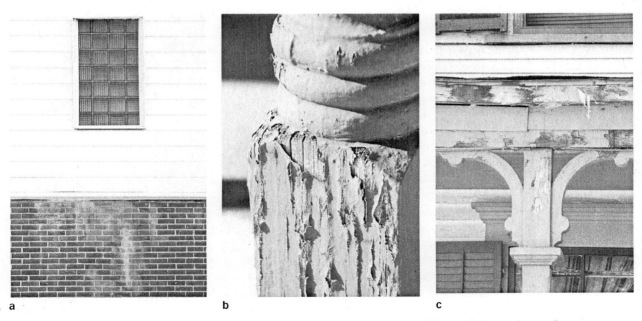

a   b   c

Figure 18-7   Paint failures are typically described by their appearance: (a) chalking, (b) alligatoring and flaking, and (c) peeling.

apart through several cycles of inadequately prepared repaintings until the paint fails completely. Weakly bound paint will become chalk-like in texture. Paint that has aged and become brittle will break and rupture as daily thermal and moisture cycles cause the substrate to expand and contract. This process causes the paint to look much like the skin of an alligator and hence is called "alligatoring." Interlayer failure occurs when two layers of incompatible paint may not bond and slough off over time. Moisture migrating out of the material will cause blistering. When the blister ruptures, the paint dries and can easily flake.

Overpainting

One of the most common issues in painted decoration is layers of paint covering the original decoration. Not only can subsequent layers of paint cause stresses in the earlier layers as more weight is added to the surface, but original details can be obscured or obliterated by the addition of later layers of paint or even clear finishes. This commonly occurs during updating when previous faux finishes or clear finishes are painted to create a new appearance. This practice may have its origins in the fact that early interiors lit by nonelectric lamps and heated with fireplaces accumulated dirt and soot on any exposed surface. The practice may also have developed as aesthetic tastes changed. Whatever the cause, the result is that original features, such as stenciling, faux finishes, and murals, have been routinely lost in the name of progress.

Soiling

Coatings and clear finishes can become soiled through exposure to air pollution, smoke and cooking fumes, and grime created by the natural oils on fingers and hands. When an inappropriate cleaning method is used, the coating can fail. Modern cleansers containing harsh chemicals can destroy the surface finish or cause it to become clouded, darkened, or otherwise degraded.

Exterior paint can be cleaned with a low-pressure water wash and mild detergents. Interior paint and clear finishes can be cleaned using mild soaps and nonabrasive cleansers. As with any cleaning project, a small test section in a hidden location should be used to assess the effectiveness and effect of the cleaning process. Several tests may be needed to find the best process. Unfortunately, even people with the best intentions can do irreparable damage when they use inappropriate cleaning methods and products on historic finishes. For example, a volunteer at a small historic church attempting to clean a pew used a modern cleaning product that removed the original historic graining on a prominent portion of the pew before realizing what was happening.

Some cleaning processes may seem counterintuitive in that the cleaning product may be marketed as made for wood surfaces but, in reality, it is the finish on the wood that is being cleaned rather than the wood itself. For instance, tests revealed that when vegetable oil soap was used on a polyurethane-finished hardwood floor, the long-term result was a buildup of the dried vegetable oil, which degraded the appearance of the floor and then hampered efforts to refinish it.

Lead

Lead was a primary component in paint for many centuries. Long valued for its durability, lead paint became unpopular when its effects on health were fully understood. Research revealed that long-term exposure to lead can cause learning and hearing disabilities in humans, especially young children. In 1978, lead was banned for use in house paint. Unfortunately, that date makes it extremely likely that lead paint was used in houses from the historic period described in this book.

Testing kits are available to confirm the presence of lead. These kits include a solution that is applied to the suspected surface and change color (e.g., purple or black) in the presence of lead. For multiple layers of paint, locate a hidden spot, scratch the paint through to the layers beneath the surface, and then test for lead. Alternatively, ultraviolet light can be used to detect lead (Carden 1991). Various paints will fluoresce in different tones that can be used to indicate the presence of lead. There are also portable hand-held devices that use x-ray fluorescence (XRF) technology to identify the existence of lead paint below the visible surface.

While removal of the lead is not required, the presence of all lead-based materials, including paint, must be disclosed to any potential buyer or renter. In lieu of complete removal, the painted surfaces that contain lead must be stabilized, and all loose and flaking paint must be removed following lead abatement guidelines available from the local county health department or building department. Typical interior paints are not sufficient to encapsulate lead paint over the long term. There are, however, epoxy-based paints for this purpose when necessary.

Recommended Treatments

When the historic and artistic merits of a particular finish of a decorative treatment are significant, employing a consultant who specializes in paint and finishes is imperative. The number of professionals who can perform the appropriate analysis and provide accurate findings and recommendations on how to proceed is limited. This is especially true for murals, faux finishes, trompe l'oeil, and stencils. The best experts often have a long waiting list and may be unavailable in the short term. The SHPO and other prominent state or national preservation organizations maintain directories of paint consultants. When suitable experts have been identified, ask them for a portfolio of their project work and references from previous clients. When possible, visit former projects to view the results.

Other painting tasks, such as painting the overall exterior, may be completed by local contractors based on their qualifications and pricing. Simply going with the lowest bid may not be the best method, since there are usually other factors concerning preservation sensitivity, awareness of local requirements regarding lead paint abatement, and so forth. One important issue is that it is only necessary to address paints and finishes that have become unstable. Conservation means retaining the original material in place. Restoration and repair involve the removal of *only* loose, damaged, and unconsolidated paint or finish.

Complete removal and replacement of all paint or finishes, including intact finishes that remain otherwise well bonded to the surface, is not recommended. When limited removal is warranted, it should be done with the gentlest methods possible — hand scraping and sanding. Using a heat gun or a heat plate can speed up the stripping process, but one must be careful not to singe or char the underlying substrate. There are also a variety of chemical strippers that can remove the finish, ranging from extremely benign soy- and citrus-based strippers to particularly nasty chemicals that contain methylene chloride (MC), acetone toluene methanol (ATM), N-methyl pyrrilodone (NMP), dibasic esters (DBEs), pure sodium hydroxide, or ammonium hydroxide. The more powerful chemical strippers all come with specific health and safety concerns that should be investigated before using them. By no means should aggressive stripping methods like sandblasting, open-flame blowtorching, or mechanical power strippers be used that could irreversibly damage the underlying substrate.

Concerns about health and sustainability have led to the increased use of low volatile organic compound (VOC) paints and coatings. VOCs consist of chemicals that evaporate under normal room conditions. They are found in paints, paint strippers, cleaning products, carpets, and other building materials. As the VOC off-gases, it can cause both short- and long-term health problems. As these compounds accumulate, they reduce the indoor air quality (IAQ) by contributing to what is known as "sick building syndrome" in inadequately ventilated buildings.

Conservation

Paint and finished surface treatments may be conserved as long as the sources of decay are identified and either removed or mitigated. Many of the safeguards and recommendations given for the surfaces described earlier in this book apply to these finishes as well—remove moisture sources, limit access for abrasive contact, provide protective screening materials, protect surfaces during construction, and so on. Any conservation treatment should be reversible and should allow for appropriate elimination or prevention of trapped moisture.

In interior spaces, where light (especially ultraviolet light from the sun or fluorescent lighting) can cause color to fade, sunlight and fluorescent light need to be minimized. Period-appropriate drapes, shades, and other shading devices should be considered in mitigating light-related decay. Inappropriate and noncompatible cleaning methods should be eliminated or replaced by methods that are benign. Harsh chemicals and abrasive cleansers are commonly the cause of rapid short-term decay of finishes. This problem usually can be quickly identified and corrected, as long as the maintenance staff is adequately trained or supervised.

Restoration

A common goal has been to accurately match the colors of a specific time period. Since removal of overlying layers of paint to entirely reveal a particular layer is not advised, the primary goal is to identify the chronology of paint layers or establish a reasonable representation of the color used during the period of interest. There are two approaches to paint color analysis—scientific and historical (Carden 1991; Moss and Winkler 1986; Perrault 1978).

In the scientific method, there are three ways to identify paint coloration: scraping, cratering (also known as "surface polishing"), and extracting, all of which cause nominal damage to the historic fabric. Scraping is done by simply scraping away layers of paint with a scalpel or hobby knife and solvents to reveal the successive layers of paint. Cratering is done by removing layers of paint with sandpaper and mineral oil in concentric circles. The paint is usually sanded with a piece of sandpaper down to the original surface material. Then, as each layer is revealed in larger concentric circles, the circle is expanded to reveal the next layer, leaving the rings formed by earlier sanding intact. Eventually, the circles are expanded to reveal all the layers of paint (see Figure 18-8). These methods have several drawbacks, including alteration of the surface color while scraping and disruption of the overall surface.

In the extracting approach, a small, discreet sample of paint and its wood or masonry surface substrate are cut from the surface. On painted metal surfaces, the extraction method leaves the metal in place and only the paint layer portion of the sample is taken to the lab. The sample is then mounted in a clear acrylic suspension and left to cure. The acrylic block is then sanded on one side, using finer and finer grit sandpaper to ultimately reveal the cross section of the sample. This

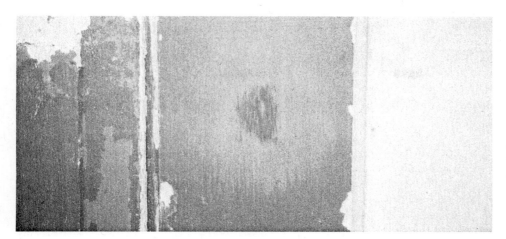

Figure 18-8 Carefully sanding away layers of overpainting is a means of identifying earlier paint colors. This is often done in a circular pattern, as shown here.

sample is then secured to a wax or clay mount and placed under a microscope. The layers are viewed through the microscope eyepiece (see Figure 18-9). In this manner, the coloration and number of layers can be catalogued. When several samples from assorted painted surfaces in a room are treated using either approach, a chromochronology (color sequence) can be created to recognize color palettes and determine when "new" construction occurred since later construction will not share all the layers of the earliest construction.

Care must be taken when analyzing colors, since linseed oil in the paint or dirt on the surface or even ambient lighting may cause a color shift. Linseed oil yellows with age. Soot and nicotine can darken and shift colors to darker or brownish hues. Some layers may be a primer rather than a finished surface. Ambient lighting can enhance or diminish certain colors, depending on the type of lighting used. Full-spectrum light is best when color matching to avoid color shifts.

To standardize color designations instead of using manufacturers' multiple and arbitrary names for the same color, the Munsell and Ostwald systems have been developed that enable color specification through scientific designation. When a color has been designated in either of these systems, it can then be ordered from a

Figure 18-9 Several layers of paint are evident in this sample. The dark spot along the top of the paint sample is a piece of dirt embedded in the second layer of paint.

manufacturer or local paint distributor. Although the Munsell and Ostwald systems measure the same thing, their terminology and designation method are slightly different. The Munsell system is based on hue, value, and chroma (saturation level). The Ostwald system is based on hue, saturation, and brightness.

In the historical approach, paint manufacturers created specific palettes based on paint color research done in the early twentieth century for Colonial Williamsburg and later for other architectural periods that can be used to establish a possible color scheme found in the targeted restoration period (Moss and Winkler 1986; Phillips and Weiss 1975; Schweitzer 2002). Although the sampling done using the scientific method may identify earlier combinations more precisely, the historical method uses the reproduction palette instead and provides appropriate color matching usually at a significantly lower cost.

Repair Strategies

Much like conservation, the repair of murals, trompe l'oeil, and stenciling may require an experienced decorative painting conservator to complete repairs appropriately. Such repairs should then be protected from damage for the duration of the project.

Paint and clear finish repairs consist of applying a compatible finish on the damaged surface. Critical to any repair is surface preparation. While it is not necessary to remove all layers of finish, the surface should be cleared of loose, flaking, and peeling paint or finish. If necessary, sharp edges can be scraped or sanded down. Interior paint and finishes can be reapplied as needed. For exterior locations, bare surfaces should be primed before applying new paint.

For lead paint, it may be too costly to strip the woodwork; instead, encapsulation may be appropriate. Some woodwork, such as a door, may cost-effectively be demounted and chemically stripped off-site. However, the wood may be damaged in the process of demounting or the stripping solvents may adversely affect the wood grain and glues used to construct the woodwork. When considering paint stripping off-site, go to the facility where the stripping will be performed to verify the process and materials used.

Lead paint may also be removed using a lead abatement contractor. Dust and effluent created by the stripping process must be carefully contained and disposed of properly following local health department guidelines. This on-site removal is most likely to be the more costly alternative in terms of time and labor costs, so budget accordingly.

Replacement Strategies

Completely stripping and replacing finishes is not recommended. The accumulated effects of age, wear, use, and environment have helped form a patina that provides character and authenticity. When a portion of the original finish has been lost, the processes for applying glazes and other clear finishes should be those used in refinishing and repairing finishes on furniture. In these situations, the assistance of a craftsperson or finishes consultant is of the utmost importance to match the tone and sheen of the remaining original finish. The final finish that matches the existing finish may result from a trial-and-error process that combines traditional finish recipes with modern commercial mixtures. For example, in rehabilitation work where it is often necessary to infill missing or damaged woodwork, it is common to apply samples of a test finish formulation to a separate piece of wood of a similar species to match the tone and sheen. Since age, unknown initial ingredients, and long-term darkening or fading may have altered the original finish, mixing several finish coloring additives to match the original finish exactly may be necessary.

Samples of the chromochronology stored in off-site locations may become lost or unavailable. If removing finishes is necessary, then it is customary to leave an undisturbed portion of the finish that may serve as the physical on-site evidence of the previous accumulations of finishes. If future methods and processes emerge that can provide more accurate results than current technology, this undisturbed portion can be used for future investigation(s).

References and Suggested Readings

Albee, Peggy A. 1984. A study of historic paint colors and the effects of environmental exposures on their colors and their pigments. *Bulletin of the Association for Preservation Technology*, 16(3/4): 3–25.

Alderson, Caroline. 1984. Re-creating a 19th century paint palette. *Bulletin of the Association for Preservation Technology*, 16(1): 47–56.

Allen, Sam. 1994. *Classic finishing techniques.* New York: Sterling.

Association for Preservation Technology. 1969. Paint color research and house painting practices. *Newsletter of the Association for Preservation Technology*, 1(2): 5–20.

Aurisicchio, S., A. Finizio, and G. Pierattini. 1983. Real-time holographic interferometry for painting conservation and restoration. *Bulletin of the Association for Preservation Technology*, 15(2): 11–16.

Brown, Ann Eckert. 2003. *American wall stenciling 1790–1840.* Hanover, NH: University of New England Press.

Calloway, Stephen, ed. 1996. *The elements of style: A practical encyclopedia of interior architectural details from 1485 to the present.* Rev. ed. New York: Simon & Schuster.

Carden, Marie L. 1991. Use of ultraviolet light as an aid to pigment identification. *APT Bulletin*, 23(3): 26–37.

Cirker, Blanche. 1996. *Victorian house designs in full color: 75 plates from "Scientific American—Architects and Builders Edition," 1885–1894.* Mineola, NY: Dover Publications.

Coffin, Margaret. 1986. *Borders and scrolls: Early American brush-stroke wall painting 1790–1820.* Albany, NY: Albany Institute of History and Art.

Cummings, Abbott Lowell. 1979. *The framed houses of Massachusetts Bay 1625–1725.* Cambridge, MA: Harvard University Press.

Downs, Arthur Channing, Jr. 1974. The introduction of American zinc paints, ca. 1850. *Bulletin of the Association for Preservation Technology*, 6(2): 36–37.

Dresdner, Michael. 1992. *The woodfinishing book.* Newtown, CT: Taunton Press.

Finkelstein, Pierre. 1997. *The art of faux: The complete sourcebook of decorative painted finishes.* New York: Watson-Guptill Publications.

Fisher, Charles E. and Hugh C. Miller, eds. 1998. *Caring for your historic house: Preserving and maintaining: Structural systems, roofs, masonry, plaster, wallpapers, paint, mechanical and electrical systems, windows, woodwork, flooring, landscapes.* New York: Harry N. Abrams.

Johnson, Ed. 1983. *Old house woodwork restoration: How to restore doors, windows, walls, stairs and decorative trim to their original beauty.* Englewood Cliffs, NJ: Prentice Hall.

Little, Nina Fletcher. 1989. *American decorative wall painting 1700–1850.* New York: E. P. Dutton.

Livingston, Dennis. 1997. *Maintaining a lead safe home: A do-it-yourself manual for home-owners, property managers, and contractors.* Baltimore: Community Resources.

Marx, Ina Brouseau, Allen Marx, and Robert Marx. 1991. *Professional painted finishes: A guide to the art and business of decorative painting.* New York: Watson-Guptill Publications.

Matero, Frank G. and Joel C. Snodgrass. 1992. Understanding regional painting traditions: The New Orleans exterior finishes study. *APT Bulletin*, 24(1/2): 36–52.

Moss, Roger W. and Gail Caskey Winkler. 1986. *Victorian interior decoration: American interiors 1830–1900*. New York: Henry Holt and Company.

Moss, Roger, ed.1994. *Paint in America: The colors of historic buildings*. Washington, DC: Preservation Press.

Newton, Roger Hale. 1943. On the tradition of polychromy and paint of the American dwelling from colonial to present times. *The Journal of the American Society of Architectural Historians*, 3(3): 21–25, 43.

Park, Sharon C. and Douglas C. Hicks. 1995. *Appropriate methods for reducing lead-paint hazards in historic housing*. Preservation Brief No. 37. Washington, DC: United States Department of the Interior.

Penn, Theodore Zuk. 1984. Decorative and protective finishes, 1750–1850: Materials, process, and craft. *Bulletin of the Association for Preservation Technology*, 16(1): 3–46.

Perrault, Carole L. 1978. Techniques employed at the North Atlantic Historic Preservation Center for the sampling and analysis of historic architectural paints and finishes. *Bulletin of the Association for Preservation Technology*, 10(2): 6–46.

Phillips, Morgan W. 1983. Acrylic paints for restoration: Three test applications. *Bulletin of the Association for Preservation Technology*, 15(1): 2–11.

—— and Norman R. Weiss. 1975. Some notes on paint research and reproduction. *Bulletin of the Association for Preservation Technology*, 7(4): 14–19.

Richey, Jim. 2000. *Finishing: Methods of work*. Newtown, CT: Taunton Press.

Schweitzer, Robert. 2002. *Bungalow colors: Exteriors*. Salt Lake City, UT: Gibbs Smith, Publisher.

Shivers, Natalie. 1990. *Walls and moldings: How to care for old and historic wood and plaster*. Washington, DC: Preservation Press.

United States Environmental Protection Agency. 1998. *Lead in your home: A parent's reference guide*. Washington, DC: Government Printing Office.

Weeks, Kay D. and David W. Look. 1982. *Exterior paint problems on historic woodwork*. Preservation Brief No.10. Washington, DC: United States Department of the Interior.

Welsh, Frank S. 1982. Paint analysis. *Bulletin of the Association for Preservation Technology*, 14(4): 29–30.

—— and Charles L. Granquist. 1983. Restoration of the exterior sanded paint at Monticello. *Bulletin of the Association for Preservation Technology*, 15(2): 2–10.

PART V

Special Topics

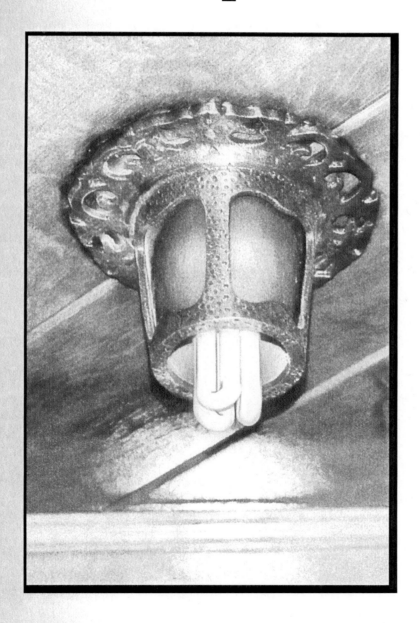

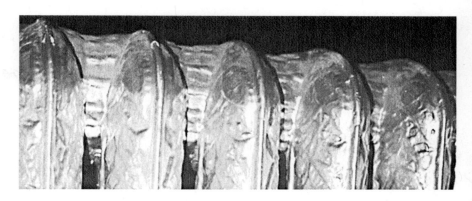

19 Heating, Ventilating, and Air-Conditioning Systems

OVERVIEW

Modern heating, ventilating, and air-conditioning (HVAC) systems consist of an integrated group of components that make buildings safe and comfortable to inhabit and operate. Early buildings relied on natural opportunities for sunlight and ventilation to enhance comfort. Where these opportunities were lacking, occupants frequently devised artificial means, such as fireplaces, stoves, and fans, to enhance or replicate the natural phenomena. Many modern HVAC systems are the result of enhancements of those earlier heating or cooling efforts.

In early construction, sunlight and natural ventilation were provided directly by operable windows and overhead skylights. The sun provided passive solar heat, which was controlled by movable awnings, porches, and shutters. Buildings were constructed to reflect the climate. These devices were smaller and more compact in cold climates and bigger and more open in warm, humid climates. Prior to the development of the skeletal framing systems of the late nineteenth century and the lightweight curtain wall systems of the twentieth century, heavier materials (e.g., stone, masonry, timber, and logs) provided a thermal mass that tempered the immediate extreme effects of the sun and the outdoor temperature.

The HVAC systems introduced in the past four centuries developed from enhancements of earlier methods. For example, heating by a central fireplace in a compact seventeenth-century New England colonial hall and parlor house had evolved into central heating using a fan and a ductwork distribution system by the early twentieth century. Many systems in use today are based on similar transformations over time.

What are typically seen inside a building are the terminal devices of each system. These devices (e.g., air registers, diffusers, and radiators) often are character-defining features that are an important part of the preservationists' responsibility. What is not seen is the infrastructure that produces or transports heating, ventilation, and air-conditioning media.

Unfortunately, these devices are vulnerable to wear and tear, removal, and insensitive alterations that reduce the visual integrity of a historic space. In contrast, the infrastructure serving these devices is often much of the original construction. Lastly are the units that produce the service media. While most buildings do not or no longer generate their own power or water resources on-site, many do produce heating, ventilating, and air-conditioning using generating devices, such as furnaces, boilers, chillers, and fan systems, that are generally not part of the public building space.

As technology evolved and improvements were introduced, many devices and systems were updated or retrofitted with new components to either lower operating costs or use less nonrenewable energy resources, such as coal, oil, and natural gas. The late twentieth century saw the advent of digital technologies that were added as an overlay control system for various HVAC systems. This trend has been particularly strong since the energy crises of the 1970s and the emerging late-twentieth-century demand for sustainable buildings. These controls typically monitor HVAC equipment and automatically adjust computer-based controls that can schedule specific operations, such as temperature settings, air flow, and other energy-saving activities.

A rehabilitation project will probably involve introducing or expanding modern service systems. Methods for integrating modern services often require creative approaches to managing the visual intrusions that their installation can create. A close survey may reveal paths for distribution networks that can be concealed from public view. These paths include closets (particularly if they are vertically aligned on multiple floors), cavity walls, or an unused horizontal plenum or vertical chase created as part of the original construction. Similarly, products can be selected that minimize the visual impact of the terminal devices due to their size and color. If need be, these devices can often be faux finished to conceal them more effectively.

HEATING, VENTILATING, AND AIR-CONDITIONING SYSTEMS

Modern thermal comfort requirements have produced a wide variety of systems to provide heated, tempered, and cooled air that are commonly referred to as "heating, ventilating, and air-conditioning" (HVAC) systems. These systems can be designed to keep temperatures and moisture conditions within a specific narrow range as determined by the user's needs. As these systems evolved, the range of thermal variance that occupants encountered also diminished. Occupants of buildings built before the advent of modern systems adapted to wider variations throughout the day and year as part of the cycle of daily life.

By comparison, modern systems provide more uniform comfort levels to which the occupant has become accustomed, although an occupant's comfort level varies based on gender, health, activity level, and metabolism. Throughout the second half of the twentieth century, as more opportunities for thermal comfort were introduced in new buildings, a growing expectation of comparable comfort developed among the users of older buildings. This expectation led to adaptations to existing systems or the installation of separate package HVAC units, especially for cooling.

Systems for residential and small commercial spaces have long been the simplest configuration due to the smaller buildings they typically served. A single-family house is typically the simplest unit of thermal control. As building types increase in size, the size and complexity of their HVAC systems increase as well. While small- and large-scale systems use the same basic principles, the larger systems tend to be more complex because they include a multitude of individual thermostatically controlled zones and a variety of performance requirements for temperature, humidity, fresh air, and odor control.

Demands for energy conservation in the late twentieth century produced increasingly complex control systems to ensure that energy use was minimized, if not optimized. HVAC, lighting, and electrical energy conservation standards, such as the American Society of Heating, Refrigerating, and Air-Conditioning Engineers (ASHRAE) ASHRAE 90 series and the Model Energy Code, were introduced and are still used to define energy performance parameters of residential and nonresidential buildings. In this same period, conservation also meant the adaptation of heating or cooling generating equipment or conversion to different fuel sources. In some instances, the mechanical equipment was removed and replaced; in others, new units were installed throughout the building as needed.

Historic Development

In vernacular buildings, the fireplace was the primary source of heating, while windows provided ventilation. Throughout time, architectural form has also been a factor in enhancing comfort. The high ceilings that allowed daylight penetration deeper into an interior space also allowed warm air to rise above the occupied space, particularly where winters were mild and summers were hot and humid (see Figure 19-1). In colder regions, where winters were long and cold, spaces were more compact and had lower ceilings to conserve heat. Larger windows that admitted daylight also admitted passive solar sunlight that could warm a space during the cooler parts

Figure 19-1 The main house at the San Francisco Plantation in Saint John the Baptist Parish, Louisiana, features numerous strategies to increase ventilation and provide shade from the sun, which increase occupant comfort, including (among other things) high ceilings, deep wraparound porches with open railings, and a central hall for horizontal and vertical air circulation.

of the year. In warmer climates, windows were shaded with awnings or louvered shutters that admitted cooling breezes while blocking the ultraviolet energy that caused a space to overheat.

In the twentieth century, as thermal comfort expectations, especially for cooling, shifted with the enhancement of mechanically operated HVAC systems, familiarity with and the use of the natural thermal comfort strategies used in many vernacular buildings faded. However, designers of notable sustainable buildings of the late twentieth century have begun to reintroduce these strategies with growing acclaim.

Heating

Since the human body does not adapt well to cold environments, heating systems were the first to emerge in common usage. Primitive societies used open fires to warm themselves and cook their food. Over time, fireplaces and, subsequently, free-standing stoves were used to provide central heating to buildings. In 1820, the first warm-air central heating systems came into use in which the heat was generated in the basement and fed into enclosed vertical chases that allowed heated air to rise and flow into occupied spaces (Ierley 1999b). One of the earliest large-scale systems for heating and ventilating was located at the United States Capitol. Planning for this system began in 1855, and it has been noted as the first complete system of its type installed in the United States (Donaldson and Nagenast 1994).

Early warm-air central heating systems relied on natural convection of air that allowed it to rise when heated or drop when cooled. In this process, warmed air migrated up to the living spaces. In the earliest systems, the increased air pressure created as the heated air entered the space caused the stale cooler air to be displaced to the outside atmosphere. A gravity air return system was introduced in 1915 (Konzo 1992, 128) that, instead of venting to the atmosphere, allowed the cooler air in the room to flow back down into the furnace. At this point, although fan-powered systems were in use in larger nonresidential buildings, the high cost and limited availability of electricity made it too costly to install them in individual residential buildings. However, after World War II, forced-air furnace systems became more common in residential construction (Ierley 1999a; Konzo 1992).

Steam heating came into use in the second half of the nineteenth century. The construction of large central steam-generating plants in the later 1870s and 1880s fostered the expanded use of steam in commercial buildings (Donaldson and Nagenast 1994). As boiler technology improved, smaller boilers were developed for residential buildings. In 1874, William Baldwin introduced the "Bundy," one of the earliest American radiators known (Oliver 1956). By the late nineteenth century, a combination of a steam boiler and a network of distributed cast iron radiators (see Figure 19-2) had been developed for residential applications modeled after the larger systems used in commercial heating. In these early systems, boilers provided steam via a one-pipe system to the radiators. As the steam circulated through the radiators, it condensed back into water. This condensate flowed back down the sloping pipes to the boiler, where it was reheated to steam. This system did not use pumps but instead relied on the thermal convection of the steam-condensate system. A later two-pipe system evolved that incorporated a separate return line to the boiler for the condensate. Although more efficient, this system was only used in public commercial buildings or large private residences because of its higher cost.

The early twentieth century saw the introduction of hot-water radiation systems that incorporated principles similar to those of steam radiation systems. Gravity hot-

Figure 19-2 Radiators were made from cast iron that was embellished with ornamental images to make them more decorative. Unfortunately, inappropriate paint can act as an insulator and decrease the efficiency of the radiator.

water systems circulated heated water through the radiators, and the cooled water was returned to the boiler for reheating. This system was quieter and more effective than the steam systems. With the later reductions in electricity costs, small pumps were added to create a forced hot-water system that competed very effectively with forced warm-air systems (Konzo 1992). In 1922, the fin-tube convector radiation system was introduced (Donaldson and Nagenast 1994). It became the forerunner of the baseboard radiation systems used in all building types today. The fin-tube was a combination of thin metal fins attached to a tube or small-diameter pipe. The fins provided a greater exposed surface that was needed to dissipate heat from the heating media. The fin-tube radiation system was much smaller than the earlier cast iron radiators and therefore took up less living space. Later versions also included electric heating sources.

Ventilating

Ventilation enhanced indoor air quality by providing sufficient fresh air to overcome odor problems or to remove stale air from a building. Early buildings relied on operable windows and louvers linked directly to the outdoors to provide fresh air through wind-blown breezes (see Figure 19-3). The introduction of the electric fan in the late nineteenth century was a major step forward in ventilation efforts.

Historically, ventilation air was tempered by a heating system to offset cold outdoor air temperatures but generally no method for cooling warm outdoor air was provided. When cooling was included, the system was known as an "air-conditioning system." Specifically, ventilation was and still is provided by a fan drawing outdoor

Figure19-3 Louvered shutters and panels allowed ventilation while maintaining privacy.

air into the building, which provides positive pressure. This pressurization can be enhanced by the use of exhaust fans that draw air from a space (e.g., a bathroom) and force it out of the building. This process creates a negative pressure that draws air from adjoining spaces through louvers in and gaps around a door or window. In this fashion, fresh outdoor air from an adjoining positively pressurized space displaces the fouled air that is removed by the exhaust fan.

In private buildings, such as single-family residences, ventilation often occurred by happenstance. Gaps around windows and doors, openings in fireplace flues, and operable windows and doors allowed fresh air to enter the building and served as the traditional means of ventilation, since residential heating and cooling systems typically ran independently of each other. Ventilation was due to natural air flow, and there were no requirements for continuous mechanically powered ventilation. However, the International Mechanical Code and ASHRAE ventilation standards set forth by ASHRAE 62 state that commercial, institutional, and industrial buildings must be mechanically ventilated when occupied. In addition, local and state ordinances may restrict or ban smoking and dictate a minimum distance between a fresh-air inlet and an airborne contamination source, such as an exhaust fan or a chimney.

The resulting operational dynamics of all-air systems, where heating and cooling are provided by air only, have led to the integration of heating, ventilating, and cooling into one system. Outdoor air is continuously drawn into the building and is tempered so that even when heating or cooling is not required, fresh air circulates throughout the building.

Air-water and all-water systems that use hot water in some form of radiation system for heating must also provide some form of ventilation. Air-water systems rely on centralized fan systems to provide ventilation and (if desired) air-conditioning. All-water systems, typically found in many noncentrally air-conditioned buildings, such as schools and motels, are smaller units that draw air directly from the outdoors into a space.

Air-Conditioning

Early efforts to cool air are considered minimally effective by modern comfort standards but were considered adequate for their day. Like heating and ventilating, cooling needs were met by architectural adaptations to create opportunities to moderate overheated spaces (see Figure 19-4). Buildings were built to include belvederes and towers (the precursors of the solar chimneys now popular in sustainable buildings) that acted as siphons to draw air up through the building. High ceilings were prominent in humid locations. Doors were topped by transoms and, when combined with the lowered top sash of double-hung windows, allowed hot air trapped at the top of a space to escape. Although none of these features directly cooled the air, the heat relief came from the sensation of cooling as moving air passed over the occupants. In arid regions, the use of thermally massive materials, like stone and masonry, delayed the radiant energy from the sun by several hours and, when combined with nighttime ventilation, drew heat out of the building to offset the daily heat gains.

One method for cooling was to harvest ice from frozen lakes and rivers and store it in underground vaults for use in the summer. By the 1860s, mechanical refrigeration machines had been developed to make ice year round. Daniel Livingston Holden, recognized as the first person in the United States to pursue mechanical refrigeration, supervised the construction of such a system in San Antonio, Texas, in 1865 (Donaldson and Nagenast 1994). Because mechanical refrigeration systems

Figure 19-4 Many architectural features of buildings, such as those of the Honolulu House in Marshall, Michigan, allowed heat to be removed from occupied spaces, thus providing a sense of coolness to the occupants.

were still not widely available, the introduction of the electric fan in the 1880s prompted early efforts to cool spaces by blowing air through a large room-sized chamber filled with blocks of ice to cool it. Ducts and openings to the space(s) being cooled allowed this cooled air to enter and moderately lower the room's temperature.

To melt into water (i.e., change state from solid to liquid), ice absorbs heat from the surrounding air, which in turn cools the air. In arid regions with low humidity, evaporating water does the same thing as it changes to vapor (evaporates). By extension, these same principles apply to mechanical refrigeration. Refrigerants will change state from a liquid to a gas and back again when exposed to controlled pressures inside an enclosed coil and the resulting temperature differences with either the liquid or gas outside the coil. In the process, the refrigerant will absorb or reject heat. The absorption of heat cools whatever liquid (changing water into ice) or gas (lowering air temperature) is in contact with the heat exchanger coil.

In climates with low humidity, evaporative coolers were used to cool air. These systems basically were created by drawing air through or across a wetted material. The evaporative cooling reduced the temperature but raised the humidity level of the air. The earliest systems used natural convection, but later systems used an electric fan to move the air.

By the 1890s, mechanical refrigeration systems were becoming increasingly available. These machines were the precursors of what eventually became air-conditioning. An interesting side note relates to how chilling capacity is quantified: Since ice was previously ordered by the ton, the amount of heat removed by the melting ice was measured in tons. This unit reference for cooling and refrigeration remains in use today.

True air-conditioning was invented to overcome problems caused by humidity. In 1902, Willis Carrier developed the first system to condition the air (hence, air-conditioning) by removing humidity and simultaneously cooling the air. This system was installed in a printing plant in Brooklyn, New York, to control the dimensions of the paper and the stability of the inks used in printing. The use of air-conditioning was initially limited to industrial applications but soon found its way into theaters by the 1920s. The cost of air-conditioning limited its residential use to the richest households until after World War II. As with so many other building-related products and systems, advances made during the war helped improve air-conditioning equipment and lower its cost. By the 1950s, air-conditioning had gained an increasingly large residential market that continues to grow today (Ackerman 2002; Cooper 1998).

Along with the ever-increasing size of commercial buildings constructed in the mid-twentieth century came the increased use of electric fluorescent lighting that introduced substantial heat gains along with those generated by the occupants. Air-conditioning became the norm, and as it was installed in older buildings, ceilings were lowered, operable windows and transoms were sealed, and the occupants became totally reliant on air-conditioning to maintain their comfort.

Distribution Systems

The most cost-effective systems for larger commercial, institutional, and industrial buildings (e.g., those with multiple floors and/or large square footage) are those that incorporate the concept of a central mechanical room and distributed fan or equipment rooms. The hot water or steam used for heating and the chilled water used for cooling are generated, respectively, by boilers and chilling equipment in a central mechanical room located in the service area of the building. Piping located in vertically aligned shafts transports the resulting heating and cooling media to the distributed fan and equipment rooms. In an all-air system, the hot water or steam and the chilled water are commonly piped to a heating and a cooling coil, respectively, in an air-handling unit that delivers warm or cool air to the space via ductwork. In all-water systems, hot water is pumped through a distribution network of pipes to the terminal device (e.g., radiator, fin-tube convector). Older steam heating systems followed this same practice. In a water system, air for ventilation is handled separately. In some cases, air-water systems were installed that provided cooling and ventilation air through the air-handling unit and heating to the radiation devices located along the walls with exposure to exterior thermal conditions.

In some instances, the heating and cooling media are generated off-site and delivered directly into the building distribution system rather than having a central mechanical equipment room. This process can occur on a campus of institutional or industrial buildings or within a municipally operated heating or cooling network. Many major cities operate these networks, especially steam distribution systems, in their central business districts.

Mechanical HVAC systems are a significant part of the infrastructure of a building. The boilers, chillers, pumps, fans, and other auxiliary equipment are typically located in publicly inaccessible service areas. As modern upgrades were made, original components were often traded for "better" ones. Over time, mechanical rooms evolved as old equipment was replaced or abandoned in place. As space allowed, new equipment was installed as needed. In general, little attention has been paid to preserving historical mechanical spaces since these areas are not typically under public scrutiny for retaining character-defining features.

Typical Problems

Aside from normal wear and tear, care must be taken to perform appropriate periodic maintenance on all systems as recommended by the manufacturer. Deferred maintenance and stopgap measures reduce the useful life of any mechanical equipment. Like an automobile, mechanical equipment should have a periodic (e.g., daily, weekly, monthly) schedule of maintenance. Failure to adhere to this schedule can cause significant decline in performance over the long term, which can be seen in the increasing fuel usage that compounds the operating cost problems.

To increase efficiency and shield workers from contact with hot surfaces, many historical and older heating systems included asbestos insulation on boilers and piping. As mentioned in earlier chapters, asbestos in a bonded condition (e.g., asbestos cement boards and panels) may be safe so long as it is not friable. Unfortunately, this safety is compromised when these materials are broken or cut. In addition, some asbestos may be secured only by a canvas covering that, if disturbed, can allow asbestos fibers to become airborne and possibly inhaled.

The major problem in many rehabilitation projects is the difficulty of running ductwork and piping through the building to access previously unconditioned or underconditioned portions of the building or installing mechanical equipment to better service remote portions of the building. The author was involved in a systems investigation in the mid-1980s for a major museum originally constructed in the 1930s with two additions in the 1960s. Beyond the centralized mechanical space and the expected major fan rooms serving the original building and the additions, subsequent installations of HVAC equipment to solve a variety of individual problems created more than thirty additional mechanical equipment service locations.

This increase in equipment space and distribution networks, which is typical of rehabilitated historic buildings as well, caused numerous logistical problems in servicing the building, as well as reducing the occupied spaces to allow space for the HVAC equipment. Evidence that the HVAC system is no longer adequate (or perhaps never was adequate by today's standards) is the piecemeal installation of numerous separate mechanical devices, such as room window air-conditioners and other small packaged cooling devices.

Recommended Treatments

The HVAC equipment should be maintained following practices recommended by the manufacturer. This maintenance includes periodic inspections to identify problems and maintenance needs and a system to ensure that maintenance activities are scheduled and performed in a timely manner. Following this practice will extend the useful life of the equipment and should optimize its efficiency. Terminal devices in public spaces should be inspected for wear and tear and repaired as necessary. Finishes should be cleaned and protected as needed. During any construction, radiators and other terminal or control devices (e.g., grilles, registers, louvers, or thermostats) in the space should be protected from potential construction-related damage. Any chimneys that are expected to be reused should be inspected for deterioration in the lining within the chimney stack and repaired as needed. Reused air distribution systems should be checked for deterioration, and their interiors should be inspected for obstructions and contaminants and then cleaned as needed.

The decision to rehabilitate or adaptively use a building must take into account the effects of rehabilitating, expanding, or modernizing the HVAC systems. This process may include reconfiguring the mechanical rooms, as well as reconfiguring or installing systems that enhance the overall thermal performance of the HVAC system. The need to install new HVAC services often requires the routing of ductwork and piping discreetly through the existing building. Typically, space for these distribution networks may be found in less obtrusive areas, such as closets, wall cavities, or smaller, less important rooms. While lowering ceilings (or a portion thereof) and furring out vertical chases in more prominent spaces is certainly possible, this practice is not recommended by the *Standards*.

References and Suggested Readings

Ackerman, Marsha E. 2002. *Cool comfort: America's romance with air-conditioning*. Washington, DC: Smithsonian Institution Press.

Brown, Dennis. 1996. Alternatives to modern air-conditioning systems: Using natural ventilation and other techniques. *APT Bulletin*, 27(3): 46–49.

Calloway, Stephen, ed. 1996. *The elements of style: A practical encyclopedia of interior architectural details from 1485 to the present*. Rev. ed. New York: Simon & Schuster.

Clark, Clifford, Jr. 1986. *The American family home: 1860–1960*. Chapel Hill: University of North Carolina Press.

Cooper, Gail. 1998. *Air-conditioning in America: Engineers and the controlled environment, 1900–1960*. Baltimore: Johns Hopkins University Press.

De Mare, Eric, ed. 1954. *New ways of servicing buildings*. London: Architectural Press.

Donaldson, Barry and Bernerd Nagenast. 1994. *Heat and cold: Mastering the great indoors*. Atlanta: American Society of Heating, Refrigerating, and Air-Conditioning Engineers.

Elliott, Cecil D. 1992. *Technics and architecture: The development of materials and systems for buildings*. Cambridge, MA: MIT Press.

Fisher, Charles E. and Hugh C. Miller, eds. 1998. *Caring for your historic house: Preserving and maintaining: Structural systems, roofs, masonry, plaster, wallpapers, paint, mechan-*

ical and electrical systems, windows, woodwork, flooring, landscapes. New York: Harry N. Abrams.

Fitch, James Marston with William Bobenhausen, 1999. *American building: The environmental forces that shape it.* New York: Oxford University Press.

Harland, Edward. 1999. *Eco-renovation: The ecological home improvement guide.* White River Junction, VT: Chelsea Green Publishing Company.

Ierley, Merritt. 1999a. *Open house: A guided tour of the American home 1637–present.* New York: Henry Holt and Company.

———. 1999b. *The comforts of home.* New York: Clarkson Potter Publishers.

Jandl, H. Ward. 1998. *Rehabilitating interiors in historic buildings.* Preservation Brief No. 18. Washington, DC: United States Department of the Interior.

Kay, Gersil Newmark. 1989. Mechanical/electrical systems in older buildings. *APT Bulletin,* 21(3/4): 5–7.

———. 1992. *Mechanical and electrical systems for buildings: Profitable tips for professionals, practical tips for the preservationists.* New York: McGraw-Hill Book Company.

Konzo, Seichi with Marylee MacDonald. 1992. *The quiet indoor revolution.* Champaign, IL: Small Homes Council–Building Research Council.

Lechner, Norbert. 2001. *Heating, cooling, lighting: Design methods for architects* 2nd ed. New York: John Wiley & Sons.

Marcus, Alan I. and Howard P. Segal. 1989. *Technology in America: A brief history.* San Diego, CA: Harcourt Brace Jovanovich Publishers.

Oliver, John W. 1956. *History of American technology.* New York: Ronald Press.

Park, Sharon C. 1991. *Heating, ventilating, and cooling historic buildings: Problems and recommended approaches.* Preservation Brief No. 24. Washington, DC: United States Department of the Interior.

Pursell, Carroll, ed. 2005. *A companion to American technology.* Malden, MA: Blackwell Publishing.

Randl, Chad. 2005. *The use of awnings on historic buildings: Repair, replacement and new design.* Preservation Brief No. 44. Washington, DC: United States Department of the Interior.

Rose, William B. 2005. *Water in buildings: An architect's guide to moisture and mold.* New York: John Wiley & Sons.

Smith, Baird M. 1978. *Conserving energy in historic buildings.* Preservation Brief No. 3. Washington, DC: United States Department of the Interior.

Stein, Benjamin, John S. Reynolds, Walter T. Grondzik, and Alison G. Kwok. 2006. *Mechanical and electrical equipment for buildings.* 10th ed. New York: John Wiley & Sons.

CHAPTER 20 Building Service Systems

OVERVIEW

Building service systems include plumbing, vertical transportation, and fire protection systems. As technology grew, enhancements to building services transformed the way daily activities were performed. An example of this is plumbing. Today, it is taken for granted that water will be supplied to a building, heated and distributed where needed, and then drained away. However, in early settlements, water had to be carried in a bucket from a stream or pond to the fireplace, where it was heated as needed. Hygiene was provided by bathing, perhaps, in that same stream or pond or washing using water in a bucket or tub, as bathrooms did not exist. Outhouses, set apart from the house, were used for privacy. The search for labor-saving opportunities and concerns for public health, modesty, and societal expectations led to the modern conveniences of indoor plumbing systems that are enjoyed today. Similar evolution characterized each building service in common use.

Since only 1 percent of the houses in the United States in 1921 had both electricity and plumbing (Stilgoe 2006), a true restoration of a residential building to a period before that time would technically include the removal of these now standard services. In commercial, institutional, and industrial buildings, removing and not replacing measures mandated by the ADA is illegal, while including them compromises the definition of restoration. So, most projects are actually a rehabilitation that involves introducing or expanding modern service systems. Methods for integrating modern services often require creative approaches to managing the visual intrusions that their installation can create.

365

The building service system elements that are most noticeable are plumbing fixtures, elevator cars, doors, and call panels. Less obvious are the control and monitoring devices, such as thermostats and equipment necessary for fire and life safety (e.g., fire alarm pull stations, fire hose cabinets, emergency egress lighting, and sprinkler heads). What is usually not seen is the service infrastructure that distributes power or water to these elements.

When new services need to be introduced and distributed throughout the building, seek paths for distribution networks that can be concealed from public view. Some products allow for color selection that can diminish the visual intrusion of control panels or protective enclosures. In products in which the original manufacturer's finish cannot be repainted later, such as covers for concealed sprinkler heads, color selection sensitivity is essential.

PLUMBING SYSTEMS

Water has long been used as a source of sustenance, a means of mechanical power, and a means of disposing of unwanted materials. The development of the colonies in North America was based on access to water. Plumbing systems consist of supply, drainage, and vent piping networks (see Figure 20-1). These networks provide fresh water, remove soiled water, and vent unwanted sewer gases from the building. Piping has been made from wood, lead, cast iron, steel, galvanized steel, copper, and assorted plastic polymers. The joints at the fittings have been sealed with lead, various types of solder, and modern glues. Connections also have been made using threaded connections on fittings. State and municipal jurisdictions introduced plumbing codes in the late nineteenth and early twentieth centuries. These codes were consolidated into the National Plumbing Code in 1947 and adapted by the International Plumbing Code at the turn of the twenty-first century.

Firgure 20-1 Typical water drainage and supply systems (from *Building Control Systems*, 2nd ed., Copyright © 1993 by John Wiley & Sons, Inc. Reprinted with permission of John Wiley & Sons, Inc.).

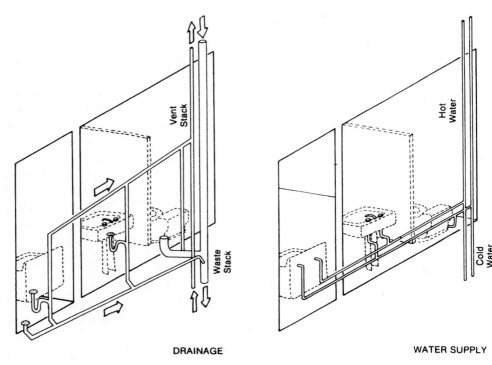

DRAINAGE   WATER SUPPLY

Figure 20-2 Cisterns that collect runoff from the roof could be used to provide water for domestic purposes. When the cistern was located above the place where the water was used, water flowed under the pressure of its own weight by gravity without the need for manual labor or a pump.

Supply piping ends at a faucet or a connection to a plumbing fixture. Openings to drainage systems occur as drains (sink or floor) and cleanouts. Immediately after the drain inlet is a trap that retains a small portion of water that prevents unwanted sewer gases from entering the occupied space. Cleanouts are closed by a threaded cover plate and are used to access the drain line when removing clogs. Vent piping is connected to the drainage piping at a point above which no soiled water can enter it and extends upward through the roof. Vent piping allows sewer gases to escape to the outside atmosphere rather than into the occupied space. Vent piping also breaks any potential vacuum that might siphon or draw out water from the trap.

Traps are a P- or U-shaped assembly of pipes that trap water at the bottom of the curved portion of the assembly. P-traps were used when a drain line penetrated the wall to connect with the building drainage system. U-traps separate the building drain from the building sewer line. S-traps, now illegal, were used when the penetration went through the floor instead of the wall. Another outdated and illegal trap is a drum trap, which used a straight length of piping connected to a drum-shaped container (hence the name) located beneath the floor.

Early water resources involved manual labor to collect and transport water to the building. Collection systems evolved that included retention ponds, cisterns, wells, and pumps. Since water flows downhill, cisterns could be used to collect runoff from roofs and also, if located above the place where the water was being used, could supply water without the aid of a pump (see Figure 20-2).

Historic Development

Similarly, the hydrostatic pressure (e.g., the force due to the difference in elevation or the effects of compression) in underground aquifers allowed water to flow upward when the overlying soil was penetrated. Unfortunately, as communities began to develop and grow congested, multiple users reduced the water pressure. Thus, the modern reservoir system came into being; water is collected and stored at an elevation above the community. The water towers located throughout the country are examples of this system.

As communities grew, the demand for denser land use prompted an increase in building height. Initially, this was limited to four or five stories due to water pressure. Steam-powered pumps and hydraulic rams developed in the early nineteenth century could pump water into tanks located above the limits of natural water pressure. However, the late-nineteenth-century advent of electrical power for pumping systems, along with the introduction of the electric elevator, led to the development of the modern high-rise. Plumbing became more sophisticated, with a variety of supply systems that relied on pumps and storage tanks located in separate mechanical rooms located at regular intervals in a high-rise or on the roof. Meanwhile, low-rise commercial, industrial, and residential buildings continued to use the traditional systems fed by the municipal water suppliers. In less congested areas without municipal water systems, it is still possible to find systems drawing water from naturally occurring aquifers that is pumped into the building.

Waste drainage systems have also evolved. The earliest forms were usually a ditch or a simple outhouse. Concerns for sanitary conditions and privacy led to their placement away from the house. Eventually, indoor plumbing systems took advantage of the opportunity for running water to remove wastes. This practice often meant that soiled water drained into a local river or lake. Although the technology for the modern sewer was known, based on earlier Roman civil works, the modern sewer was not built in many cities until after the Civil War. Drainage piping inside the building was largely cast iron or galvanized steel (when it became available in the late nineteenth century) through the late twentieth century, when polymer-based products such as acrylonitrile butadiene styrene (ABS) and polyvinyl chloride (PVC) were introduced (Stein, Reynolds, Grondzik, and Kwok 2006).

Early settlements used hollowed logs to form crude piping systems. By the late eighteenth century, cast iron was in use. The first modern large-scale public water supply systems were constructed in the middle third of the nineteenth century. At this time, sinks were often not connected to a plumbing system. The water supply often ended at a faucet in the yard from which water was carried into the building. By the 1860s, plumbing had begun to move indoors and consisted of simple water supply piping. Domestic hot water was provided by a heating coil and tank arrangement attached to the rear of large stoves. Supply piping inside the building included lead, copper, brass, or galvanized steel pipe as they became available. Modern polymer-based supply piping and tubing accepted by the International Plumbing Code at the end of the twentieth century included chlorinated polyvinyl chloride (CPVC), polybutylene (PB), cross-linked polyethylene (PEX), and cross-linked polyethylene/cross-linked aluminum/cross-linked polyethylene (PEX-AL-PEX) (Stein, Reynolds, Grondzik, and Kwok 2006).

The first stages of the modern bath appeared in the 1860s, when the bathtub was located in a separate bathroom for privacy. By corresponding necessities and proximity to supply piping, the water closet (which was first made in large numbers in the 1840s and is considered an early version of the modern toilet) was also located in the bathroom. Sinks remained located elsewhere. Waste drainage at first involved the use of pans and privy pits. In the 1890s, after the completion of municipal sewer systems that could carry away soiled water, new vitreous china water closets were introduced that were largely equivalent to the present-day toilet. Still, the inclusion of bathrooms was slow to develop. By 1900, only 15 percent of American homes had a complete bathroom, and by 1940, this number had increased to only 55 percent (Ierley 1999).

As tastes changed, plumbing fixtures varied considerably. Materials ranged from stone to vitreous china and porcelain enamel–coated cast iron to modern fiberglass and other polymers. Sinks and toilets have had the widest variety of uses, ranging from ornamental versions used in private bathrooms to simple utility versions used in kitchens and service areas. Bathtubs evolved from portable metal tubs to the magnificent porcelain enamel–coated cast iron claw–footed behemoths of the 1880s to the streamlined vitreous china products used in the early twentieth century. Bathing customs have also seen the rise and decline of bidets, needle or steam cage showers, and sitz baths. Colors have also varied. White is perhaps the most common, but the aesthetics of the early to mid-twentieth century saw acceptance of a rainbow of offerings from fixture manufacturers.

The late twentieth century saw a rise in concern for water conservation that led to new plumbing fixtures (some digitally controlled) in the marketplace. Low-flush toilets and low-flow faucets with infrared detecting controls were introduced. This water conservation effort has often resulted in the replacement of older toilets, which may use up to 5 or more gallons per flush, with modern ones using less than 2 gallons. Composting toilets have also been introduced that eliminate flushing altogether.

Typical Problems

Plumbing fixtures are vulnerable to architectural and aesthetic fashions. In residential buildings, bathrooms and kitchens are among the two most common spaces to be updated. In other buildings, fixtures may be changed as a building is periodically updated or undergoes a conservation retrofit. When this happens, original fixtures are often replaced with those of contemporary merit. This merit may be aesthetic, as is often the case, or it may be functional in that the older fixtures may use excessive amounts of water by today's standards.

Faucets and fixture accessories were often made from porcelain or brass finished with nickel, chrome, or other plating metals. Their surfaces were subject to wear from harsh or abrasive cleaning methods. Internally, moving parts eventually wear out due to both the friction created when moved and the properties of the water passing through them. These items can be removed and reconditioned. Replacements may also be available from reproduction manufacturers and salvage yards.

The materials used for piping can lead to several problems, not the least of which is related to the use of lead for piping and as an ingredient in solder and packing materials used to seal joints. Lead piping was commonly used because it was malleable and could be bent or curved slightly to get past obstructions. Some communities commonly used lead piping to connect the street distribution main to individual buildings. Lead piping was used inside buildings as well. Due to its relatively low melting temperature, lead was used to seal joints in metal piping. Lead is dangerous since it can leach out into the water and contaminate it.

A problem that occurs in such ferrous metals as steel or cast iron piping is corrosion. This corrosion results from contact with air and water but also may be accelerated by the extreme pH of the water itself. Even mildly acidic or alkaline conditions will cause corrosion. Another problematic form of corrosion is galvanic action that occurs when two dissimilar metals are in contact with one another. The corrosive actions just described may lead to clogged pipes. As corrosion occurs, the cross-sectional area of the pipe shrinks and materials collect on the roughened interior surface.

On drainage and vent systems, vents may become clogged with debris from nests or other small objects and not work properly, while underground sewer and drainage lines from the building may weaken or fail due to corrosion and may be crushed by the weight of soil or other materials bearing down on them. A common cause of drainage problems is the failure of the joints in clay tile or cast iron pipe or the holes caused by corrosion, which enable the penetration of tree roots that are seeking the moisture within the pipe. In addition to the clogs created by tree roots are obstructions created by insoluble objects and materials caught within the pipe. Grease, kitchen and bathroom wastes, and other debris also form clogs that reduce water flow (see Figure 20-3). Methods used to remove obstructions may damage the pipe. Caustic cleaners can further erode the interior of corroded pipes, and air pressure blowout methods can cause weakened pipes to crack or even burst. Care should be taken when selecting the method for clearing obstructions.

On the supply side, pipes can often burst due to freezing temperatures since as water freezes, it expands. If the pipe is constricted or the temperatures are low enough, the pressure that builds up at the point where freezing occurs will rupture the pipe or the fitting. This will occur in an unheated attic or an inadequately insulated wall or floor cavity adjacent to the outdoors. Piping for both supply and wastewater systems was frequently added in exposed locations within the house or was installed under raised floors. It is still possible to see drainage systems added in plain view since there was no economical way to install it wholly within the construction of the existing building.

Recommended Treatments

The first and most important treatment is to locate and repair any leaking pipes and fittings and to correct any malfunctioning fixtures. When leaks are located inside enclosed floor, ceiling, or wall cavities, they may be identified by the staining and swelling

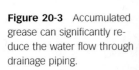

Figure 20-3 Accumulated grease can significantly reduce the water flow through drainage piping.

of surface materials as well as by peeling paint or wallpaper. Likewise, waste and vent lines should be inspected to locate obstructions and deterioration inside the pipe. The obstructions should be cleared and the pipe repaired or replaced as needed.

When a fixture has been damaged or its surfaces have been marred by deterioration, staining, or pitting, the surfaces may be repaired or reconditioned. This work is typically done off-site. Any reconditioned and repaired fixtures should be stored safely until they are reinstalled on-site. Modern manufacturers make period reproductions to replace missing or damaged fixtures. Replacement fixtures and accessories may also be obtained from salvage yards.

A common concern is water quality, particularly in light of the uses of lead in plumbing supply system construction. Likewise, hard water can aggravate health problems of the occupants and leave stains on plumbing fixtures. Two ways to deal with these issues are to install an individual filter on each fixture or to install a single central filtration system. The first option requires periodic multiple changes of the filters. The second option can reduce contamination problems caused before the water enters the building but not those from within the piping system itself, such as lead that leaches out of the pipes and the solder.

Installing new plumbing may be required to bring a building up to modern standards. Unlike ductwork, supply lines are relatively small in diameter and can readily fit in ceiling and wall cavities. Vertically oriented piping can be routed through closets or other service rooms. Waste lines pose a problem because they must be moderately sloped (e.g., a $1/4$-inch drop/foot of run) to accommodate flow requirements. Fixtures must be located so that waste lines can remain within the enclosed horizontal floor/ceiling cavity and not protrude into occupied spaces below.

ELEVATORS

Elevators are a specific character-defining feature found in many older commercial, institutional, and industrial buildings. Particularly unique are the open-cab (car) systems used in early office and industrial buildings and the highly ornate doors and interior finishes found in office, residential, and retail buildings. Their retention is recommended in the *Standards*.

Historic Development

While an elevator is a simple concept that has been used for millennia in lifting people and goods from one level to another, the safe operation of the elevator had long been a cause of concern. In 1853, Elisha Otis invented a safety mechanism that immediately stopped an elevator from descent when the cable(s) supporting it broke. The use of the elevator in American buildings can be traced to a series of innovations introduced throughout the nineteenth century that included the hydraulic elevator (1878), the electric elevator (1889), and the escalator (1899) (Otis Elevator Company 2007).

Modern elevator systems are composed of an elevator car, cables, guide rails, a shaft, a motor, safety equipment, control panels, and either a counter-weight (for traction systems) or a hydraulically powered ram (for hydraulic systems). Early systems designed as freight elevators were not particularly ornamental. They were often manually controlled by a human operator, and had open cars (cages or railed platforms) and sometimes an open shaft. As the use of passenger elevators grew, the expected amount of enclosure increased in reaction to safety concerns, passenger comfort, and fire and smoke control.

By the early twentieth century, elevator doors and interior walls had often become quite elaborate and ornate. Enhancements to controls led to the replacement of elevator operators by the mid-twentieth century. Today, the gearless traction system is used in both low- and high-rise buildings, while the hydraulic elevator is used only in low-rise buildings.

Typical Problems

Older elevators are slow and noisy by modern standards, so even if they are in perfect working order, there may be complaints when occupants compare them to those found in modern buildings. In addition, there may not be a sufficient number of them to meet existing passenger demands or expected new demands when the project is completed. Many older elevators were upgraded to meet the requirements of the ADA. The ADA specifically addressed issues of control access, operations, and labeling. In addition, there are concerns about whether elevator cabs are large enough to accommodate wheelchairs, emergency medical stretchers and other equipment. In some instances, upgrades have met the requirements of the law but may have compromised the historic appearance of the elevator cab or doors.

Recommended Treatments

The most important treatment is to inspect the elevator for safety defects such as control problems, worn-out equipment, and inadequate emergency communication and braking systems. When the defects have been corrected, the feasibility of reusing the existing elevator can be assessed. If the assessment reveals that the elevator cannot meet the expected demand, then there may be an opportunity to upgrade it or install a new elevator elsewhere in the building without disturbing the configuration of the existing elevator. The reuse and upgrading of existing elevators are covered under the provisions of the American Society of Mechanical Engineers (ASME) standard A17.3, *Safety Code for Existing Elevators and Escalators*, which outlines specific requirements for the safe reuse of elevators and escalators.

In light of the earlier construction of some elevators and the possible adaptive use of a building, updating the controls and safety features of existing elevators may be necessary. New elevator systems may also be needed to meet expected future demand. Adding elevators in this manner is expensive and needs to be done with caution so that historically significant spaces and finishes are not needlessly destroyed. One way to do this is to locate new elevator shafts in service spaces adjoining the original elevator lobby or in secondary spaces near historically significant interior spaces.

FIRE PROTECTION SYSTEMS

Fire protection systems are an important aspect of modern buildings. Many of the early building codes in the United States were developed in reaction to disastrous fires and the safety issues they revealed. Many projects in historic buildings routinely include the upgrading of fire protection systems, if they exist at all, to meet the requirements for expected use after the project is completed. This is particularly important when a building's change to a higher level of use is anticipated.

Historic Development

Early fire sprinkler systems were used in nineteenth-century commercial buildings whose owners paid for them based on the savings generated by reduced insurance policy costs. In 1852, the first perforated piping systems used for fire protection were installed in Lowell, Massachusetts, in the Plant of the Proprietors of the Locks and Canals on the Merrimack River. In 1872, Philip W. Pratt received the first patent for a sprinkler system. In 1874, Henry S. Parmelee obtained a patent for a sprinkler head (Grant, 1996).

Concern for fire safety expanded the use of sprinkler systems to schools, commercial buildings, and retail buildings in the twentieth century. New construction concealed piping in walls and ceiling cavities, but existing construction often had these systems installed in plain view.

Sprinkler heads controlled the distribution pattern and flow of water. Sprinklers were fully open or were controlled by a fusible material holding the sprinkler closed. When the fusible material melted, the sprinkler opened and water was sprayed outward. Sprinkler heads were laid out in a uniform pattern to ensure full coverage in a space. Spacing between sprinkler heads was determined by the fire code hazard level assigned to a space—light hazard, ordinary hazard, or extra hazard.

The orientation of the sprinkler varied. In fully exposed systems, sprinkler heads were mounted either in a pendant orientation below the supply piping or in an upright position above this piping. The upright orientation provided protection due to accidental damage from occupants below. For spaces with dropped ceilings, sprinklers were mounted in the pendant orientation as fully exposed below, semirecessed within, or fully recessed above the ceiling. Lastly, sidewall orientations were used in a vertical wall.

The two common types of sprinkler systems were wet pipe and dry pipe. Wet pipe systems were used in heated locations. Dry pipe systems were used in unheated spaces or where there was concern for water damage. Wet pipe systems were the simplest and most common. In this system, a sprinkler head was activated by the heat of the fire and only dispersed water through the activated sprinkler head. In a dry pipe system, the pipe was filled with compressed air or nitrogen. When the heat of the fire activated the system, the pressure was relieved and water flowed into the pipe and through the activated sprinkler head only.

There were two dry pipe systems: a preaction system and a deluge system. In a preaction system, water was held back by a preaction valve that responded to the activation of a heat or smoke detector that was more sensitive than the sprinkler head. When smoke or heat was detected, the preaction valve opened and let water into the pipe, whereupon the systems acted like a wet pipe system. When the water flowed in, an alarm sounded to notify occupants to evacuate before the fire activated any sprinkler heads. The deluge system had open sprinkler heads and used the more sensitive sensor but allowed immediate water flow through all sprinkler heads simultaneously. The deluge system was used in extra-hazard spaces like airplane hangars (Stein, Reynolds, Grondzik, and Kwok 2006).

Typical Problems

Fire sprinkler systems are tested periodically to ensure that they function properly. A test valve located at the farthest point from the main is opened, and water flow and quality are evaluated. Reports of this test are filed with the fire marshal's office so that a record of performance issues can be available for confirming potential problems. More commonly, problems can stem from accidental damage or intentional vandalism to the sprinkler heads, alarm pull stations, fire extinguishers, and hose cabinets. Other problems arise when the building's use is changed and, in turn, causes the hazard level to become higher or if the fire safety code has changed since the sprinkler system was first installed. In these cases, the original system may not be adequate to protect occupants in the proposed use and the system may need to be expanded, upgraded, or replaced.

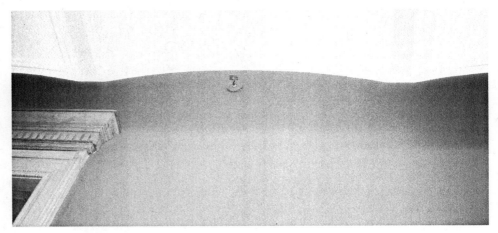

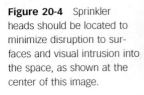

Figure 20-4 Sprinkler heads should be located to minimize disruption to surfaces and visual intrusion into the space, as shown at the center of this image.

Recommended Treatments

The IBC and the IEBC enable the fire marshal to require specific reports to substantiate that the system and any deviations from contemporary construction practices comply with the code. The design team should work with local code officials to avoid disruption of historic finishes. Fire protection equipment, such as piping and sprinkler heads, should be located to minimize visual intrusion (see Figure 20-4). Like the other building systems, new fire safety systems should be installed to minimize intrusion into historic spaces by instead routing mains and branches through nonpublic service areas.

References and Suggested Readings

Calloway, Stephen, ed. 1996. *The elements of style: A practical encyclopedia of interior architectural details from 1485 to the present.* Rev. ed. New York: Simon & Schuster.

Clark, Clifford, Jr. 1986. *The American family home: 1860–1960.* Chapel Hill: University of North Carolina Press.

De Mare, Eric, ed. 1954. *New ways of servicing buildings.* London: Architectural Press.

Elliott, Cecil D. 1992. *Technics and architecture: The development of materials and systems for buildings.* Cambridge, MA: MIT Press.

Fisher, Charles E. and Hugh C. Miller, eds.1998. *Caring for your historic house: Preserving and maintaining: Structural systems, roofs, masonry, plaster, wallpapers, paint, mechanical and electrical systems, windows, woodwork, flooring, landscapes.* New York: Harry N. Abrams.

Grant, Casey Cavenaugh.1996 The birth of NFPA. *NFPA Journal.* http://www.nfpa.org/itemDetail.asp?categoryID=500&itemID=18020&URL=About%20U s/History.

Gray, Lee E. 2002. *A history of the passenger elevator in the nineteenth century.* Mobile, AL: Elevator World.

Harrison, Molly. 1972. *The kitchen in history.* New York: Charles Scribner's Sons.

Ierley, Merritt. 1999. *The comforts of home.* New York: Clarkson Potter Publishers.

Jandl, H. Ward. 1998. *Rehabilitating interiors in historic buildings.* Preservation Brief No.18. Washington, DC: United States Department of the Interior.

Kaplan, Marilyn. 1989. *Replicating historic elevator enclosures.* Preservation Tech Notes: Mechanical Systems Number 1. Washington, DC: National Park Service.

Kay, Gersil Newmark. 1989. Mechanical/electrical systems in older buildings. *APT Bulletin,* 21(3/4): 5–7.

———, 1992. *Mechanical and electrical systems for buildings: Profitable tips for professionals, practical tips for the preservationists.* New York: McGraw-Hill Book Company.

Labine, Clem and Carolyn Flaherty, eds. 1983. *The Old-House Journal compendium.* Woodstock, NY: Overlook Press.

Lupton, Ellen and J. Abbott Miller. 1992. *The bathroom, the kitchen, and the aesthetics of water: A process of elimination.* Cambridge, MA: MIT List Visual Arts Center.

Marcus, Alan I. and Howard P. Segal. 1989. *Technology in America: A brief history.* San Diego, CA: Harcourt Brace Jovanovich Publishers.

Oliver, John W. 1956. *History of American technology.* New York: Ronald Press.

Otis Elevator Company. 2007. About Otis. http://www.otis.com/cp/ categorydetails/1,2239, CLI1_CPI1_RES1,00.html.

Peterson, Charles. 1976. *Building early America.* Radnor, PA: Chilton Book Company.

Plante, Ellen M. 1995. *The American kitchen: 1700 to the present.* New York: Facts on File.

Pursell, Carroll, ed. 2005. *A companion to American technology.* Malden, MA: Blackwell Publishing.

Stein, Benjamin, John S. Reynolds, Walter T. Grondzik, and Alison G. Kwok. 2006. *Mechanical and electrical equipment for buildings.* 10th ed. New York: John Wiley & Sons.

Stilgoe, John. 2006. Ignore drain traps at your peril. Boston Globe. July 16. http://www.boston. com/news/local/massachusetts/articles/2006/07/16/ignore_drain_traps_ at_your_peril/

Winkler, Gail Caskey. 1989. *The well-appointed bath: Authentic plans and fixtures from the 1900's.* Washington, DC: National Trust for Historic Preservation.

Wright, Lawrence. 1960. *Clean and decent: The fascinating history of the bathroom and the water closet.* London: Routledge and Kegan Paul.

CHAPTER 21 Lighting and
Electrical Systems

OVERVIEW

Daylight provided the original source of illumination. As civilized societies emerged, so too did torches, candles, and lamps as sources of light. The initial primary fuel sources were locally available natural fuels, such as reeds and woody plant materials. As technology improved, tallow, fats, oils from plants and animals, and subsequently petroleum-based liquids and gases were used well into the early twentieth century.

Historic period lighting fixtures are one of the most noticeable character-defining features of a building. So too, but less immediately obvious in most cases, are light switches and, to a lesser extent, electrical outlets. The service infrastructure that distributes electricity is generally concealed. As with other building service systems, the restoration of lighting to a specific period may technically be possible for the various historic lighting systems, but modern building codes have fire safety precautions regarding open flames (such as when gas lighting or candles are used) and emergency lighting requirements for egress.

Just as indoor plumbing was not prevalent until well after 1921 in many residential buildings, electric lighting was not available before 1879. A true restoration of a residential building to a period before that time would technically include the unlikely removal of electric lighting. Instead, what commonly occurs is the inclusion of electrified period fixtures with lamps that imitate the light output of a candle or other open flame. A restoration to a period before 1879 is actually a rehabilitation that retains electric lighting and service systems. A project that restores a building to a period after the date when the building had electric lighting installed is a true restoration.

As technology evolved and improvements were introduced, many devices and systems were updated or retrofit with new components, mostly due to changing aesthetic and architectural fashions, fire and life safety concerns, and ADA requirements. The late twentieth century also saw rising concern for energy conservation that brought digital controls into the marketplace. For example, lighting controls can now monitor occupancy and turn lights off when a space is empty for a given length of time, and computer-based controls can schedule the turning of lights on or off, depending on the time of day and exterior daylighting conditions.

Caution is advised when installing modern fixtures and controls, especially those that will be in plain sight. If installed insensitively, individually or together they can can create a jarring visual intrusion that compromises the visual integrity of a historically significant space. This is also true when new services need to be introduced and distributed throughout the building. Following the same strategies described for integrating HVAC and other system distribution networks, it is important to seek paths for fixtures and controls that can be concealed from public view.

LIGHTING

Lighting is available in two forms, natural and artificial. The building itself can take advantage of natural lighting by integrating features, such as large windows and high ceilings, that allow daylight to penetrate interior spaces. Artificial lighting (i.e., light generated from nonsolar sources) came into use as people struggled to see in the dark. While a variety of fuels for artificial light has been used over time, contemporary lighting relies on electrical sources.

Historic Development

From the earliest colonial settlements to the present, lighting systems have had several overlapping stages. The noted historian Roger W. Moss, in *Lighting for Historic Buildings* (1988), classified these periods as follows:

1620–1850: Candleholders

1783–1839: Whale-oil, lard-oil, and burning-fluid fixtures

1854–1934: Kerosene fixtures

1817–1907: Gas lighting

1879–present: Electric lighting

New lighting technologies typically first became available in urban areas. Gas power was used throughout most of the nineteenth century for street lighting (see Figure 21-1) and for domestic lighting in the homes of the rich. It was later eclipsed by electric lighting systems of the late nineteenth century in the larger cities along the East and West Coasts and in the upper Midwest. Nonelectric lighting systems persisted well into the 1930s in many rural areas until several programs, such as the Tennessee Valley Authority and the Rural Electrification Administration, were established to increase the standard of living in these areas.

As electrical service became more widely available, many nonelectric lighting fixtures were converted to electricity. During this transition period, some manufacturers made fixtures that used both gas and electricity. Gas fixtures can be identified by the turnkey located on the armature of the fixture that was used to adjust the gas flow rate and thus the flame brightness. Evidence of the existence of gas fixtures can be found within walls or ceiling cavities where gas piping is discovered often abandoned in place or reused as an electrical wiring conduit.

Figure 21-1 An example of a gas-powered street light.

The first electrical lighting, known as "arc lighting," was based on the arc that was made by electricity passing between the ends of two electrodes. This lighting was used in lighthouses and in some streetlights after the mid-nineteenth century. In 1879, Thomas Edison invented the incandescent light bulb using a carbon filament and spent years improving it, as well as creating the power distribution systems needed to serve individual buildings. Various derivative and alternative forms of incandescent lighting were used into the mid-twentieth century.

In 1937, the General Electric Company introduced the fluorescent lamp at the World's Fair in New York City and then manufactured it for sale in 1938. Instead of a filament, the lamp used electrical current to excite a phosphorus coating inside a hollow tube. Fluorescent lighting transformed the way commercial lighting systems were laid out and led to an increasing departure from traditional daylighting methods after World War II. Atria and light shafts were considered obsolete, and daylighting itself became perceived as a source of glare and visual discomfort. Eventually, this change caused growing reliance on a uniform distribution of hundreds of fluorescent light fixtures across the ceilings of the work space. In many buildings, this led to the lowering of ceilings as part of the installation of recessed fixtures and the infill and reduction of window glazing originally designed to admit daylight.

Neon lighting was first used for signage in the United States in 1923 at a Packard automobile dealership in Los Angeles. By the late twentieth century, it had become a significant part of advertising signage both on buildings and on freestanding signs. Neon signs consisted of a series of gas-filled tubes connected to an electrical source. Red light was produced when neon was excited by an electrical current. The use of neon was later augmented by argon, krypton, and phosphors to obtain different colors.

Typical Problems

The most common problem with historic lighting systems has been damage to or removal of original fixtures due to conversion from candles or gas to electricity, misuse, normal wear and tear, vandalism, inappropriate or insensitive upgrades to wiring, changing aesthetic tastes, or theft. Weather exposure, fire, and water damage from leaks are other important factors. The glass portions of a fixture are vulnerable to breakage, and the fixture is vulnerable to electrical failures due to moisture in the wiring system. The light fixture may suffer from a short circuit of its internal wiring as the wiring ages. The wiring to and within the fixture may deteriorate, become corroded, or loosen. In any event, the light fixture may not work properly. In all cases, if a short circuit is suspected, power to the fixture must be turned off before servicing it.

A common problem is damage that occurs during the project. Light fixtures of historical interest should be protected from damage. If repair or reconditioning is needed, then remove the fixture, complete the repairs, and store the fixture off-site until it needs to be reinstalled. If a fixture is not expected to be removed, it should be enclosed in a protective covering. Portable temporary lighting can be used during construction.

In the late twentieth century, as the cost of generating electricity increased and the environmental cost of building new power plants became more relevant, many modifications to existing lighting systems were made. Many building operators attempted to replace incandescent lighting with various forms of fluorescent lighting. In some instances, this meant that older fixtures were removed; in others, lamps in existing fixtures were replaced with supposedly equivalent fluorescent lamps. One unfortunate result has been the proliferation of compact fluorescent lamps that neither fit in the fixture nor distribute light the same way (see Figure 21-2). Fluorescent fixtures themselves were also delamped to reduce energy costs.

Figure 21-2 Compact fluorescent replacement lamps were designed for contemporary light fixtures. When installed in older fixtures, they may compromise the visual integrity of the fixture. Care must be taken to match the dimensions of the replacement lamp with those of the original.

The relamping and delamping activities of the late twentieth century caused many fixtures to be removed or replaced. While this may have lowered energy consumption, the damaging effect on historic integrity was significant as original lighting systems were summarily replaced with modern systems, often with insufficient regard for their visual appearance. Demand for reproduction lighting fixtures that could accommodate new, efficient lighting products led to more historically compatible lamps by the early twenty-first century.

Recommended Treatments

The energy crises of the 1970s led to standards and energy codes for electrical systems that include lighting and other energy-consuming uses. These standards have been incorporated into the *International Energy Conservation Code (IECC)* that supplements the *IBC*. Many jurisdictions have either adopted these codes directly or adapted them to meet more stringent local requirements. These adaptations can be at the state level (e.g., California Title 24) or at the municipal level (e.g., New York City's High Performance Building Guidelines). Concern for reducing operating expenses may still prompt the adaptation of older lighting systems to better meet modern requirements.

An increasing number of craftspeople are engaged in the repair, retrofit, and upgrading of existing light fixtures. Salvage operations, reproduction manufacturers, and custom fabricators can be consulted in locating or re-creating missing fixtures. Existing fixtures can be restored or rehabilitated, and reproduction fixtures can be used when evidence exists to confirm the appearance of earlier fixtures. As with many of the previously discussed aspects of replacing missing features, any review agencies overseeing the project should be consulted in selecting the replacement.

Ironically, many new sustainable buildings are being designed to include daylighting. Older and historic buildings, with lowered ceilings and enclosed windows can take advantage of this opportunity by removing their late-twentieth-century installations. This approach must address the problems mitigated by earlier interventions, which typically include heat loss and heat gain through glazing and stratification of air due to higher ceilings (e.g., warm air rises to the highest part of the space, which is good for summer cooling but not for winter heating). Historic windows can be upgraded or augmented (e.g., interior storm windows) to incorporate performance features comparable to those of many modern windows or the windows can be refitted with glazing units (e.g., laminated glass, simulated divided lights) that match the appearance of earlier windows while performing at twenty-first-century levels.

ELECTRICAL POWER

Long before the use of electricity, many emerging civilizations had discovered how to use muscle power (e.g., human, mule, or ox) to produce the motive force needed to get work done. Water power came next, with the use of vertically oriented water wheels and horizontally oriented turbines connected via a system of belts to machinery in mills to provide the motive force to perform cutting, grinding, and other operations more easily than using muscle power alone. Wind power (e.g., windmills have been used to grind grain and pump water) has also been used throughout the past 400 years, although on a much smaller scale. Steam power came into use in the late eighteenth and early nineteenth centuries and fueled the Industrial Revolution in Europe and America. The first steam-powered equipment used in America was a pump in 1755 that drained the main shaft at the Schuyler family copper mine in Bergen County, New Jersey (Ierley 1999).

The early- and mid-nineteenth-century efforts of scientists and engineers like Alessandro Volta, Charles Coulomb, Joseph Henry, Michael Faraday, and Thomas Davenport in exploring magnetic fields and electromagnetic induction led to electromagnetic devices that converted mechanical energy into an electrical current (Ierley 1999; Marcus and Segal 1989). This led to the development of electromagnetic dynamos that replaced batteries as sources of power. These dynamos were the precursors of the early power generating systems. Some of the earliest uses for electrical power included telegraph systems, electroplating, and arc lighting systems.

Historic Development

The late nineteenth century witnessed tremendous changes as electrical power and light transformed the American way of life. Although many of the earliest power supply and distribution systems have been modernized or replaced altogether, this period saw a variety of electrical systems introduced that formed the foundations of modern technology.

Power Production and Distribution

The earliest power generators used an electromagnetic dynamo system that produced direct current (DC). This current is similar to that of the modern battery and is what Thomas Edison used to fuel his first electric lamp in 1879. Edison was an early advocate of DC. However, the distribution of DC was limited by the ability of a network to maintain voltage over long distances. This eventually prompted Edison's conversion to alternating current (AC), which can be transported over much greater distances with only nominal losses. While AC is the predominant form of electricity available today, DC is used for elevators and for emergency lighting and communications systems, either directly from battery sources or converted from an AC source.

The earliest systems were often constructed for single locations, such as factories and large private residences, and power was generated on-site. For general use, however, experimentation and economic considerations led to the recognition that AC power could best be generated in a large single facility or an interconnected network of facilities and transported at higher voltages via a power cable grid to the site. While transported at rates of thousands of volts, the power was transformed (using a step-down transformer) to levels acceptable by modern appliances and lighting. The standard voltage for the most common usage is 110 or 220 volts, but there are certain types of commercial, institutional, and industrial equipment, motors, and lighting that can also use 208-, 277-, or 480-volt systems.

Today electricity is produced in power plants across the country using nonrenewable coal-fired, natural gas, and nuclear fuel sources. Hydroelectric power produced by flowing water (essentially an updated version of the water-based motive force systems used in the earliest settlement period) provides a renewable source, but at the environmental cost of the loss of flooded ecosystem habitats. In the late twentieth century, environmental concerns led to the introduction of additional renewable energy sources, such as photovoltaic panels, wind turbines, and biomass fuels, which can generate electricity for on-site use or be supplied to the existing power grid.

Much debate has arisen over the use of photovoltaic systems, which generate electricity from the sun, on historic buildings. While still relatively expensive, these systems provide renewable energy and will become a sustaining force in many new buildings. Their use on historic buildings is problematic in that they usually consist of flat panels or have been integrated into common construction materials like shingles and spandrel glass. The latter form can be rather inconspicuous (Johnson 1997), but the former two tend to be visually obtrusive and sometimes materially incompatible with historic roofing. One solution has been to allow them on nonprimary elevations and roofs that cannot be seen from the street. This debate can only broaden as the availability of photovoltaics increases and their price goes down.

Electrical Distribution Systems

The earliest wiring in electrical distribution systems consisted of two exposed metal wire conductors that were initially secured in place by wooden cleats nailed directly to the wall. Many original electrical wiring systems were installed as a surface-mounted configuration in existing buildings rather than being concealed by removing and replacing surface finishes. Wiring for buildings built after the widespread use of electrical power became common was installed in attics, wall cavities, and crawl spaces that were covered by finished surface materials.

Unfortunately, direct contact with the combustible materials found in typical building construction led to fires as the wires heated up when used or were close enough together to permit arcing or short-circuiting between them. To reduce the risk of fire and short-circuiting, a (glazed or unglazed) ceramic porcelain knob and tube system was introduced and used wherever the conductors respectively changed direction or penetrated a combustible material such as wood framing (see Figure 21-3). Over time, wiring was improved by enclosing it in a variety of materials, including various forms of treated paper, braided cloth, rubber, and, by the 1970s, plastics.

Modern electrical codes, such as the National Electric Code (NEC or ANSI/NFPA 70), require specific types of wiring for specific locations as a safeguard against fire and electrocution. The most common wiring in residential buildings today consists of two conductors and a ground wire enclosed but separated in a plastic covering. This wiring is known as "nonmetallic sheathed cable" (NEC types NM and NMC, commonly referred to as Romex, a registered trademark of the Southwire Company).

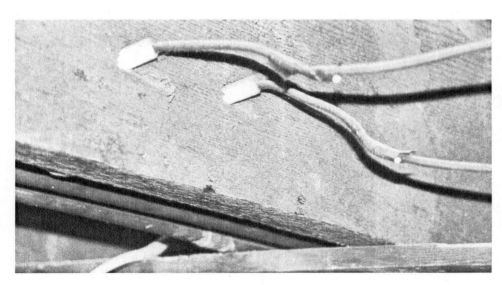

Figure 21-3 Knob and tube wiring was widely used in early wiring systems and is still in service today. Many local electrical codes do not require the wiring to be updated unless the circuits it serves are being altered. These systems do not provide a grounding wire that is required for many twentieth-century appliances and electronic equipment.

As electrical systems get larger and more complicated, such as those in commercial, institutional, and industrial buildings, power is supplied by flexible armored cable (NEC type AC, commonly referred to as BX, a registered trademark of the General Electric Corporation). In even larger systems, the power is distributed by cabling in conduits or by busducts. Busducts consist of copper plates in a housing to which a cable runout can be connected. An example of a small busduct is the track used in surface-mounted track lighting.

The electrical loads in a building are divided into circuits that serve specific equipment or spaces. Many early circuits were designed for much smaller loads since there were fewer electrically powered items than there are today. Early electrical service for many houses was extremely limited by today's standards. It often consisted of one or perhaps two circuits that were each typically rated to draw a maximum of 25 amps. As the use of electrical appliances grew in the late nineteenth and throughout the twentieth centuries, the number of circuits and loads steadily increased. Modern residential service now consists of twenty or more circuits drawing a total of more than 200 amps. Similar growth in the demand for services has affected other building types as well given the increased use of electric equipment throughout the past century.

For safety, the amount of electricity flowing into a building and subsequently to each circuit is controlled by a service panel. Originally, a fuse box, often located in an attic or another service space, performed this duty and used fuses that, when overheated by excessive current draw, would burn out and stop all flow into the circuit. Later, as electrical systems grew larger, distribution panels were developed with

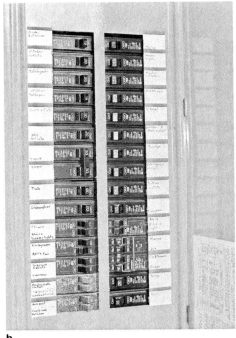

a   b

Figure 21-4   Early electrical service was controlled by fuse boxes (a), which eventually were replaced by circuit breaker systems (b). In modern construction, the number of circuits has significantly increased and the range of equipment used has expanded to require a variety of voltages.

circuit breakers that would trip to an off position but could be reset when the loads were reduced (see Figure 21-4).

A complete inspection of the electrical system should be performed to identify problems and system deficiencies. Inspection and assessment are performed by a licensed electrician or an electrical engineering consultant. Electrical wiring systems create many safety concerns, primarily combustion and electrocution. Problems arise from exposed conductors, faulty or missing ground wires, compromised protective coverings (e.g., deteriorated by age or gnawed on by animals), and non-code-compliant alterations and additions to the circuits. These can usually be initially identified by "dead" switches and circuits or inoperative fixtures. Where connections and wiring are exposed, check for melted or burned wires, connectors, or protective coverings. Unfortunately, most of the problems will lie concealed within walls, and locating them may require removal of surface finishes.

Modern electrical codes, such as the NEC or ANSI/NFPA 70, require specific types of wiring for specific locations as a safeguard against fire and electrocution. The NEC has evolved significantly over time, with existing systems being "grandfathered" (allowed to remain although not in accordance with the new code). This has commonly created a "spaghetti" of successive generations of wiring as persons completing later alterations simply bypassed (or not) existing wiring and abandoned (fully, partially, or not) the wiring for existing circuits in place.

Another problem is electrical capacity. The increased demand for electricity eventually led to the extension of existing circuits, the installation of separate new circuits, and the use of unsafe methods to defeat the safety measures incorporated into the service panel. For example, a fuse could be bypassed by inserting a copper penny into the base of the fuse. This permitted more current than the rated amperage allowed for the circuit and stopped fuses from burning out, but it also significantly increased the probability of fire as the wires overheated. Similarly, installation of new circuits and alterations to existing circuits to add capacity by using undersized wiring or not working in accordance with code requirements resulted in unsafe conditions. As stated above, the clues to these situations include melted or burned wiring covers, exposed connections, and inadequate grounding. Even when done in accordance with codes, the multiple alterations to and additions of various new circuits and their wiring pose a potential safety problem when it becomes necessary to identify active or inactive circuits throughout the building.

Typical Problems

Traditionally, little has been done to encourage the preservation of electrical service systems since they are largely unseen by the general public and may be in violation of modern codes. Many suspicious construction practices of earlier times may result in unsafe conditions, particularly in older and historic buildings that predate modern codes. Whether caused by an amateur or an (unlicensed) electrical contractor, code defects need to be corrected as mandated by local building code officials and the fire marshal. While a new use as a house museum may change the status of a previously residential building to include safety measures (e.g., life safety and security systems) for public buildings, most private residences may not be required to add these features. Based on the scope of the work, buildings historically used by the public may need to be brought up to modern codes.

If expanded service is required, it may be necessary to install completely new service panels and distribution systems to meet those new demands. Existing systems

Recommended Treatments

that are compliant with current codes typically do not have to be modified if they are not part of the proposed alterations. However, the installation of the new electrical system needs to take into account how the distribution network will affect the appearance of the spaces being served. Electrical panels can be located in closets or service areas away from public view. Wiring (AC, NM, NMC) may be routed through cavity walls or concealed behind baseboards and other trim pieces. Larger distribution systems, such as conduit, cable trays, and busducts, should be installed in accordance with local codes while mitigating the loss of historic integrity of the public spaces. Overall, the intent should be to minimize the visual impact of the new service and its effect on room finishes.

References and Suggested Readings

Association for Preservation Technology. 2000. APT/AIC Guidelines for light and lighting in historic buildings that house collections. *APT Bulletin*, 31(1): 11.

Calloway, Stephen, ed. 1996. *The elements of style: A practical encyclopedia of interior architectural details from 1485 to the present*. Rev. ed. New York: Simon & Schuster.

Cauldwell, Rex. 1996. *Wiring a house*. Newtown, CT: Taunton Press.

Clark, Clifford, Jr. 1986. *The American family home: 1860–1960*. Chapel Hill: University of North Carolina Press.

Craft, Meg Loew and M. Nicole Miller. 2000. Controlling daylight in historic structures: A focus on interior methods. *APT Bulletin*, 31(1): 53–59.

De Mare, Eric, ed. 1954. *New ways of servicing buildings*. London: Architectural Press.

Ehrenkranz, Florence and Lydia Inman. 1958. *Equipment in the home: Appliances, wiring and lighting, kitchen planning*. New York: Harper & Row Publishers.

Elliott, Cecil D. 1992. *Technics and architecture: The development of materials and systems for buildings*. Cambridge, MA: MIT Press.

Fisher, Charles E. and Hugh C. Miller, eds. 1998. *Caring for your historic house: Preserving and maintaining: Structural systems, roofs, masonry, plaster, wallpapers, paint, mechanical and electrical systems, windows, woodwork, flooring, landscapes*. New York: Harry N. Abrams.

Ierley, Merritt. 1999. *The comforts of home*. New York: Clarkson Potter Publishers.

Johnson, Elizabeth. 1997. *The Thoreau Center for sustainability: A model public–private partnership*. Washington, DC: National Trust for Historic Preservation.

Kay, Gersil Newmark . 1992. *Mechanical and electrical systems for buildings: Profitable tips for professionals, practical tips for the preservationists*. New York: McGraw-Hill Book Company.

Lechner, Norbert. 2001. *Heating, cooling, lighting: Design methods for architects*. 2nd ed. New York: John Wiley & Sons.

Luckiesh, M. 1920. *Lighting the home*. New York: Century Company.

———. 1925. *Lighting fixtures and lighting effects*. New York: McGraw-Hill Book Company.

Marcus, Alan I. and Howard P. Segal. 1989. *Technology in America: A brief history*. San Diego, CA: Harcourt Brace Jovanovich Publishers.

Maril, Nadja. 1989. *American lighting: 1840–1940*. West Chester, PA: Schiffer Publishing.

Moss, Roger W. 1988. *Lighting for historic buildings*. Washington, DC: Preservation Press.

Myers, Denys Peter. 1978. *Gaslighting in America: A guide for historic preservation*. Washington, DC: United States Department of the Interior.

Oliver, John W. 1956. *History of American technology*. New York: Ronald Press.

Peterson, Charles. 1976. *Building early America*. Radnor, PA: Chilton Books.

Phillips, Derek. 1997. *Lighting historic buildings*. New York: McGraw-Hill Book Company.

Plante, Ellen M. 1995. *The American kitchen: 1700 to the present*. New York: Facts on File.

Pursell, Carroll, ed. 2005. *A companion to American technology*. Malden, MA: Blackwell Publishing.

Randl, Chad. 2005. *The use of awnings on historic buildings: Repair, replacement and new design*. Preservation Brief No. 44. Washington, DC: United States Department of the Interior.

Shapiro, David E. 1998. *Old electrical wiring maintenance and retrofit*. New York: McGraw-Hill Book Company.

Stein, Benjamin, John S. Reynolds, Walter T. Grondzik, and Alison G. Kwok. 2006. *Mechanical and electrical equipment for buildings*. 10th ed. New York: John Wiley & Sons.

Taylor, Thomas H., Jr. 2000. Lighting historic house museums. *APT Bulletin*, 31(1): 7.

Thuro, Catherine. 1999. *Oil lamps: The kerosene era in North America*. Paducah, KY: Collector Books.

CHAPTER 22 Sustainability

Sustainability has been described as how well a process or product meets social, environmental, and economic needs. In contrast, the extraction-depletion approach used for most of the past four centuries has encouraged the pursuit of the short-term gains of expanding the built environment. While this action has long been viewed as socially and economically positive, the unfortunate result has also been the long-term degradation of the natural environment. The historic tradition of this model has been that there was always more somewhere else—more land, more energy resources, and more opportunity to exploit them.

Stewardship of the built environment is a sustainability philosophy that recognizes the natural and built environments as two complementary subsystems of the larger ecosystem. By understanding the closed system of forces, the preservation and reuse of the built environment is viewed as a sustainability strategy. An increasing amount of the built environment is underused throughout the country as suburban sprawl continues to drain resources into an increasingly unsustainable model of living. Reusing the built environment reverses this process.

A number of preservationists have long understood the sustainable aspects of reusing the built environment. Conversely, advocates of sustainability understand the need for the conservation of resources to promote a healthier ecosystem but have often overlooked the contributions that preservation and reuse of the built environment can make. Long-term benefits of both preservation and sustainable design present a strong case for adopting the mantra "reduce, reuse, and recycle" that can be interpreted at multiple parallel levels. From a sustainability perspective, this practice can be seen simply as reducing consumption by reusing resources or recycling them back into the consumer market for such things as paper, glass, plastic, or just about any consumer good or product.

A SUSTAINABLE PHILOSOPHY: STEWARDSHIP OF THE BUILT ENVIRONMENT

PRESERVATION AND SUSTAINABLE DESIGN SYNERGIES

Synergies arise when growth pressures are redirected back into the existing built environment rather than the continued depletion of open land at the suburban perimeter. The reinvigoration of inner-city and first-tier suburbs can take advantage of existing municipal service infrastructure, reduce atmospheric pollution from commuters, and provide the benefits of living in a physically diverse neighborhood that already has many of the features that suburban neo-traditional developments strive to achieve.

This broadened perception of preservation as sustainability has been overlooked by many but has been increasingly supported by economic and life-quality research findings from the NTHP Main Street Program efforts nationwide. Nearly every successful community revitalization project has succeeded when preservation of the built environment has been integrated into the planning goals of those communities (Rypkema 2006). Many of the projects that have been less than successful removed the character-defining features of the community (e.g., the buildings and the sense of continuity that accompanies them).

GLOBAL GOALS OF SUSTAINABILITY

In the final decade of the twentieth century, growing awareness of the need for sustainable design began to make an impact on design practices. In the past thirty years, many efforts have been made to understand the root causes of *un*sustainable design. Many of the imperatives that have been produced can be interpreted to include the preservation and reuse of the built environment as a primary tenet of sustainability. One such dictum (Mendler, Odell, and Lazarus 2006, 3) that provides an insight into the parallel nature of preservation and sustainable design is:

- Waste Nothing
  - Reduce construction, remodeling, and building operating waste.
  - Take a "less is more" approach to resource use.
  - Design for flexibility and long-term use.
  - Encourage resource reuse; avoid use of scarce materials.

- Adapt to Place
  - Use indigenous strategies.
  - Strive for diversity.

- Fit Form to Function

- Use "Free" Resources
  - Use renewable energy and materials resources.
  - Use materials and resources available locally and in abundance.

- Optimize Rather Than Maximize
  - Seek synergistic solutions.
  - Reduce reliance on mechanical systems.

- Create a Livable Environment
  - Protect sensitive and endangered ecosystems.
  - Support restoration of degraded natural systems.

- Promote development of pedestrian-friendly, mixed-use communities.
  - Create healthy environments free of toxic materials.
  - Provide daylight and views, and direct connections to nature.
  - Provide for personal control.
  - Create opportunities for personal expression.
  - Seek opportunities to improve social equity.

Consider how the goals of a historic preservation or reuse project can respond positively to any of these imperatives. Some affirmative responses can consist of the direct application of construction materials and practices. Others can focus on what the completed project will contribute to the occupants and the community.

Given the overarching idea that preservation and reuse are a form of sustainable design on their own merits, a question often still arises: what can be done to make the reused building more sustainable? That question involves a search for strategies that can be employed within an existing building to (1) make it more resource efficient and minimize the use of energy and water and (2) make it a healthier and more comfortable place.

SUSTAINABLE PRESERVATION PRACTICES

Low-Technology Approaches

Low-technology approaches refer to materials and constructions that usually do not have significant mechanical or electrical components or control systems. Such items often are manually controlled (e.g., shutters, awnings, and windows). Low technology also refers to a stationary aspect of the building or site (e.g., roof color, porch overhang, extra thermal insulation, and shade trees) that, once in place, may require no user input except maintenance.

Some aspects of existing buildings have already contributed to sustainability through their low-technology solutions that are part of the original construction. To understand some of the sustainable benefits afforded by preservation and reuse of the built environment, recall that before the early twentieth century, the construction of many buildings reflected their local climate and local materials (Lechner 2001). With the long-term use of modern HVAC and lighting systems, familiarity with a number of the low-technology practices that had commonly been incorporated into historic buildings faded. A number of these approaches have been described separately in the development of construction practices and building usage discussed earlier in this book; they are summarized below.

Architectural Form as Environmental Control
Take advantage of existing climate-based design features.
Include daylighting.
Reuse passive solar control and orientation.
Reuse passive natural cooling and ventilation.

Materials
Use locally available materials.
Recycle/salvage existing materials.
Use thermally massive building materials.
Retain the embodied energy.

Health and Safety
Remove contaminants and toxins.
Use clean water and air.

Architectural form as an environmental control system involves the use of porches, balconies, awnings, shutters, and mature deciduous trees for sun control, coniferous trees for wind control, towers and belvederes as thermal chimneys, skylights for light and ventilation, operable windows, climate-appropriate room volumes (e.g., expansive and open in hot climates, compact and closed in cold climates), sunspaces, and a host of other features (Lechner 2001).

The reuse of the building involves locally available, recycled, or salvaged material, and the building materials themselves may have been locally produced for the original construction. The use of thermally massive materials that can moderate daily temperature extremes began to fade with the introduction of light-weight curtain wall and veneer masonry systems. To appreciate the difference, contrast the thermal comfort of an attic and the thermal comfort of a basement on a hot summer afternoon. The last aspect of materials is the embodied energy of existing materials. Embodied energy is the sum of all the energy used to create, transport, and install the material in the building (Jackson 2005b). Discarding a material also means wasting the embodied energy within it.

Last, from a health and safety perspective, the benefits of removing toxins and contaminants from the building, including the water and air within the building, are self-evident. Toxins are removed from the greater community as well with the reduction of commuting time and the increased opportunities to walk or take mass transit, which will increase air quality in the long term.

Not surprisingly, contemporary projects that are winning recognition for innovative thinking have borrowed directly from or conceptually derived from these same practices.

High-Technology Systems

Many high-technology systems are HVAC or other building service systems configured to meet specific application needs. For example, a system for a laboratory is designed to perform specific functions to optimize the use of the laboratory. Meanwhile, a simple forced-air furnace is common in many residential applications. Between these two extremes, there are choices for fuel source, air filtration, ventilation, humidification, and indoor air quality options.

How well the proposed system will meet the expected performance requirements, and the extent of physical and visible disruption created by the installation of the system, are the primary selection criteria for many system designs. However, beyond that, other important selection criteria should include (1) the initial cost to purchase and install the system, (2) the projected costs for operation and maintenance, and (3) the expected useful life of the system, assuming proper operating and maintenance practices are followed. Through a process known as "life-cycle cost analysis," the overall cost of the system can be computed and compared with that of alternative systems. In this manner, a system that costs more initially but has lower lifetime operating costs can be equitably compared to a less expensive system that has higher lifetime operating costs. In many cases, systems with higher first costs have lower life-cycle costs because of improved operating and maintenance efficiencies. Selection based solely on first costs is shortsighted but unfortunately is the more common practice.

Most high-technology systems consist of an assembly of integrated mechanical and electrical components that work together to optimize energy consumption, reduce labor, or generally improve the quality of life. The more sophisticated the operating controls, the more likely it is that they will require specific training or expertise to make them operate properly. In recognition of this constraint, part of sustainable design has been the use of commissioning. Commissioning occurs when the manufacturer's representative, the engineer, or another person familiar with the control and operation of the equipment provides training and instructions to the operating "staff," which can range from the homeowner to a staff of building operations personnel. As part of the start-up and testing of each building system, the operating staff is shown how to control and operate it. Depending on the complexity of the systems involved, training options include on-site hands-on training sessions, preparing an operations manual describing each system and its requirements for operation and maintenance, or simply providing a file folder with the operating instruction for each piece of equipment.

Many sustainability strategies use low-technology solutions that must involve changes in manufacturing technology. As in the early 1970s, before the preservation industry became a well-established market segment, sustainable products are still in their infancy even though they are using timeless concepts. As demand for such things as low-VOC paints and coatings, higher recycled-content materials, sustainable flooring (e.g., linoleum), and sustainably harvested woods begins to grow, manufacturers will make more efforts to produce and market them. From a historic preservation technology perspective, these products will face scrutiny concerning their compatibility with the historic fabric, but given the economic forces of the market, they will increasingly find their way into the preservation and reuse marketplace.

Similar to the demand for environmentally sustainable products, increasing scrutiny of construction processes should lead to changes in the way demolition wastes in preservation projects are handled. The abatement of lead, asbestos, and other hazardous toxins and contaminants will be closely monitored for compliance with disposal regulations. Although a nationwide network of salvage operations for historic architectural materials already exists, a growing number of building product recycling centers are emerging to serve the needs of the contemporary market. And lastly, the economic feasibility of separating demolition wastes by material types (e.g., wood, metals, gypsum, and other building materials) will become greater as increasing demand for postconsumer recycled content grows and thus creates the loop back to the supply side of the process.

In the United States, the organizations that have developed sustainability-oriented programs include the United States Department of Energy, the United States Environmental Protection Agency, the National Association of Home Builders, and the American Institute of Architects (see websites in Appendix B). One of the more successful sustainable design programs is Leadership in Energy and Environmental Design (LEED), created by the United States Green Buildings Council, which reflects the growing concern to design and construct more environmentally friendly buildings. While it was originally oriented toward new construction, users of LEED are successfully adapting its strategies to existing buildings.

Like the *Secretary of the Interior's Standards for the Treatment of Historic Properties*, LEED is a voluntary standard. LEED is intended to provide verifiable parameters that quantify the environmental sensitivity of a building. The original program was designed for application to new building construction, to the dismay of a number of preservationists, who felt that it overlooked the sustainable aspects of preservation and adaptive use of existing buildings (Carroon, Roberts, and Simonsen 2004). The program has been expanding and revising its standards to accommodate these six target market segments:

New Construction (LEED-NC)

Core and Shell Development (LEED-CS)

Existing Buildings (LEED-EB)

Neighborhood Development (LEED-ND)

Sustainable Preservation Practices: The Loop Closes

SUSTAINABILITY-ORIENTED PROGRAMS

Commercial Interiors (LEED-CI)

Homes (LEED-H)

In defining the standards, LEED has also created a scoring system that quantifies how well a building performs with regard to a holistic view of sustainability. The scoring system defines six areas (i.e., sustainable sites, water efficiency, energy and atmosphere, materials and resources, indoor environmental quality, and innovation and design process) and provides a sequence of credits that can earn points in each area. Points can be earned only after meeting certain prerequisites in some categories. The points vary between the areas; some list as few as 5 possible points (water efficiency), while others have as many as 17 (energy and atmosphere). A maximum of 69 points can be earned.

The design team for each project must have at least one LEED-accredited professional (LEED-AP) to ensure that the standards are properly interpreted and the work complies with the standards. To obtain a LEED-AP designation, a person may take an exam in one of the six target market segments listed above. As a project's scope is defined, the potential cost and design impact of each credit are evaluated by the owner, the LEED-AP consultant, and the design team. As the credits for each area are tallied, the success of the project is quantified by the total number of points earned and is designated as platinum (52–69 points), gold (39–51 points), silver (33–38 points), or certified (26–32 points) (see Figure 22-1).

While the process is voluntary, an increasing number of cities and government agencies are requiring that the design team, if not actually use LEED directly on the project, then at least emulate the principles of LEED. At the time of this writing, the program has been plagued by the resistance of some persons (e.g., owners, architects, engineers, contractors, manufacturers, and public officials) to the extra effort and cost of tracking information flows through the project and confirming that all materials used meet the standards. This resistance, however, should fade as the design, construction, and manufacturing communities develop information tracking systems that facilitate the data requirements of LEED. As with the historic preservation incentives of the past few decades, a number of funding incentives are emerging that will help promote adoption of the LEED principles.

Figure 22-1 The reused Fuller Paint Company Building in Salt Lake City was among the first rehabilitation projects in the country to earn a LEED gold rating and simultaneously qualify for federal preservation tax credits.

G. H. SCHETTLER HOUSE REHABILITATION

THIS CASE STUDY DEMONSTRATES how the preservation and reuse aspects of rehabilitating buildings are a means of fostering sustainable design practices. The study also demonstrates how traditional and affordable technologies can be combined with advanced technologies to revitalize existing buildings. While this project is at the scale of a single building, the processes and goals are similar to those of much larger projects involving other building types. The difference is a matter of increased scale. In reviewing this case study, recall and reflect on how it interprets the global goals described earlier in this chapter.

The rehabilitation of the G. H. Schettler House (Young 2004) illustrates the processes used to meet the owners' concurrent historic preservation and environmental conservation goals for a privately owned residential building. The G. H. Schettler House was built in Salt Lake City in 1904 (see Figure 22-2). This single-family house, typical of housing found in this historic district, was one-half mile from downtown and was served by three local bus routes. The ten-room house was approximately 2,500 square feet.

The house was converted to five apartments in 1936. When it was purchased in 1994, most of the infrastructure for the apartments remained and the building service systems were incompatible with twenty-first-century needs. Although the house was heated by small furnaces, there was no central heating system. Cooling was provided by an evaporative cooler that only served a portion of the first floor. Utilities were substandard. However, the overall structural system was in good condition. Much of the original woodwork and plaster remained, although covered by many layers of paint and wallpaper. The 1970s-era asphalt roofing and aluminum siding concealed a roof structure and wood siding that were relatively intact.

The context, location, and amenities of the house included many features that sustainable projects seek. This project did, however, have one feature that other projects did not: it reused an existing building rather than constructing a new one in the suburbs or replacing an existing building in an older neighborhood. The features of the rehabilitation are:

- Architectural
  - Brick walls serve as thermal mass.

a   b

Figure 22-2   The G. H. Schettler House before (a) and after (b) rehabilitation.

- Ceiling height allows summer heat to rise above occupied space.
- Transoms over doors allow nighttime convective cooling.
- Double-hung windows allow nighttime convective cooling.
- Large windows permit passive winter heating and year-round daylighting.
- Natural ventilation allows warm air to rise out of the stairwell skylight.

- Thermal Control/Efficiency
  - Programmable thermostats employ day/night setbacks.
  - Thermostats control individual zone dampers for each floor.
  - Attic and crawl space insulation exceeds energy code requirements.
  - Weather stripping has been replaced.
  - Storm windows have been refurbished.
  - Light-colored roof shingles reflect summer heat.
  - Attic vent fan relieves excessive heat.

- Electrical/Lighting
  - Daylighting is used extensively.
  - Task lighting is used to meet visual needs.
  - Programmable timers are used for exterior lighting.
  - Security lighting is a two-stage, motion sensor–activated system.
  - The new wiring system meets the requirements of twenty-first-century living.

- Environmental
  - Recycled content (i.e., the house itself) is approximately 85.9 percent.
  - Replacement wooden simulated divided light windows on the second floor meet fire code egress requirements and reduce sound transmission and heat loss/gain.
  - Nighttime convective cooling is used throughout the house.
  - Natural materials are used throughout the house.
  - No threatened tree species or old-growth trees were used.
  - A warm-air furnace provides high-efficiency electrostatic air filtration.
  - Cooking and domestic hot water use natural gas.
  - A whole-house water filtration system removes potential hazards.
  - A water softener provides treated water.
  - Domestic hot water tank insulation improves fuel efficiency.
  - Lead-bearing materials have been abated.
  - Lowflow plumbing fixtures have been installed.

- Socioeconomic
  - Retaining the building maintains continuity of the historic urban fabric.
  - Reuse demonstrates economic feasibility to others.
  - The house reuses existing public utilities.
  - Financing of the project capitalizes on tax incentives.
  - Location of the house on three bus routes promotes the use of mass transit.
  - The location is within walking distance of downtown.

The rehabilitation shows that reusing existing residential buildings can meet both sustainability and preservation goals that can have a far-reaching impact on the stewardship of the built environment. This project succeeds in many ways. First, it models a successful process. Second, the modifications reduce energy consumption while increasing thermal comfort. Third, the project shows how to address environmental contamination. Fourth, the rehabilitation demonstrates how building reuse can reduce material streams by recycling the existing house and reducing the environmental impact caused by producing new building materials. Fifth, the project demonstrates that rehabilitating a house in an existing neighborhood, even in a historic district, is financially competitive with new house construction in the suburbs.

The homeowners used a process referred to as "project definition" (Mendler, Odell, and Lazarus 2006) to establish the overall vision for the project in interviewing potential project team members. All design team members, contractors, and consultants were selected based on their demonstrated sensitivity to and expertise in preservation. The overall process can be used as a model for other similar or larger projects. While relatively well known to members of the design and construction community, the steps followed are outlined here to provide insights into combining historic preservation design review with the overall process:

- Condition assessment: A building inspector completed a thorough investigation of the conditions during the prepurchase phase with a specific focus on the roof, structure, and service utilities. A walk-through visual inspection was conducted by a preservation consultant to identify historic character-defining elements and establish a baseline for the rehabilitation work.

- Programming: The homeowners developed performance requirements to set the tone for the project's definition of the philosophical goals. The philosophical goals were "to rehabilitate the house and retain its historic character while providing a home that meets the environmental demands of twenty-first-century urban living."

- Schematic design development: After several interviews, the architect was hired to develop schematic designs using the programming information provided by the homeowners. Since this was a tax credit project, the homeowners consulted the SHPO to confirm that the work would satisfy the *standards* for rehabilitation. Input was also sought from the city planning department since the house is a contributing building in Salt Lake City's historic district where any work affecting the exterior of the building had to receive a "Certificate of Appropriateness."

- Schematic design review: Alternatives were reviewed by the homeowners, and a final design was selected based on modifications suggested by the city and state regulating agencies, as well as to get the project within the budget defined by the homeowners.

- Construction document development: The construction documents were developed. At this point, home refinancing was confirmed.

- Contractor selection: A short list of qualified contractors was developed and the plans were sent out for bids. The contractor was selected based on the bid, an interview, reference checks, and philosophical compatibility with the project goals.

- Construction: Demolition was extensive: the roof was completely replaced, the decayed plaster on the first and second floors was removed, and the first floor was reconfigured back to the pre-1936 layout. New building service systems were installed throughout. The 1970s-era aluminum siding and aluminum porch elements were replaced with wooden elements based on photographs and existing physical evidence.

The project definition developed in the preprogram phase enabled the home-owners, the architect, and the contractor to reach a common understanding of the project goals and led to a smoother realization of the finished work.

There was no central HVAC system, and thermal comfort was extremely poor prior to the rehabilitation. Temperature extremes on the second floor made only the first floor inhabitable. Recognition of the contribution of the house itself to environmental control needs was based in part on the energy-conserving principles incorporated into the original construction. Thermal upgrades reduced the average heating bill by more than 25 percent even though the project increased the thermally comfortable and usable living space by 50 percent. Since there was no air-conditioning before the rehabilitation, there is no baseline for comparison. However, in the first three cooling seasons, the air-conditioning system was needed only in July and August. A calculation of the heating and cooling loads revealed load reductions of approximately 37 and 23 percent, respectively.

On-site tests revealed two forms of lead in this house—lead piping and lead-based paint. The main water line was made of lead, and the supply lines used lead-base solder in the joints. All plumbing was replaced. A whole-house water filtration system was installed to remove water-borne lead and other impurities. The lead paint was stripped from interior surfaces. Test samples of linoleum, ceiling tile, and insulation verified that no asbestos was present.

A significant aspect of the project was to demonstrate how reusing a building is more resource-sensitive than constructing a new building. Despite virtually gutting the second floor and removing a significant portion of the deteriorated plaster on the first floor, a comparative analysis showed that the project created approximately 47.3 tons of material streams for new materials to and demolition materials from the building. If, instead of rehabilitating the house, the owners had constructed an identical house in the suburbs, this amount would have been 182.4 tons. A second alternative illustrating the teardown trend in older neighborhoods showed that if the house had been torn down and rebuilt, the material stream would have increased to 351.8 tons (This number may actually be low since the replacement house is usually much bigger than the original.) So, this analysis showed that building a new house in the suburbs would have generated nearly four times as great a material stream as the rehabilitation project and a teardown and rebuild project would have generated more than seven times as much. These calculations clearly demonstrate that rehabilitating older housing stock is the most resource-sensitive sustainable alternative.

The cost of the project was $84.42/square foot (sf) without tax credits and $67.97/sf with them. Including the initial purchase price of the property, the overall costs without and with tax credits were $135.58 and $119.13/sf, respectively. These numbers were well below the costs for new-house construction in the local suburbs at the time (2000) and show that it is possible to rehabilitate an older home, even when following preservation standards, and be financially competitive.

Overall, these five factors confirm that the project has met the goal "to rehabilitate the house and retain its historic character while providing a home that meets the environmental demands of twenty-first-century urban living." The historic preservation technology principles included the integration of compatible processes and products to remove the lead hazards and the multiple layers of "updating" projects and then restore the interior finishes, woodwork, and ornamental features of the

a   b

Figure 22-3   The north parlor of the G. H. Schettler House before (a) and after (b) rehabilitation.

house (see Figure 22-3). The project won preservation awards from the Utah Heritage Foundation and the Salt Lake City Historic Landmarks Commission.

What further measures can be taken? The answer is complex and is complicated by an assortment of myths about preservation that have emerged where the promotion of processes, products, or materials that in the long run are insensitive and sometimes incompatible with preservation-oriented goals have collided with the realities of building science and economics (Sedovic and Gotthelf 2005).

Take the case of replacing original wood-sash single-paned windows with vinyl double-paned windows. As noted in Chapter 10, while the energy performance of the replacement *glass* may be better, the majority of the heat loss comes from the infiltration of air in the gap between the window and the wall surface and the gap between the sashes and the window frame. This gap can easily be eliminated by installing weather stripping and caulking far less expensively than replacing the entire window. This saves a character-defining feature; retains the embodied energy of the original window; avoids consuming new materials for the new window and the embodied energy needed to make, transport, and install the new window; and reduces landfill wastes by not discarding the original window.

The potential missed energy savings opportunity could be offset by installing additional attic or roof insulation. The lower additional cost of adding the weather stripping, caulking, and insulation is recovered more quickly in energy cost savings than the cost of the replacement window. In the long term, this will also probably save more energy, and therefore is both sustainable and preservation sensitive. This concept of trading off one solution for another that is more or equally sustainable and preservation friendly is the key to success in creatively interpreting the performance requirement of a system rather than fixating on a single-minded solution. This approach is referred to as the "whole-building concept" or "whole-systems" thinking. The building is a system that must meet certain performance requirements. As most projects are limited by a budget, selections based on performance become important factors when looking at long-term savings and costs. The whole-building trade-off approach is necessary when trying to resolve sustainability questions, code compliance issues, and almost any other preservation issue as well. (Note: This last

**CONCLUDING
COMMENTS**

statement provides an insight not only into sustainability but also into how modern practices, materials, and expectations must be adapted to the preservation and reuse of buildings, and not the other way around.)

The objective is not to accept a single-minded approach or prescriptive imperative blindly but to understand the problem from a historic preservation technology perspective and solve it in the context of modern design and construction practices. Consult with other design team members to confirm or dispute the appropriate solution instead of proceeding with a single-minded approach that in fact may be shortsighted in the long term. Appropriate use of the collective expertise of the design team is part of what makes a project successful.

The sustainability principles and goals in use today already include aspects of preservation and reuse of the built environment. Using historic preservation technology to further these principles and goals is an emerging phenomenon across the country. The growing availability of sustainable design processes and products can be readily adapted to historic preservation technology to accept those that are compatible with preservation, restoration, rehabilitation, and reconstruction. The key will be understanding the materials, construction practices, and remediation methods of historic preservation technology and then effectively applying the insights that understanding provides in making appropriate choices.

References and Suggested Readings

Carroon, Jeanne, Allen Roberts, and Soren Simonsen. 2004. "Green" design and historic buildings. *Forum News*, X(3):1–2, 6.

Cavallo, James. 2005. Capturing energy-efficiency opportunities in historic houses. *APT Bulletin*, 36(4): 19–23.

Cazayoux, Edward Jon. 2003. *A manual for the environmental and climatic responsive restoration and renovation of older houses in Louisiana*. Baton Rouge: Louisiana Department of Natural Resources.

Elefante, Carl. 2005. Historic preservation and sustainable development: Lots to learn, lots to teach. *APT Bulletin*, 36(4): 53.

Fitch, James Marston with William Bobenhausen. 1999. *American building: The environmental forces that shape it*. New York: Oxford University Press.

International Code Council. 2006. *International Energy Conservation Code*. Country Club Hills, IL: International Code Council.

Jackson, Mike. 2005a. Building a culture that sustains design. *APT Bulletin*, 36(4): 2.

———. 2005b. Embodied energy and historic preservation: A needed reassessment. *APT Bulletin*, 36(4): 47–52.

Kibert, Charles J. 2005. *Sustainable construction: Green building design and delivery*. New York: John Wiley & Sons.

Lechner, Norbert. 2001. *Heating, cooling, lighting: Design methods for architects*. 2nd ed. New York: John Wiley & Sons.

McInnis, Maggie and Ilene R. Tyler. 2005. The greening of the Samuel T. Dana Building: A classroom and laboratory for sustainable design. *APT Bulletin*, 36(4): 39–45.

McMahon, Edward T. and A. Elizabeth Watson. 1992. *In search of collaboration: Historic preservation and the environmental movement*. Information Series No. 71. Washington, DC: National Trust for Historic Preservation.

Mendler, Sandra, William Odell, and Mary Ann Lazarus. 2006. *The HO+K guidebook to sustainable design*. 2nd ed. New York: John Wiley & Sons.

Merryman, Helena. 2005. Structural materials in historic restoration: Environmental issues and greener strategies. *APT Bulletin* 36(4): 31–38.

National Audubon Society and Croxton Collaborative Architects. 1994. *Audubon house*. New York: John Wiley & Sons.

Powter, Andrew and Susan Ross. 2005. Integrating environmental and cultural sustainability for heritage properties. *APT Bulletin*, 36(4): 5–11.

Rose, William B. 2006. Should walls of historic buildings be insulated? *APT Bulletin*, 36(4): 13–18.

Rypkema, Donovan D. 2006. Economics, sustainability, and historic preservation. *Forum Journal*, 20 (2). http://forum.nthp.org/subNTHP/displayNews.asp? lib_ID=921.

Sedovic, Walter and Jill H. Gotthelf. 2005. What replacement windows can't replace: The real cost of removing historic windows. *APT Bulletin*, 36(4): 25–29.

Sedway Cooke Associates. 1983. *Retrofit right: How to make your old house energy efficient.* Oakland, CA: City of Oakland Planning Department.

Venolia, Carol and Kelly Lerner. 2006. *Natural remodeling for the not-so-green-house.* Asheville, NC: Lark Books.

Young, Robert A. 2003. Fort Stephen A. Douglas: Adaptive reuse for a community of scholars. In *Protecting our diverse heritage: The role of parks, protected areas, and cultural sites*, eds. David Harmon, Bruce M. Kilgore, and Gay E. Vietzke, 205–209. Hancock, MI: George Wright Society.

———. 2004. Stewardship of the built environment: The emerging synergies from sustainability and historic preservation. In *Archipelagos: Outposts of the Americas enclaves amidst technology*, eds. Robert Alexander Gonzalez and Marilys Rebeca Nepomechie, 35–50. Washington, DC: Association of Collegiate Schools of Architecture.

Appendix A

Secretary of the Interior's Standards for the Treatment of Historic Properties

OVERVIEW

The *Secretary of the Interior Standards* recognize four separate treatment approaches (preservation, rehabilitation, restoration, and reconstruction) to any construction project involving a historic building. Accordingly, standards and guidelines have been developed to assist in determining which course of action should be pursued and in defining which technologies are appropriate to use. The National Park Service administers the enforcement of these standards, and the sections below have been drawn directly from their website at http://www.cr.nps.gov/hps/tps/standguide/index.htm.

PRESERVATION

Standards for Preservation

1. A property will be used as it was historically, or be given a new use that maximizes the retention of distinctive materials, features, spaces, and spatial relationships. Where a treatment and use have not been identified, a property will be protected and, if necessary, stabilized until additional work may be undertaken.
2. The historic character of a property will be retained and preserved. The replacement of intact or repairable historic materials or alteration of features, spaces, and spatial relationships that characterize a property will be avoided.
3. Each property will be recognized as a physical record of its time, place, and use. Work needed to stabilize, consolidate, and conserve existing historic materials and features will be physically and visually compatible, identifiable upon close inspection, and properly documented for future research.
4. Changes to a property that have acquired historic significance in their own right will be retained and preserved.
5. Distinctive materials, features, finishes, and construction techniques or examples of craftsmanship that characterize a property will be preserved.
6. The existing condition of historic features will be evaluated to determine the appropriate level of intervention needed. Where the severity of deterioration requires repair or limited replacement of a distinctive feature, the new material will match the old in composition, design, color, and texture.
7. Chemical or physical treatments, if appropriate, will be undertaken using the gentlest means possible. Treatments that cause damage to historic materials will not be used.

8. Archeological resources will be protected and preserved in place. If such resources must be disturbed, mitigation measures will be undertaken.

GUIDELINES FOR PRESERVATION AS A PHILOSOPHICAL TREATMENT APPROACH

Introduction

When the property's distinctive materials, features, and spaces are essentially intact and thus convey the historic significance without extensive repair or replacement; when depiction at a particular period of time is not appropriate; and when a continuing or new use does not require additions or extensive alterations, *Preservation* may be considered as a treatment. Prior to undertaking work, a documentation plan for *Preservation* should be developed.

In *Preservation*, the options for replacement are less extensive than in the treatment, *Rehabilitation*. This is because it is assumed at the outset that building materials and character-defining features are essentially intact (e.g., that more historic fabric has survived, unchanged over time). The expressed goal of the *Standards for Preservation and Guidelines for Preserving Historic Buildings* is retention of the building's existing form, features and detailing. This may be as simple as basic maintenance of existing materials and features or may involve preparing a historic structure report, undertaking laboratory testing, such as paint and mortar analysis, and hiring conservators to perform sensitive work, such as reconstituting interior finishes. Protection, maintenance, and repair are emphasized while replacement is minimized.

Identify, Retain, and Preserve Historic Materials and Features

The guidance for the treatment *Preservation* begins with recommendations to identify the form and detailing of those architectural materials and features that are important in defining the building's historic character and which must be retained in order to preserve that character. Therefore, guidance on *identifying, retaining, and preserving* character-defining features is always given first. The character of a historic building may be defined by the form and detailing of exterior materials, such as masonry, wood, and metal; exterior features, such as roofs, porches, and windows; interior materials, such as plaster and paint; and interior features, such as moldings and stairways, room configuration and spatial relationships, as well as structural and mechanical systems; and the building's site and setting.

Stabilize Deteriorated Historic Materials and Features

Deteriorated portions of a historic building may need to be protected through preliminary stabilization measures until additional work can be undertaken. *Stabilizing* may include structural reinforcement, weatherization, or correcting unsafe conditions. Temporary stabilization should always be carried out in such a manner that it detracts as little as possible from the historic building's appearance. Although it may not be necessary in every preservation project, stabilization is nonetheless an integral part of the treatment *Preservation*; it is equally applicable, if circumstances warrant, for the other treatments.

Protect and Maintain Historic Materials and Features

After identifying those materials and features that are important and must be retained in the process of *Preservation* work, then *protecting and maintaining* them are addressed. Protection generally involves the least degree of intervention and is preparatory to other work. For example, protection includes the maintenance of historic materials through treatments, such as rust removal, caulking, limited paint removal, and re-application of protective coatings; the cyclical cleaning of roof gutter systems; or installation of fencing, alarm systems and other temporary protective measures. Although a historic building will usually require more extensive work, an overall evaluation of its physical condition should always begin at this level.

Repair (Stabilize, Consolidate, and Conserve) Historic Materials and Features

Next, when the physical condition of character-defining materials and features requires additional work, *repairing* by *stabilizing, consolidating, and conserving* is recommended. *Preservation* strives to retain existing materials and features while employing as little new material as possible. Consequently, guidance for repairing a historic material, such as masonry, again begins with the least degree of intervention possible, such as strengthening fragile materials through consolidation, when appropriate, and repointing with mortar of an appropriate strength. Repairing masonry as well as wood and architectural metal features may also include patching, splicing, or otherwise reinforcing them by using recognized preservation methods. Similarly, within the treatment *Preservation*, portions of a historic structural system could be reinforced using contemporary materials, such as steel rods. All work should be physically and visually compatible, identifiable upon close inspection and documented for future research.

Limited Replacement in Kind of Extensively Deteriorated Portions of Historic Features

If repair by stabilization, consolidation, and conservation proves inadequate, the next level of intervention involves the *limited replacement in kind* of extensively deteriorated or missing parts of features when there are surviving prototypes (e.g., brackets, dentils, steps, plaster, or portions of slate or tile roofing). The replacement material needs to match the old both physically and visually (e.g., wood with wood). Thus, with the exception of hidden structural reinforcement and new mechanical system components, substitute materials are not appropriate in the treatment *Preservation*. Again, it is important that all new material be identified and properly documented for future research. If prominent features are missing, such as an interior staircase, exterior cornice, or a roof dormer, then a *Rehabilitation* or *Restoration* treatment may be more appropriate.

Energy Efficiency/Accessibility Considerations/Health and Safety Code Considerations

These sections of the *Preservation* guidance address work done to meet accessibility requirements and health and safety code requirements; or limited retrofitting measures to improve energy efficiency. Although this work is quite often an important aspect of preservation projects, it is usually not part of the overall process of protecting, stabilizing, conserving, or repairing character-defining features; rather, such work is assessed for its potential negative impact on the building's historic character. For this reason, particular care must be taken not to obscure, damage, or destroy character-defining materials or features in the process of undertaking work to meet code and energy requirements.

REHABILITATION

Standards for Rehabilitation

1. A property will be used as it was historically or be given a new use that requires minimal change to its distinctive materials, features, spaces, and spatial relationships.
2. The historic character of a property will be retained and preserved. The removal of distinctive materials or alteration of features, spaces, and spatial relationships that characterize a property will be avoided.
3. Each property will be recognized as a physical record of its time, place, and use. Changes that create a false sense of historical development, such as adding conjectural features or elements from other historic properties, will not be undertaken.
4. Changes to a property that have acquired historic significance in their own right will be retained and preserved.
5. Distinctive materials, features, finishes, and construction techniques or examples of craftsmanship that characterize a property will be preserved.
6. Deteriorated historic features will be repaired rather than replaced. Where the severity of deterioration requires replacement of a distinctive feature, the new feature will match the old in design, color, texture, and, where possible, materials. Replacement of missing features will be substantiated by documentary and physical evidence.
7. Chemical or physical treatments, if appropriate, will be undertaken using the gentlest means possible. Treatments that cause damage to historic materials will not be used.
8. Archeological resources will be protected and preserved in place. If such resources must be disturbed, mitigation measures will be undertaken.
9. New additions, exterior alterations, or related new construction will not destroy historic materials, features, and spatial relationships that characterize the property. The new work shall be differentiated from the old and will be compatible with the historic materials, features, size, scale and proportion, and massing to protect the integrity of the property and its environment.
10. New additions and adjacent or related new construction will be undertaken in such a manner that, if removed in the future, the essential form and integrity of the historic property and its environment would be unimpaired.

GUIDELINES FOR REHABILITATION AS A PHILOSOPHICAL TREATMENT APPROACH

Introduction

When repair and replacement of deteriorated features are necessary; when alterations or additions to the property are planned for a new or continued use; and when its depiction at a particular period of time is not appropriate, *Rehabilitation* may be considered as a treatment. Prior to undertaking work, a documentation plan for *Rehabilitation* should be developed. In *Rehabilitation*, historic building materials and character-defining features are protected and maintained as they are in the treatment *Preservation*; however, an assumption is made prior to work that existing historic fabric has become damaged or deteriorated over time and, as a result, more repair and replacement will be required. Thus, latitude is given in the *Standards for Rehabilitation and Guidelines for Rehabilitation* to replace extensively deteriorated, damaged, or missing features using either traditional or substitute materials. Of the four treatments, only Rehabilitation includes an opportunity to make possible an efficient contemporary use through alterations and additions.

Identify, Retain, and Preserve Historic Materials and Features

Like *Preservation*, guidance for the treatment *Rehabilitation* begins with recommendations to identify the form and detailing of those architectural materials and features that are important in defining the building's historic character and which must be retained in order to preserve that character. Therefore, guidance on *identifying, retaining, and preserving* character-defining features is always given first. The character of a historic building may be defined by the form and detailing of exterior materials, such as masonry, wood, and metal; exterior features, such as roofs, porches, and windows; interior materials, such as plaster and paint; and interior features, such as moldings and stairways, room configuration and spatial relationships, as well as structural and mechanical systems.

Protect and Maintain Historic Materials and Features

After identifying those materials and features that are important and must be retained in the process of *Rehabilitation* work, then protecting and maintaining them are addressed. Protection generally involves the least degree of intervention and is preparatory to other work. For example, protection includes the maintenance of historic material through treatments, such as rust removal, caulking, limited paint removal, and re-application of protective coatings; the cyclical cleaning of roof gutter systems; or installation of fencing, alarm systems and other temporary protective measures. Although a historic building will usually require more extensive work, an overall evaluation of its physical condition should always begin at this level.

Repair Historic Materials and Features

Next, when the physical condition of character-defining materials and features warrants additional work, *repairing* is recommended. *Rehabilitation* guidance for the repair of historic materials, such as masonry, wood, and architectural metals, again begins with the least degree of intervention possible, such as patching, piecing-in, splicing, consolidating, or otherwise reinforcing or upgrading them according to recognized preservation methods. Repairing also includes the limited replacement in kind—or with compatible substitute material—of extensively deteriorated or missing parts of features when there are surviving prototypes (e.g., brackets, dentils, steps, plaster, or portions of slate or tile roofing). Although using the same kind of material is always the preferred option, substitute material is acceptable if the form and design as well as the substitute material itself convey the visual appearance of the remaining parts of the feature and finish.

Replace Deteriorated Historic Materials and Features

Following repair in the hierarchy, *Rehabilitation* guidance is provided for *replacing* an entire character-defining feature with new material because the level of deterioration or damage of materials precludes repair (e.g., an exterior cornice; an interior staircase; or a complete porch or storefront). If the essential form and detailing are still evident so that the physical evidence can be used to re-establish the feature as an integral part of the rehabilitation, then its replacement is appropriate. Like the guidance for repair, the preferred option is always replacement of the entire feature in kind, that is, with the same material. Because this approach may not always be technically or economically feasible, provisions are made to consider the use of a compatible substitute material. It should be noted that, while the National Park Service guidelines recommend the replacement of an entire character-defining feature that is extensively deteriorated, they never recommend removal and replacement with new material of a feature that—although damaged or deteriorated—could reasonably be repaired and thus preserved.

Design for the Replacement of Missing Historic Features

When an entire interior or exterior feature is missing (e.g., an entrance, or cast iron facade, or a principal staircase), it no longer plays a role in physically defining the historic character of the building unless it can be accurately recovered in form and detailing through the process of carefully documenting the historical appearance. Although accepting the loss is one possibility, where an important architectural feature is missing, its replacement is always recommended in the *Rehabilitation* guidelines as the first, or preferred, course of action. Thus, if adequate historical, pictorial, and physical documentation exists so that the feature may be accurately reproduced, and if it is desirable to reestablish the feature as part of the building's historical appearance, then designing and constructing a new feature based on such information is appropriate. However, a second acceptable option for the replacement feature is a new design that is compatible with the remaining character-defining features of the historic building. The new design should always take into account the size, scale, and material of the historic building itself and, most importantly, should be clearly differentiated so that a false historical appearance is not created.

Alterations/Additions for the New Use

Some exterior and interior alterations to a historic building are generally needed to assure its continued use, but it is most important that such alterations do not radically change, obscure, or destroy character-defining spaces, materials, features, or finishes. Alterations may include providing additional parking space on an existing historic building site; cutting new entrances or windows on secondary elevations; inserting an additional floor; installing an entirely new mechanical system; or creating an atrium or light well. Alteration may also include the selective removal of buildings or other features of the environment or building site that are intrusive and therefore detract from the overall historic character. The construction of an exterior addition to a historic building may seem to be essential for the new use, but it is emphasized in the *Rehabilitation* guidelines that such new additions should be avoided, if possible, and considered only after it is determined that those needs cannot be met by altering secondary (e.g., non-character-defining interior spaces). If, after a thorough evaluation of interior solutions, an exterior addition is still judged to be the only viable alterative, it should be designed and constructed to be clearly differentiated from the historic building and so that the character-defining features are not radically changed, obscured, damaged, or destroyed. Additions and alterations to historic buildings are referenced within specific sections of the *Rehabilitation* guidelines, such as Site, Roofs, Structural Systems, etc., but are addressed in detail in New Additions to Historic Buildings.

Energy Efficiency/Accessibility Considerations/Health and Safety Code Considerations

These sections of the guidance address work done to meet accessibility requirements and health and safety code requirements, or retrofitting measures to improve energy efficiency. Although this work is quite often an important aspect of *Rehabilitation* projects, it is usually not a part of the overall process of protecting or repairing character-defining features; rather, such work is assessed for its potential negative impact on the building's historic character. For this reason, particular care must be taken not to radically change, obscure, damage, or destroy character-defining materials or features in the process of meeting code and energy requirements.

RESTORATION

Standards for Restoration

1. A property will be used as it was historically or be given a new use which reflects the property's restoration period.
2. Materials and features from the restoration period will be retained and preserved. The removal of materials or alteration of features, spaces, and spatial relationships that characterize the period will not be undertaken.
3. Each property will be recognized as a physical record of its time, place, and use. Work needed to stabilize, consolidate and conserve materials and features from the restoration period will be physically and visually compatible, identifiable upon close inspection, and properly documented for future research.
4. Materials, features, spaces, and finishes that characterize other historical periods will be documented prior to their alteration or removal.
5. Distinctive materials, features, finishes, and construction techniques or examples of craftsmanship that characterize the restoration period will be preserved.
6. Deteriorated features from the restoration period will be repaired rather than replaced. Where the severity of deterioration requires replacement of a distinctive feature, the new feature will match the old in design, color, texture, and, where possible, materials.
7. Replacement of missing features from the restoration period will be substantiated by documentary and physical evidence. A false sense of history will not be created by adding conjectural features, features from other properties, or by combining features that never existed together historically.
8. Chemical or physical treatments, if appropriate, will be undertaken using the gentlest means possible. Treatments that cause damage to historic materials will not be used.

9. Archeological resources affected by a project will be protected and preserved in place. If such resources must be disturbed, mitigation measures will be undertaken.
10. Designs that were never executed historically will not be constructed.

GUIDELINES FOR RESTORATION AS A PHILOSOPHICAL TREATMENT APPROACH

Introduction

When the property's design, architectural, or historical significance during a particular period of time outweighs the potential loss of extant materials, features, spaces, and finishes that characterize other historical periods; when there is substantial physical and documentary evidence for the work; and when contemporary alterations and additions are not planned, *Restoration* may be considered as a treatment. Prior to undertaking work, a particular period of time (e.g., the restoration period) should be selected and justified, and a documentation plan for *Restoration* developed. Rather than maintaining and preserving a building as it has evolved over time, the expressed goal of the *Standards for Restoration and Guidelines for Restoring Historic Buildings* is to make the building appear as it did at a particular—and most significant—time in its history. First, those materials and features from the "restoration period" are identified, based on thorough historical research. Next, features from the restoration period are maintained, protected, repaired (e.g., stabilized, consolidated, and conserved), and replaced, if necessary. As opposed to other treatments, the scope of work in *Restoration* can include removal of features from other periods; missing features from the restoration period may be replaced, based on documentary and physical evidence, using traditional materials or compatible substitute materials. The final guidance emphasizes that only those designs that can be documented as having been built should be re-created in a restoration project.

Identify, Retain, and Preserve Materials and Features from the Restoration Period

The guidance for the treatment *Restoration* begins with recommendations to identify the form and detailing of those existing architectural materials and features that are significant to the restoration period as established by historical research and documentation. Thus, guidance on *identifying, retaining, and preserving features from the restoration period* is always given first. The historic building's appearance may be defined by the form and detailing of its exterior materials, such as masonry, wood, and metal; exterior features, such as roofs, porches, and windows; interior materials, such as plaster and paint; and interior features, such as moldings and stairways, room configuration and spatial relationships, as well as structural and mechanical systems; and the building's site and setting.

Protect and Maintain Materials and Features from the Restoration Period

After identifying those existing materials and features from the restoration period that must be retained in the process of *Restoration* work, then *protecting and main-*

taining them is addressed. Protection generally involves the least degree of intervention and is preparatory to other work. For example, protection includes the maintenance of historic material through treatments, such as rust removal, caulking, limited paint removal, and re-application of protective coatings; the cyclical cleaning of roof gutter systems; or installation of fencing, alarm systems and other temporary protective measures. Although a historic building will usually require more extensive work, an overall evaluation of its physical condition should always begin at this level.

Repair (Stabilize, Consolidate, and Conserve) Materials and Features from the Restoration Period

Next, when the physical condition of restoration period features requires additional work, *repairing* by *stabilizing, consolidating, and conserving* is recommended. *Restoration* guidance focuses upon the preservation of those materials and features that are significant to the period. Consequently, guidance for repairing a historic material, such as masonry, again begins with the least degree of intervention possible, such as strengthening fragile materials through consolidation, when appropriate, and repointing with mortar of an appropriate strength. Repairing masonry as well as wood and architectural metals includes patching, splicing, or otherwise reinforcing them using recognized preservation methods. Similarly, portions of a historic structural system could be reinforced using contemporary material, such as steel rods. In *Restoration*, repair may also include the limited replacement in kind—or with compatible substitute material—of extensively deteriorated or missing parts of existing features when there are surviving prototypes to use as a model. Examples could include terra-cotta brackets, wood balusters, or cast iron fencing.

Replace Extensively Deteriorated Features from the Restoration Period

In *Restoration*, *replacing* an entire feature from the restoration period (e.g., a cornice, balustrade, column, or stairway) that is too deteriorated to repair may be appropriate. Together with documentary evidence, the form and detailing of the historic feature should be used as a model for the replacement. Using the same kind of material is preferred; however, compatible substitute material may be considered. All new work should be unobtrusively dated to guide future research and treatment. If documentary and physical evidence are not available to provide an accurate recreation of missing features, the treatment *Rehabilitation* might be a better overall approach to project work.

Remove Existing Features from Other Historic Periods

Most buildings represent continuing occupancies and change over time, but in *Restoration*, the goal is to depict the building as it appeared at the most significant time in its history. Thus, work is included to remove or alter existing historic features that do not represent the restoration period. This could include features such as windows, entrances and doors, roof dormers, or landscape features. Prior to altering or removing materials, features, spaces, and finishes that characterize other historical periods, they should be documented to guide future research and treatment.

Re-create Missing Features from the Restoration Period

Most *Restoration* projects involve re-creating features that were significant to the building at a particular time, but are now missing. Examples could include a stone balustrade, a porch, or cast iron storefront. Each missing feature should be substantiated by documentary and physical evidence. Without sufficient documentation for these "re-creations," an accurate depiction cannot be achieved. Combining features that never existed together historically can also create a false sense of history. Using traditional materials to depict lost features is always the preferred approach; however, using compatible substitute material is an acceptable alternative in *Restoration* because, as emphasized, the goal of this treatment is to replicate the "appearance" of the historic building at a particular time, not to retain and preserve all historic materials as they have evolved over time. If documentary and physical evidence are not available to provide an accurate re-creation of missing features, the treatment *Rehabilitation* might be a better overall approach to project work.

Energy Efficiency/Accessibility Considerations/Health and Safety Code Considerations

These sections of the *Restoration* guid[elines] address work done to meet accessibility requirements and health and safety code requirements; or limited retrofitting measures to improve energy efficiency. Although this work is quite often an important aspect of restoration projects, it is usually not part of the overall process of protecting, stabilizing, conserving, or repairing features from the restoration period; rather, such work is assessed for its potential negative impact on the building's historic appearance. For this reason, particular care must be taken not to obscure, damage, or destroy historic materials or features from the restoration period in the process of undertaking work to meet code and energy requirements.

RECONSTRUCTION

Standards for Reconstruction

1. *Reconstruction* will be used to depict vanished or nonsurviving portions of a property when documentary and physical evidence is available to permit accurate reconstruction with minimal conjecture, and such reconstruction is essential to the public understanding of the property.
2. *Reconstruction* of a landscape, building, structure, or object in its historic location will be preceded by a thorough archeological investigation to identify and evaluate those features and artifacts which are essential to an accurate reconstruction. If such resources must be disturbed, mitigation measures will be undertaken.
3. *Reconstruction* will include measures to preserve any remaining historic materials, features, and spatial relationships.
4. *Reconstruction* will be based on the accurate duplication of historic features and elements substantiated by documentary or physical evidence rather than on conjectural designs or the availability of different features from other historic properties. A reconstructed property will re-create the appearance of the non-surviving historic property in materials, design, color, and texture.

5. A reconstruction will be clearly identified as a contemporary re-creation.
6. Designs that were never executed historically will not be constructed.

GUIDELINES FOR RECONSTRUCTION AS A PHILOSOPHICAL TREATMENT APPROACH

Introduction

When a contemporary depiction is required to understand and interpret a property's historic value (including the re-creation of missing components in a historic district or site); when no other property with the same associative value has survived; and when sufficient historical documentation exists to ensure an accurate reproduction, *Reconstruction* may be considered as a treatment. Prior to undertaking work, a documentation plan for *Reconstruction* should be developed. Whereas the treatment *Restoration* provides guidance on restoring—or re-creating—building features, the *Standards for Reconstruction and Guidelines for Reconstructing Historic Buildings* address those aspects of treatment necessary to re-create an entire nonsurviving building with new material. Much like *Restoration*, the goal is to make the building appear as it did at a particular—and most significant—time in its history. The difference is, in *Reconstruction*, there is far less extant historic material prior to treatment and, in some cases, nothing visible. Because of the potential for historical error in the absence of sound physical evidence, this treatment can be justified only rarely and, thus, is the least frequently undertaken. Documentation requirements prior to and following work are very stringent. Measures should be taken to preserve extant historic surface and subsurface material. Finally, the reconstructed building must be clearly identified as a contemporary re-creation.

Research and Document Historical Significance

Guidance for the treatment *Reconstruction* begins with *researching and documenting* the building's historical significance to ascertain that its re-creation is essential to the public understanding of the property. Often, another extant historic building on the site or in a setting can adequately explain the property, together with other interpretive aids. Justifying a reconstruction requires detailed physical and documentary evidence to minimize or eliminate conjecture and ensure that the reconstruction is as accurate as possible. Only one period of significance is generally identified; a building, as it evolved, is rarely re-created. During this important fact-finding stage, if research does not provide adequate documentation for an accurate reconstruction, other interpretive methods should be considered, such as an explanatory marker.

Investigate Archeological Resources

Investigating archeological resources is the next area of guidance in the treatment *Reconstruction*. The goal of physical research is to identify features of the building and site which are essential to an accurate re-creation and must be reconstructed while leaving those archeological resources that are not essential undisturbed. Information that is not relevant to the project should be preserved in place for future research. The archeological findings, together with archival documentation, are then used to replicate the plan of the building, together with the relationship and size of rooms, corridors, and other spaces, and spatial relationships.

Identify, Protect and Preserve Extant Historic Features

Closely aligned with archeological research, recommendations are given for *identifying, protecting, and preserving* extant features of the historic building. It is never appropriate to base a *Reconstruction* upon conjectural designs or the availability of different features from other buildings. Thus, any remaining historic materials and features, such as remnants of a foundation or chimney, and site features, such as a walkway or path, should be retained, when practicable, and incorporated into the reconstruction. The historic as well as new material should be carefully documented to guide future research and treatment.

Reconstruct Nonsurviving Building and Site

After the research and documentation phases, guidance is given for *Reconstruction* work itself. Exterior and interior features are addressed in general, always emphasizing the need for an *accurate depiction* (e.g., careful duplication of the appearance of historic interior paints, and finishes, such as stenciling, marbling, and graining). In the absence of extant historic materials, the objective in reconstruction is to re-create the appearance of the historic building for interpretive purposes. Thus, while the use of traditional materials and finishes is always preferred, in some instances, substitute materials may be used if they are able to convey the same visual appearance. Where nonvisible features of the building are concerned—such as interior structural systems or mechanical systems—it is expected that contemporary materials and technology will be employed. Re-creating the building site should be an integral aspect of project work. The initial archeological inventory of subsurface and aboveground remains is used as documentation to reconstruct landscape features, such as walks and roads, fences, benches, and fountains.

Energy Efficiency/Accessibility/Health and Safety Code Considerations

Code requirements must also be met in *Reconstruction* projects. For code purposes, a reconstructed building may be considered as essentially new construction. Guidance for these sections is thus abbreviated, and focuses on achieving design solutions that do not destroy extant historic features and materials or obscure reconstructed features.

Appendix B

Historic Preservation Technology Resources

WEBSITES

These websites can extend the reader's opportunity to learn about available preservation resources. Many of these are maintained by government agencies responsible for distribution of information both nationally and locally. The remainder includes websites maintained by groups working in the design and construction industry that are both influenced by and influence historic preservation technology practices. Several of these websites have search engines that can access information by using the terms found throughout this book.

Government Agencies

These websites link directly to government agencies that affect the use of appropriate historic preservation technology.

Advisory Council on Historic Preservation (ACHP): http://www.achp.gov

Americans with Disabilities Act (ADA): http://www.usdoj.gov/crt/ada/adahom1.htm

Environmental Protection Agency (EPA): http://www.epa.gov

Federal Emergency Management Association (FEMA): http://www.fema.gov

National Center for Preservation Training and Technology (NCPTT): http://www.ncptt.nps.gov

National Earthquake Hazard Reduction Program: http://www.nehrp.gov

National Forest Service Forest Products Laboratory: http://www.fpl.fs.fed.us

National Institute of Standards and Technology (NIST): http://www.nist.gov

National Park Service (NPS):

Heritage Documentation Programs: http://www.cr.nps.gov/hdp

Preservation Briefs: http://www.cr.nps.gov/hps/tps/briefs/presbhom.htm

Preservation Tech Notes: http://www.cr.nps.gov/hps/tps/technotes/tnhome.htm

Secretary of the Interior's Standards and Guidelines: http://www.cr.nps.gov/hps/tps/standards_guidelines.htm

National Register of Historic Places: http://www.cr.nps.gov/nr

Occupational Safety and Health Administration (OSHA): http://www.osha.gov

United States Department of Housing and Urban Development (HUD): http://www.hud.gov

Preservation Organizations

These websites are maintained by professional societies and groups engaged directly in the use of historic preservation technology. Many of these organizations maintain archives of historic preservation–related articles, sell technical publications, or can answer questions about historic preservation technology–related matters. Several offer additional resources to members.

American Association for State and Local History (AASLH): http://www.aaslh.org

Association of Living History, Farm, and Agricultural Museums (ALHFAM): http://www.alhfam.org/index.php

Historic Building Inspectors Association (HBIA): http://inspecthistoric.org

Historic New England (HNE): http://www.historicnewengland.org

International Commission on Monuments and Sites (ICOMOS): http://www.international.icomos.org/publications/index.html

International Institute for Conservation of Historic and Artistic Works (IIC): http://www.iiconservation.org/

National Conference of State Historic Preservation Officers (NCSHPO): http://www.ncshpo.org

National Main Street Center: http://www.mainstreet.org

National Trust for Historic Preservation (NTHP): http://www.nationaltrust.org

Society for Commercial Archeology (SCA): http://www.sca-roadside.org

Society for Industrial Archeology (SIA): http://www.sia-web.org

Society of Architectural Historians (SAH): http://www.sah.org

United States National Committee of the International Council on Monuments and Sites (US/ICOMOS): http://www.icomos.org/usicomos

Vernacular Architecture Forum (VAF): http://www.vernaculararchitectureforum.org

Winterthur: http://www.winterthur.org

Professional Societies and Trade Groups

These websites are maintained by professional societies and trade groups that are affiliated with the use of historic preservation technology. Many of these organizations maintain archives of historic preservation–related articles, sell technical publications, or can answer questions about historic preservation technology–related matters. Some offer additional resources to members.

Aluminum Association, Inc.: http://www.aluminum.org

American Concrete Institute (ACI): http://www.concrete.org/general/home.asp

American Institute of Architects (AIA): http://www.aia.org/hrc_default

American Institute of Steel Construction (AISC): http://www.aisc.org

American National Standards Institute (ANSI): http://www.ansi.org

American Planning Association (APA): http://www.planning.org

American Society for Nondestructive Testing (ASNT): http://www.asnt.org

American Society for Testing and Materials (ASTM): http://www.astm.org

American Society of Civil Engineers (ASCE): http://www.asce.org

American Society of Heating, Refrigerating, and Air-Conditioning Engineers (ASHRAE): http://www.ashrae.org

American Society of Home Inspectors (ASHI): http://www.ashi.org

American Society of Interior Designers (ASID): http://www.asid.org

American Society of Mechanical Engineers (ASME): http://www.asme.org

Association for Preservation Technology International (APT): http://www.apti.org

Brick Industry Association (BIA)—Technical Notes: http://www.bia.org/html/frmset_thnt.htm

Construction Specifications Institute: http://www.csinet.org/s_csi/index.asp

Copper Development Association (CDA): http://www.copper.org

Edison Electric Institute (EEI): http://www.eei.org/industry_issues/industry_overview_and_statistics/history/index.htm

Housing and Urban Development User: http://www.huduser.org

Institute for Business & Home Safety (IBHS): http://www.ibhs.org

International Code Council (ICC): http://www.iccsafe.org

International Facilities Management Association (IFMA): http://www.ifma.org

Marble Institute of America (MIA): http://www.marble-institute.com

Masonry Institute of Washington (MI): http://www.masonryinstitute.com

National Academy of Building Inspection Engineers (NABIE): http://www.nabie.org

National Association of Certified Home Inspectors (NACHI): http://www.nachi.org

National Association of Home Inspectors (NAHI): http://www.nahi.org

National Fire Protection Association (NFPA): http://www.nfpa.org

National Frame Builders Association (NFBA): http://www.nfba.org

National Precast Concrete Association (NPCA): http://www.precast.org

National Research Council Canada (NRC): http://irc.nrc-cnrc.gc.ca/index_e.html

National Roofing Contractors Association (NRCA): http://www.nrca.net

Portland Cement Association (PCA): http://www.cement.org/index.asp

Recent Past Preservation Network (RPPN): ttp://www.recentpast.org

Smart Growth Network: http://www.smartgrowth.org

Stained Glass Association of America (SGAA): http://www.stainedglass.org

Stone Foundation: http://www.stonefoundation.org/

Timber Framers Guild (TFG): http://www.tfguild.org

United States Green Buildings Council (USGBC): http://www.usgbc.org

Urban Land Institute (ULI): http://www.uli.org

Trade Journals, Publications, and Design Guides

These websites provide resources that are available directly from professional publishers related to historic preservation technology. These websites may also offer online resources.

Building Stone Magazine: http://www.buildingstonemagazine.com

Cathedral Communications: http://www.buildingconservation.com/articles.htm

Historic HomeWorks: http://www.historichomeworks.com

Masonry Advisory Council: http://www.maconline.org

Masonry Conservation Research Group:
   http://www2.rgu.ac.uk/schools/mcrg/mcrghome.htm
Masonry Construction: http://www.masonryconstruction.com
National Wood Flooring Association: http://www.woodfloors.org
Old House Web: http://www.oldhouseweb.com
Structural Building Components Magazine: http://www.sbcmag.info
Traditional Building Magazine: http://www.traditional-building.com
Traditional Roofing Magazine: http://www.traditionalroofing.com
Whole Building Design Guide (WBDG):
   http://www.wbdg.org/design/historic_pres.php
Restore Media: http://www.restoremedia.com

Educational Opportunities

These organizations provide educational opportunities related to historic preservation technology.

Campbell Center for Historic Preservation Studies: http://www.campbellcenter.org
Heritage Conservation Network: http://www.heritageconservation.net/about.htm
Historic Windsor: http://www.preservationworks.org
National Council for Preservation Education: http://www.ncpe.us
National Preservation Institute: http://www.npi.org/seminars.html
Preservation Trades Network: http://www.iptw.org/home.htm
Traditional Building Skills Institute: http://www.snow.edu/~tbsi

Glossary

abatement: removing or mitigating hazardous materials.

AC power: electricity that uses alternating current that varies in voltage through a set cycle.

acoustical tile: a tile made from clay and wood fiber that is used to reduce echoes.

adobe: a brick or block made from a combination of soil, fibrous materials (e.g., straw or dried cactus), water, and/or cactus juice that is pressed into molds to create earthen blocks that are left in the sun to bake dry.

aggregate: small stones or sand used in making concrete or plaster.

air handling unit: an assembly of components that typically includes a fan along with cooling and heating coils, humidification manifolds, and air filters.

air-water system: a mechanical system where heated or chilled water is supplied through piping to a fan coil unit that heats or cools a room and ventilation air is separately supplied to the fan coil unit.

Akoustolith: an acoustical tile invented by Raphael Guastavino and Wallace Clement Sabine.

alkali-silica reaction: the chemical action that can occur between the alkali in portland cement and the silica in the aggregates used in a concrete mixture.

all-air system: a mechanical system that uses heated or cooled air distributed through ductwork to heat or cool a room.

alligatoring: paint that has aged and become brittle will break and rupture, causing its texture to look like the skin of an alligator.

alloy: a metal created by mixing two or more metals.

all-water system: a mechanical system where heated or chilled water is supplied through piping to a fan coil unit that heats or cools a room. Alternatively, hot water for heating can be supplied to a radiator or baseboard convector.

Anaglypta: a paper-based embossed wall covering pressed between two rollers to impart a relief pattern to the material with raised elements that are hollow.

anneal: the process of cooling glass in a controlled manner to relieve internal stresses as it is being flattened.

annual rings: see *growth rings*.

apron: the vertical covering that conceals the area beneath a porch.

asbestos: a mineral fiber obtained from magnesium silicate that was used as an insulator and as a reinforcing agent in mastics, concrete, floor coverings, and numerous other building products.

asbestos abatement: the process of removing or encapsulating products that contain asbestos.

asbestos cement siding: cladding that used a mixture of asbestos and cement to make concrete that was cast into shingle and tile shapes.

asphalt siding: cladding made from bitumen-infused felt or paper. It was made in the form of shingles, sheets, and rolls that were embossed and colored to simulate other materials.

auger: a hand tool used for drilling.

balloon framing: a construction method that uses lightweight lumber instead of heavy timbers.

baluster: vertical support for a handrail.

balustrade: a railing system composed of balusters, a hand rail, and sometime a bottom rail.

base flashing: a sheet metal strip that fits under the roof cladding and along the side of a vertical projection. Base flashing is covered by counter-flashing.

baseboard: wood trim secured at the bottom (base) of a wall to protect the wall surface and conceal the joint at the edge of the flooring.

bead board: a tongue-and-groove board with a beaded edge that runs lengthwise along one edge and down the middle of the face of the board.

beam: a structural member that carries transverse loads from a joist, girder, rafter, or purlin.

bearing wall: a wall capable of supporting loads other than its own weight.

bedding plane: in geology, the bonding layer formed in sedimentary rock by fine-grained minerals deposited over a long period of time.

Bessemer process: a refinement of steelmaking achieved

by blowing air through molten iron to oxidize impurities and make steel more rapidly.

beveled: an otherwise flat material having a uniformly sloped surface such that the edge is narrower that the center of the material.

beveled siding: another name for clapboard.

blade sign: a board mounted perpendicularly to the facade containing the name of a business or store so that pedestrians could readily see it.

blind tenon: a construction method in which a tenon does not extend through the entire width of the member in which the matching mortise has been cut.

brass: a copper alloy containing zinc and small amounts of other elements.

bronze: a copper alloy containing tin.

brown coat: the second layer of plaster that was used to level out the scratch coat and provide a smooth surface for the finish coat.

building pathology: the scientific study of what causes buildings to fail.

bulkhead: the portion of a wall located below the display windows of a retail building.

calcining: a roasting method used to remove carbon dioxide from limestone and oyster shells to create calcium oxide, also known as "quicklime."

came: a narrow strip of lead cast into an H or U shape that was used to hold glass in an art or stained glass window. Cames have also been made from zinc, copper, or brass.

carbon content: the percentage of carbon found in a piece of metal, expressed as a percentage of the total weight.

carbon tetrachloride: a cleaning solvent used in dry-cleaning clothes.

carbonation: the long-term hardening effects on mortar and concrete in reaction to exposure to carbon dioxide.

cartoon: a template used in art or stained glass making. It was a diagram of the proposed window showing the outlines of the openings for each piece of glass as they would appear in the assembled window.

casement window: an operable window with a hinge on the vertical side.

cast iron: an iron alloy with a carbon content between 2.2 and 5.0 percent that has been melted and poured (cast) into molds to create specifically shaped objects.

casting: the process of placing metal or concrete into molds to form specific objects and shapes.

caulking: a resilient compound used to fill joints and prevent leaks.

chalking: disintegration of paint into fine particles.

character-defining feature: an aspect of a building, space, or material that is unique to the time period or methods used when a building was constructed.

chemical wash: a finish treatment that uses chemicals to create a desired surface patina.

chinking: materials used in log construction to fill long gaps between logs.

chip carving: removing material in small pieces using chisels, gouges, and other cutting tools.

chisel: a hand tool with a square cutting edge used to shape wood or stone.

chloride attack: a reaction in which salt ions in unwashed sand, especially in coastal areas, react with the alkali in portland cement.

chromochronology: tracing the changes of paint colors in a space over time.

circuit breaker: a device that monitors electrical current drawn into a power circuit. When the draw exceeds safety limits, the breaker shuts off the electrical circuit.

clapboard: a beveled siding that is thinner along one edge and is placed horizontally along the side of a building in parallel overlapping strips.

climate-based design: the practice of constructing buildings that enhance or mitigate the effects of local climatic forces rather than simply using larger HVAC systems.

cold paint: paint applied but not fused to a stained glass window using a kiln.

cold-rolled: metal that has been pressed between roller bars to form shapes and profiles after the steel has cooled. Cold-rolled metal sheets are generally thicker than hot-rolled metal sheets.

compact fluorescent lamps: small fluorescent lamps developed in the late twentieth century to replace less efficient incandescent lamps.

composition: also known as "compo" in woodworking, a type of putty consisting of chalk, glue, linseed oil, and resin.

compressive strength: the capacity to resist forces that push inward.

concealing pigment: materials added to paint to hide the surface being painted, including clay, whiting, various types of lead, and titanium oxide, typically in assorted shades of white.

concrete: a mixture of cement, aggregate, and water that can be placed in formwork and left to cure into a monolithic material.

conservatory: a greenhouse attached to the side of a building.

consolidant: a low-viscosity liquid that penetrates into the smallest cracks and fissures of a decayed material, where it hardens as it dries to reinforce the decayed material. A consolidant can be used alone or, more often, prior to the use of more viscous epoxies that cannot penetrate as deeply as the consolidant.

copper foil: a thin strip of copper inserted between two pieces of glass that can be used in place of a came in making or repairing an art or stained glass window.

cornice: the exterior portion of a building where the roof and wall meet or the portion of an interior wall where the wall and ceiling meet.

counter-flashing: sheet metal built into masonry that covers the upturned edges of base flashing.

counterweight system: a configuration of cords, pulleys, and weights that was used to hold an object such as a window sash or an elevator at a desired location.

crawl space: a space beneath a porch or the first floor of a building that is less than one story tall.

crazing: a interconnected network of fine cracks on a glazed surface.

cricket: flashing that protects the upper side of a chimney penetration on a steep-pitched roof.

crizzling: a network of cracks that form inside glass.

crosscut saw: a hand tool used to cut across the grain of a piece of wood.

curtain wall: an enclosure system used to cover the exterior of a skeletal framed building.

cutwork: elaborate woodwork made by using a jigsaw and/or lathe.

dado: the lower portion of a wall between the chair rail and the baseboard.

dalle de verre: literally, a "slab of glass," a form of stained glass that was revived in the twentieth century and consists of thick slabs of glass mounted in concrete or mortar-epoxy.

daubing: a rough coating of plaster.

DC power: electricity that uses a constant current rather than an alternating current.

delamination: breaking apart in layers, usually along the bedding plane.

delamping: removing an existing lamp without replacing it.

deluge system: a fire protection system where all sprinkler heads are open at the same time to allow delivery of water to all sprinkler heads simultaneously.

devitrification: the internal formation of crystals in glass that has been exposed to high heat.

dew point: the temperature at which moisture condenses out of the air.

diaphragm: in seismic upgrades, the continuous plane of reinforced material installed to withstand shearing forces generated in a seismic event.

display windows: large windows along the front and entrance to a store that allow merchandise to be viewed from the sidewalk.

double-hung window: a window with two vertically aligned sashes, which can be opened.

drainage plane: a physical gap or separation between two layers of material to interrupt moisture flow rather than permitting the moisture to move farther into the wall assembly.

drip edge: a protrusion that allows water to fall to the ground rather than continue to flow along the surface of a wall.

drop siding: a nonbeveled siding that was often cut with a distinct profile consisting of bevels, rabbets, and concave shapes.

dry pipe sprinkler: a fire protection system where the water is allowed into the piping only upon a call from a fire detection signal.

drywall: a gypsum board panel used to enclose the interior wall (and ceiling) framing without using plaster, which is considered a wet construction method.

dutchman: a piece of material spliced into a surface after the damaged portion has been removed.

efflorescence: salt crystals that form on the surface of masonry when moisture-carrying alkali evaporates.

embodied energy: the amount of energy needed to create, transport, and fabricate a material or component used in constructing a building.

embossed wall covering: a material with a raised pattern of design elements on the surface.

encaustic tile: a clay tile used for floors and walls. It was made using different colored clays inlaid on a clay tile of another color and then fired to fuse them in place.

end grain: grain located at the end of a piece of wood that can readily absorb water if not sealed.

epoxy: a viscous liquid compound applied to damaged surfaces or pressed into voids created by missing material. After curing, some epoxies can be sanded or shaped to match existing surface contours.

epoxy mortar: a mixture of epoxy, sand, and cement.

exfoliating: flaking off in fine pieces.

exhaust fan: a mechanical device used to remove foul air from a building.

expanded metal lath: sheet metal that has been slit and stretched for use as a substrate for plaster.

extruding: creating a specific shape by forcing a molten or plastic material through an opening in a die with a predetermined profile.

faience: a flat panel of glazed terra-cotta that was used as a decorative building material.

fan coil unit: an enclosed mechanical device composed of a fan that blows air across a fin tube.

fan room: a separate, publicly inaccessible room where air handling units are located.

fibrous plaster: a form of plaster that is cast with canvas or other reinforcing materials.

finish coat: the final layer of plaster that forms the exposed surface of the wall.

fired glass paint: paint composed of colored glass that was bonded to the surface of stained glass using the heat in a kiln.

fixed sash: a window with sash that does not open.

flashed glass: stained glass consisting of a layer of clear glass and a thin layer of colored glass that have been fused together.

flashing: a system of sheet metal (or other impervious) strips that are inserted across joints created by vertical penetrations in a roof. Flashing is used to keep moisture out of the joint.

flat plaster: plaster that has a flat exposed surface.

float glass: glass that is made by floating a layer of glass on a bed of molten tin and fire polishing the exposed surface.

fluorescent lighting: a system that uses electrical current to excite phosphors inside a tube to produce light.

footing: structural support located below a foundation wall or column that carries vertical loads. Individual footings each support single columns, and continuous footings support a wall.

forging: the process of creating a specific shape or object by physically striking the raw material.

freeze-thaw cycle: a process in which temperatures continuously fluctuate above and below freezing. Water expands when freezing, producing stress in moisture-laden materials that can cause those materials to eventually break apart.

fumigation: the process of applying insecticides in aerosol form.

furnace: an enclosed mechanical device used to create heated air.

fuse: an early form of circuit breaker. The fusible link inside the fuse melted when too much current was drawn by an electrical circuit.

gabled roof: a roof with a single slope on either side of the central peak.

galvanic corrosion: an electrochemical reaction that causes metal to corrode when two dissimilar metals are in contact in the presence of an electrolyte.

galvanize: to coat steel or iron with zinc.

gambrel roof: a roof with two sloped sections on either side of the central peak.

gas lighting: a lighting system that uses an open flame fueled by natural gas.

gauging plaster: a form of gypsum plaster used as a finish coat.

girder: a large beam that supports heavy loads at points along its length.

glaze: a finish that has been applied to tile to decorate and seal the surface when the tile is fired.

gouge: a chisel with a curved or V-shaped cutting blade used to shape wood.

graining: a decorative treatment in which a surface is painted to look like a specific wood by simulating the figural patterns created by the graining of that wood.

grisaille: a paint technique using varying densities of a color to form objects, textures, and shadow patterns on an otherwise clear piece of glass.

growth rings: in a tree, concentric light and dark rings indicating the growth that occurred during a growing season.

grozing: the process of chipping away the edge of flat glass to create a desired shape.

gusset plate: a flat piece of plywood or metal secured to a truss to reinforce connections.

gutter: a trough constructed along the eaves of a roof to collect and direct runoff to a downspout.

gypsum board: a gypsum plaster panel covered on two sides with water-repellent paper.

gypsum plaster: a form of plaster that includes gypsum to shorten the curing time.

hazardous waste: any material containing components that can harm human health when thrown away.

heating, ventilating, and air-conditioning (HVAC): the combination of all systems that provide heating, fresh air, and cooling.

high-carbon steel: an iron alloy with a carbon content between 0.45 and 2.0 percent.

high technology: motors, fans, and electronic controls used to actuate systems that control environmental comfort in modern buildings.

hipped roof: a roof that slopes upward from all four sides to the peak.

homogeneous vinyl tile: a resilient flooring product that contained vinyl uniformly throughout the body thickness.

horse: a device used to create linear profiles in ornamental plaster.

hot-rolled: metal that has been pressed between roller bars to form shapes and profiles while the steel is still hot from the furnace. Hot-rolled metal sheets are generally thinner than cold-rolled metal sheets.

hot-water boiler: heating equipment that produces heated water.

hydraulic concrete: concrete that cures underwater.

hydraulic elevator: an elevator system that uses a fluid-pressurized ram to move the elevator car.

hydrostatic pressure: the force exerted by the weight of water in a container.

igneous rock: a rock, such as granite, formed by volcanic action.

in situ: in place.

incandescent lighting: a system that uses electricity to heat a filament in a lamp to the point where the filament glows brightly and provides light.

inlaid: a flush-mounted piece of material that has been embedded and secured in the surface of another material.

interlayer failure: the separation created between two or more coats of incompatible paint.

japanning: a highly polished black lacquered finish.

jewels: in stained glassmaking, relatively small pieces of glass that have been cast, spun, pressed, carved, sculpted, or hammered to form a specific geometric shape or pattern.

joist: a horizontal framing member that is supported by beams, girders, or bearing walls and is used to support floor or ceiling loads.

kiln: an oven or furnace typically used to fire brick or tile, heat vitreous paint, or dry wood.

knapped: glass that has been chipped away at the edges to form a desired shape.

knob and tube wiring: an early electrical system consisting of two separate electrical conductors. A ceramic tube was used to isolate the wiring as it passed through combustible materials such as wood framing. A knob was used to secure the wiring when it changed directions so that the two conductors would not inadvertently touch.

lacquer: a coating containing resin and solvents that evaporate to leave a hard finish.

laminated floor: a floor composed of reconstituted wood pulp and glue bonded to a veneer wearing surface.

laminated glass: two layers of glass that are bonded to either side of a plastic sheet.

lateral loads: forces that impact a material along its sides.

lath: wood strips, expanded metal, or drywall used as a substrate for plaster to provide gaps for forming plaster keys.

lead abatement: the process of removing hazardous materials containing lead.

leafing: the process of applying thin metal foil to a surface.

lehr: an annealing oven that is used to control the cooling rate.

lime plaster: a mixture of lime putty, sand, and other additives that is mixed with water and used to form a flat or ornamental surface.

lime putty: quicklime (calcium oxide) that has been slaked and can then be mixed with sand and water (and other additives) to make plaster.

lime-sand mortar: a combination of water, lime, and sand that was widely used prior to the introduction of portland cement in 1871. Lime-sand mortar, also known as "lime mortar," was weaker and took longer to cure than portland cement.

Lincrusta-Walton: an embossed wall covering that was made in a manner similar to that of linoleum, except that wood pulp and paraffin wax were used instead of ground cork dust as a filler to create sheets with three-dimensional forms and patterns.

low-carbon steel: an iron alloy with a carbon content below 0.02 percent.

low-sloped roof: a roof with a pitch of less than 3:12 (a 3-inch rise in a 12-inch run).

low technology: the use of natural environmental forces rather than more advanced mechanically oriented systems.

mansard roof: a hipped roof having a double slope that is much steeper on the lower slope.

marbling: a decorative treatment in which a surface is painted to look like a specific type of marble by simulating the figural patterns created by the minerals in the marble.

marquetry: inlaid pattern of ornamental material used on floors and furniture.

material streams: the flows of various raw, finished, and demolition materials to and from a building construction site.

mechanical room: the location of the boilers and chillers that convert primary energy into heating or cooling media.

medium-carbon steel: an iron alloy with a carbon content between 0.25 and 0.45 percent.

metamorphic rock: rock that has undergone a transformation due to compressive forces in the Earth's crust, such as limestone transformed into marble.

mild steel: an iron alloy with a carbon content of 0.25 percent.

miter: a corner joint formed by cutting a 45-degree angle along the length or ends of the wood.

moisture content: the weight of water in a material expressed as a percentage of the dry weight of the material. A moisture content of 7 percent means that 7 percent of the weight of the material is due to the water within it.

molding: a linear piece of trim often cut into an ornate profile.

molding plane: a tool used to cut a decorative profile on a piece of wood.

mullion: the vertical structural member that separates individual windows in a series.

multistory porch: a porch that is two or more stories tall.

muntin: the narrow strips of wood used to hold individual panes of glass in place in a true divided light window.

natural ventilation: fresh air that enters and moves through a building.

negative mold: an object created by pressing a positive model into a bed of densely packed moist sand or using a flood mold process.

newel post: the vertical support at the end of a stairway balustrade.

nonrenewable energy: fuel drawn from carbon-based sources (e.g., petroleum, natural gas) that cannot be readily replenished economically.

notching: the method used to cut a log so that alternating logs form an interlocking joint at the corner of a log building.

novelty siding: see *drop siding*.

ocher: soil ground to a fine powder containing yellow-brown hydrated iron oxide.

one-pipe system: a heating system that provides hot water sequentially to a series of radiators.

open sheathing: boards attached to roof framing with intentional gaps of several inches between them to permit air flow to the underside of roofing shingles.

ore: raw metal taken directly from the ground. Ore is smelted to extract pure metals from it.

ornamental plaster: decorative plaster with three-dimensional features.

oxidation: the process of reacting with chemicals that contain oxygen. In ferrous metals this appears as corrosion or rust.

paint: a coating consisting of a liquid vehicle (typically linseed oil or water) and pigments for concealing and tinting.

papier-mâché: literally meaning "chewed paper," this material was composed of cellulose and glue, pressed into molds, and sealed in linseed oil to create ornamental elements.

parquet: inlaid decoration made by cutting wood into geometric forms with mitered joints to form various patterns.

patina: a change in a surface caused by aging, wear, and oxidation.

pebble dash: a finish treatment for stucco in which pebbles are thrown or troweled into the wet stucco finish coat.

pen: a single room of a log building.

photovoltaic: the process of using ultraviolet energy from the sun to create electricity.

piazza: a raised porch or verandah.

pitch: a method for defining the slope of the roof. For example, a roof with a 3:12 pitch rises 3 inches vertically in 12 inches of horizontal run.

plaster: a mixture of lime, sand, and water used to cover walls and ceilings and to create decorative design elements.

plaster key: the connection formed when plaster has been pressed through gaps in the lath and subsequently hardens.

plate glass: glass that is made by pouring molten glass onto a flat surface, letting it cool, and then polishing the glass to a uniform thickness.

plating: in stained glass, the practice of putting more than one piece of glass in a came.

pointing: the process of filling joints between masonry units with mortar.

polychlorinated biphenyls (PCBs): toxic organic chemicals formerly used as insulators in electrical transformers.

portland cement: a binder used in modern concrete manufacturing composed of a mixture of limestone, clay, and shale that is ground, burned, and then mixed with gypsum. The mixture of water, aggregate, and portland cement forms concrete.

positive model: a three-dimensional object that was carved from wood or plaster by a skilled artisan or craftsperson and served as the master copy from which casting molds were made.

post and beam: framing that uses timbers in vertical (post) and horizontal (beam) orientations.

pot-metal glass: colored or stained glass made by adding metal salts to a pot of molten glass. For example, adding gold created red glass, while adding silver created yellow glass.

precast: an object that has been cast and allowed to cure in a location other than its final installation place. Commonly used for repetitive decorative or structural elements that are separately lifted and secured in place.

pressed metal: see *stamped metal*.

primary spaces: the major public spaces of a building that significantly contribute to the historic character of the interior of a building.

priming: the process of coating bare wood surfaces with a less expensive opaque coating (e.g., primer) that seals the wood from moisture and provides a smooth, continuous bonding surface for the subsequent application of paint.

prismatic glass: a corrugated optical glass with a texture simulating prisms that is used in transoms of commercial buildings to enhance daylighting.

project definition: a construction management practice that provides a specific, clear focus for project goals.

purlins: structural members attached at right angles to the principal rafters to provide intermediate span support for the common rafters.

quarrels: small diamond-shaped panes of glass.

quarries: see *quarrels.*

radiator: an enclosed mechanical device that ejects heat into the atmosphere by both radiant and conductive heat transfer.

radon: a naturally occurring radioactive gas that is a by-product of decaying uranium deposits in soil.

rafter: an inclined structural member running from the peak of the roof to the eaves used to support roof cladding.

raised panel: a wood panel that is thicker in the middle than at the edges.

reactive aggregate: sand or stone containing chloride or sulfate ions that interact chemically with alkali in a concrete mixture.

rebar: a term commonly used to describe *reinforcing bars* used in the construction of reinforced concrete.

refrigerant: a chemical with thermodynamic properties that can absorb or reject heat when evaporated and condensed, respectively, under high temperatures or pressures.

relamping: replacing existing lamps with more energy-efficient versions.

renewable energy: fuel that is directly created from solar sources or other sustainably generated means, such as wind or water movement.

repointing: a repair process in which failing mortar is removed and replaced with new mortar.

return fan: a mechanical device that draws used air out from the occupied space and returns it to the HVAC equipment for heating or cooling.

reveal: the exposed lower face of a shingle.

rip saw: a hand tool used to cut along the grain of a piece of wood.

riser: the vertical portion of a stair step located at the rear of the tread.

rising damp: a process in which water is drawn by capillary action into masonry.

riven: split.

rock lath: perforated gypsum board used as a substrate for veneer plaster.

rolling: forcing a material through a set of rollers to create a specific shape or thickness.

rondels: round pieces of glass created by spinning or machine pressing.

rubber tile: a resilient flooring product composed of rubber, ground wood, processing oils, organic pigments, and asbestos.

Rumford tile: an acoustical tile invented by Raphael Guastavino and Wallace Clement Sabine.

rustication: 1. a process in which flat boards were beveled along the horizontal edges and had a V-shaped vertical groove cut into them at regular intervals to simulate the appearance of cut stone; 2. the process of cutting a beveled edge along the four edges of the exposed face of a stone.

saddle bar: a rod or flat bar secured into the frame of a stained glass window that was used for reinforcement and support.

sampling: the process of obtaining material samples that can be tested off-site or the process of testing specific locations on-site using nondestructive testing methods.

scagliola: a form of plaster in which granite or alabaster rock chips are mixed with pigments and plaster of Paris and then polished to imitate stone.

scaling: the flaking and peeling of concrete or mortar surfaces.

scratch coat: the first layer of plaster or stucco.

secondary spaces: locations in a building that adjoin primary spaces but have less importance and less public access.

sedimentary rock: rock, such as limestone or sandstone, that is formed by the long-term deposit of fine mineral particles.

seismic: pertaining to earthquakes.

seismic upgrades: remodeling a building so that it can better withstand an earthquake.

service porch: a small porch used for utilitarian purposes located at the back of the house.

setback: 1. a specified distance from a boundary line of a property; 2. adjusting a thermostat to use less heating or cooling.

shakes: thick wooden shingles.

shear walls: walls designed to withstand the shearing forces generated by earthquakes or wind.

shed roof: a roof with only one sloping plane.

shingles: roof cladding units made from a variety of materials (wood, clay tile, slate, stamped metal, asphalt, concrete, etc.) that were cut to uniform lengths and thicknesses. Shingles were also used as a siding material.

side gallery: a verandah consisting of two or more floors.

sill plate: a horizontal timber located at the bottom of a wood structural frame that rests directly on the foundation.

simulated divided light: a single-light, double-paned window in which spacers are placed between the glass panes in locations corresponding to simulated muntins fixed to the exterior surfaces of the glass.

single-hung window: a window with two vertically aligned sashes, only one of which is operable.

sinkage: a three-dimensional decorative strip or block of plaster that is secured into a running profile of a plaster ornament such as a cornice.

sistering: the process of reinforcing an existing structural member by attaching one or more structural members alongside it.

skirt: the vertical covering that conceals the area beneath a porch.

slaking: the process of mixing quicklime with water.

sleeping porch: porch located adjacent to a bedroom that was used for sleeping.

smelting: the process in which ore is heating to its molten state and pure metals are separated.

solarization: a process in which manganese in glass turns purple after long exposure to sunlight.

spall: chips or fragments that fall from the exposed face of masonry in reaction to compressive forces generated by movement or thermal expansion of building materials.

splicing: the process of attaching a reinforcing member to an existing member or mechanically connecting two shorter members together to form a much longer one.

sprinkler head: the fixture through which water flows from a sprinkler system.

sprinkler system: a fire protection measure consisting of piping that provides water to suppress fires in a building.

stabilization: the process of returning a material or an assembly to a state of equilibrium.

stack effect: upward movement of warmed air that occurs in rooms with high ceilings or in a tall atrium or stairwell.

stain: as a decorative coating, a paint with little to no concealing pigments.

stamped metal: sheets of metal that have been embossed with a pattern or mechanically pressed into a specific shape.

stamping: using a ram to force a sheet of metal into a specific shape or pattern.

steam boiler: mechanical device that generates steam for heating.

steel: an iron alloy with less than 2.0 percent carbon content.

steep-sloped roof: a roof having a pitch greater than 3:12 (a 3-inch rise in a 12-inch run).

stoop: a small porch or platform next to an entrance to a house.

stopped tenon: see *blind tenon*.

storefront: the portion of a building composed of entrances, windows, canopies, signage, and material finishes that emphasize the nature of the business located inside.

stratification: referring to layers.

structural sections: standard cross-sectional shapes, such as I-beams, that were developed by the steel industry to allow competitive sales.

stucco: an exterior form of plaster that uses portland cement, lime, and sand.

substrate: the layer below the exposed surface material.

sulfate attacks: a reaction that occurs when sulfates in soil or groundwater react with concrete or mortar.

sun porch: an enclosed room on the side of a building that has many windows to admit sunlight.

supply fan: a device that pushes air through ductwork and into the occupied space.

supply system: the piping used to provide water to fixtures in a building.

sustainability: the concept of evaluating processes as a combination of sociocultural, economic, and environmental forces to determine how they mitigate the depletion of natural resources.

tabby: a mixture of lime, sand, oyster shells, and water that was used as a type of concrete.

tempered glass: a high-strength glass that breaks into small pieces when shattered.

tensile strength: the maximum capacity of a material to resist tension (forces that pull).

terne: a lead alloy containing tin that was used to protect sheet steel from corrosion.

terra-cotta: a fired-clay building product that can be unglazed or decoratively glazed.

terrace: a landscaping feature in which the ground is leveled, sometimes paved, and often decorated with plants. In some cases, a terrace may be located on a roof as well.

terrazzo: a mixture of marble (or other stone chips), concrete, and any desired pigments used to create a formed-in-place floor that was ground and polished to mimic marble.

thermoplastics: a resin that will soften when heated and harden when cooled.

through tenon: a construction method in which a tenon extends through the width of the member in which the matching mortise has been cut.

timber framing: the use of wood structural members greater than 5 inches on a side to construct a structural support system.

traction elevator: an elevator system in which the car is lifted by cables and counterweights.

transom window: a window located above a door or large window that was sometimes operable to allow ventilation.

tread: the horizontal portion of a stair step.

true divided light: a multilight window with individual glass panes held in place by muntins.

truss: a combination of structural members forming a frame composed of triangular shapes.

two-pipe system: a heating system that provides a separate return pipe from each radiator. This system is a later improvement on the early one-pipe system.

umber: brown soil that was ground to a fine powder containing hydrated iron oxide.

unglazed: a tile or other fired clay material that has not been treated with glaze.

unreinforced masonry (URM): a type of construction composed of stone, brick, or block that does not include provision for lateral loads created by earthquakes or wind.

veneer: a thin layer of material covering a substrate made of other material.

veneer plaster: a surface treatment in which a thin layer (veneer) of plaster is applied to a wall made from drywall or other material.

verandah: an open covered porch that extends along the side of a house.

vinyl wall covering: a successor to wallpaper that used vinyl resins instead of paper.

visual inspection: an assessment process that uses visual cues to determine conditions.

vitreous enamel paint: translucent paint made from ground glass that is fused to a surface by heating it in a kiln or oven.

volatile organic compounds (VOCs): chemicals that evaporate at room temperature.

wainscoting: vertically mounted tongue-and-groove boards or raised paneling mounted on the dado portion of the wall to protect plaster walls.

wallboard: a family of products including various forms of gypsum board, pressed fiberboard, plywood, and other materials that were used to enclose interior and exterior framing.

wallpaper: a wood cellulose–based covering applied to walls and ceilings that included hand-painted or machine-produced scenes, patterns, and colors.

waste and vent system: the piping used to allow (waste) water to drain into a sewer while allowing sewer gases to vent to the atmosphere.

water table: 1. an ornamental ledge or drip edge at the base of an exterior wall that allows water to drip from it; 2. groundwater level.

wattle and daub: a system of infill used between timber framing to enclose a building. Sticks (wattle) were interwoven, and plaster (daub) was applied to fill in the gaps.

weatherboard: a thicker form of clapboard.

weather stripping: materials such as felts or gaskets installed around window sash or doors to prevent unwanted air from entering the occupied space.

weep hole: small, inconspicuous opening at the bottom of a wall assembly or hollow column that allows condensation and moisture to flow out from the interior of that assembly or column.

wet pipe sprinkler: a fire protection system in which the piping contains water at all times.

window putty: also known as "glazing compound," this is used along with glazier's points to hold a pane of glass in a window sash. The putty seals the joint between the glass and the rabbet along the muntin and sash.

witness mark: evidence of the presence of earlier construction such as a paint outline, discolored finish, or nail holes.

wracking: the result of forces applied to a frame that cause the connection angles to shift.

wraparound porch: a porch that covers more than the main facade of a house.

wrought iron: an iron alloy with a typical carbon content of less than 0.01 percent that has been heated and hammered (wrought) into tools, hardware, utensils, and ornamental objects.

wythe: a single thickness of brick or block.

zinc: a metal used for galvanizing sheet steel and iron, as a pigment for white paint, and as a component in various metal alloys.

Index